CARPENTRY

H. H. Siegele

Craftsman Book Company 542 Stevens Avenue Solana Beach, California 92075

Library of Congress Cataloging in Publication Data

Siegele, Herman Hugo, 1883-
Carpentry.

1. Carpentry. I. Title.
TH5604.S49 694 74-31468
ISBN 0-910460-12-4

Contents

1

Squaring and Staking Out

The Apprentice

In the beginning of his career, every apprentice carpenter must have a set of tools - - not a full set, but enough tools to keep him going without borrowing from his associates. Then as he gains in experience and in the skill of handling tools, he can add to them such tools as circumstances under which he is working might suggest. He should remember, however, that so long as he works at the trade his set of tools will never be complete - - that he must add to them constantly. Moreover, the set of tools that is suggested here should not be taken as being complete, but it will do as a list, from which he can select a beginning set:

Hammer, saw, pencil, rule, hand axe and chisels, 1-½-inch, ¾-inch, and 3/8-inch to begin with. Brace and bits, sizes as needed. Jack plane, block plane, jointer and rabbet plane. Steel square, level, wrecking bar, mortising and shingling gauges. Screw driver, nail set, carpenter's apron, flat file, saw file, saw set and saw clamp. Shingling hatchet, dividers, burnisher, oil stone and powertools, sucn as skill and radial saws, router, floor sander, drill and many other such tools that carpenters use. Contractors, as a rule, furnish the power tools necessary.

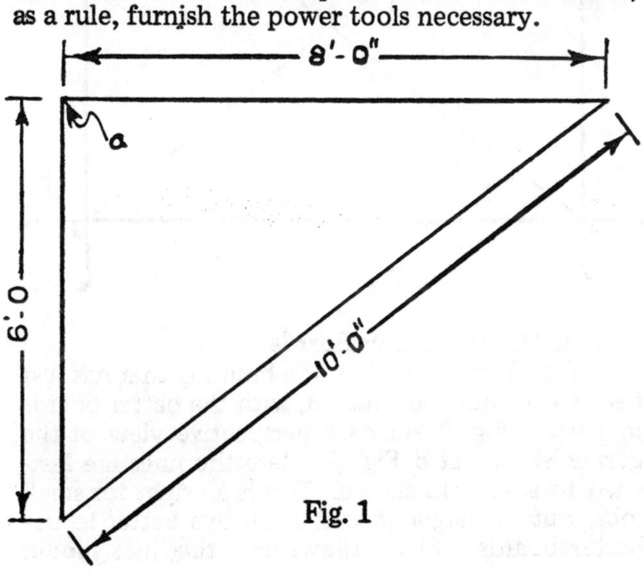

Fig. 1

Squaring Methods

Every apprentice should learn early in his experience how to square a building. This is usually done when the building is staked out. Perhaps the oldest, and the most used method of squaring a building, is the six-eight-ten method, a diagram of which is shown by Fig. 1. This triangle, whose sides are respectively, six feet, eight feet, and ten feet, makes a perfect right-angle at the point marked *a*. This can be proved by taking the sum of 6 squared, and 8 squared, and extract the square root. This would give us the hypotenuse of the triangle, or 10 feet . . . The figures, six feet, eight feet, and ten feet, are merely taken for convenience. Any other proportionate figures will give the same results. A few proportionate figures are given by Fig. 2 to show this. The three sides of the triangle we have marked *A*, *B* and *C*, respecttively, correspond with six-eight-ten. The first figures, working from the inside out, are the commonly used figures, six feet, eight feet and ten feet. The second figures, three feet, four feet and five feet, are proportionate figures, obtained by dividing the original figures by two. The third figures show the results gained by multiplying the first, or

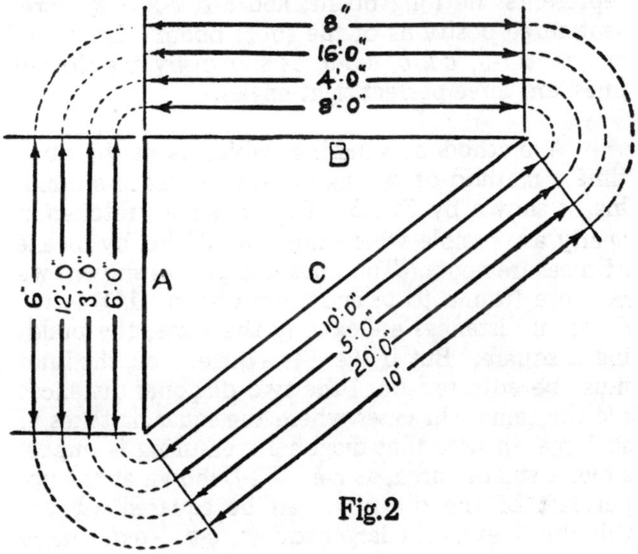

Fig.2

base figures, by two. Here we have twelve, sixteen and twenty. Multiplying the base figures by any other figure will give figures that are proportionately the same as the base figures, six-eight-ten. The last figures on the diagram show that inches can be used as well as feet; in fact, any standard of measurements can be used, provided the same standard is used on each of the three sides.

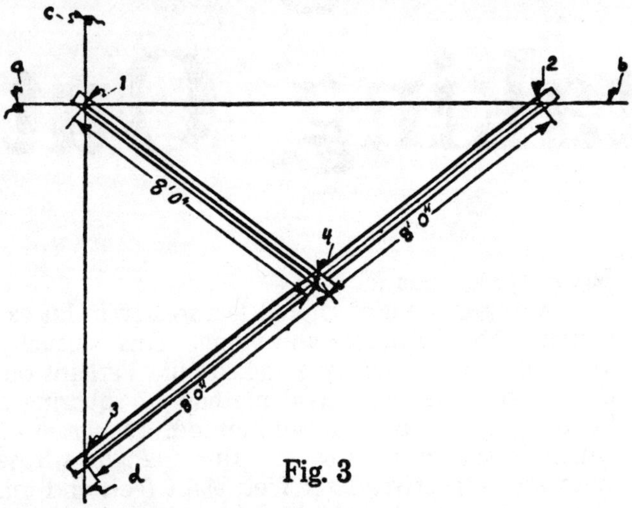

Fig. 3

Fig. 3 shows a simple method of squaring a building by means of a templet. Take a board a little over sixteen feet long, and another board a little over eight feet long, and strike a chalk line through the center of each. Now nail the short board, as shown, to the center of the long board, or to the point marked 4. Make the distance between 1 and 4, 2 and 4, 3 and 4, each 8 feet and your templet is ready for use. Stretch a line from *a* to *b*, and place point number 1 on the established corner, and let point number 2 come at any convenient place along line *a-b*. This done, stretch a line from *c* to *d*, crossing points *1* and *3*, and the angle at point *1* will be a perfect right-angle. The principle that this squaring method is based on is illustrated by the diagram shown in Fig. 4. Here the line *a-b* represents the long board; and *c-1*, *c-2*, *c-3* represent three positions of the short board. The three angles, *a-1-b*, *a-2-b*, *a-3-b*, as shown by the dotted lines, are three perfect right-angles.

A method of squaring, which is at the same time a method of testing the squareness of a building, is shown by Fig. 5. The lines are stretched as nearly as possible where they should be, by means of measurements. The lines in place, as shown, we measure from *a* to *b*, and from *c* to *d*. If the two diagonal distances are exactly the same, the building is square. But if there is a difference, the lines must be adjusted until the two diagonal distances are the same. In cases where the squaring takes in so large an area that diagonal measuring is impossible, a smaller area, as *c-e - - g-f*, shown at the upper left of the diagram, can be squared. From this the lines of the larger square, *c-b - - a-d*, can be

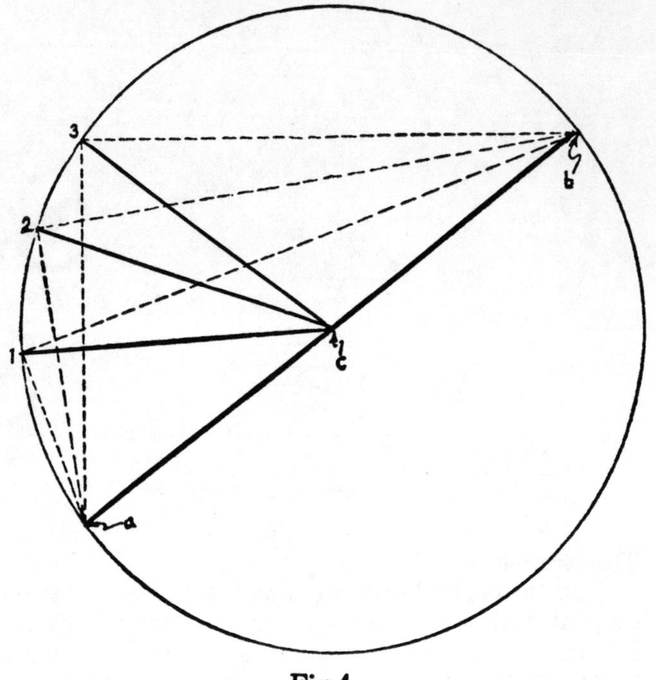

Fig.4

located by measurements, parallel to the respective lines of the small square, *c-e - - g-f*.

Surveyor's Transit

Almost every contractor who handles large jobs, owns a surveyor's transit; and whenever a surveyor's transit is available, it should be used for squaring as well as for leveling. But such instruments are not always at hand, and in those cases, any of the methods just explained will give good results.

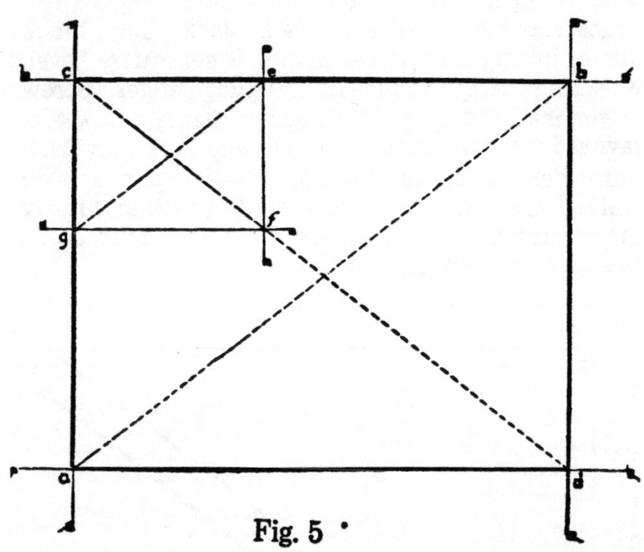

Fig. 5

Staking Out and Batter Boards

Fig. 6 shows a plan of a building that has just been staked out and squared, with the batter boards in place. Fig. 7 shows a perspective view of the corner shown at d Fig. 6. Here the lines are fastened to stakes, as shown. This is all right for small jobs, but on larger jobs it is always better to use batter boards. Fig. 8 shows how the lines should

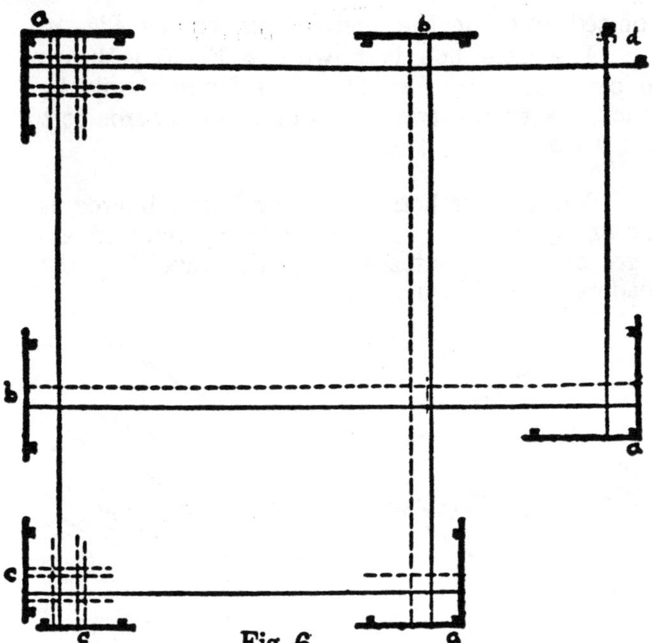

Fig. 6

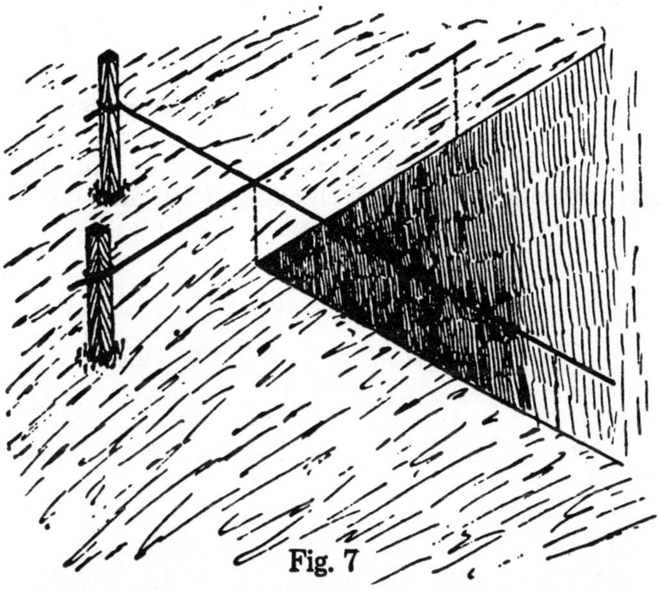
Fig. 7

be fastened to the stakes. This hitch is not only easy to make, but it is easy to unmake. Another advantage this hitch has is that it holds the line to one side of the stake, as will be seen by referring to the illustration.

Fig. 9 gives a perspective view of the corner shown at *c*, *c*, Fig. 6. It will be noticed here that each of the two boards is fastened to two stakes. This method of placing batter boards has an advantage over the method shown at the corners marked *a*, Fig. 6. The advantage is this: If one of the boards, for some reason, is forced out of place, the other will not be affected by it. The points indicated with the reference letters are (Fig. 9) speaking of both boards; *a*, the building line; *b*, the inside wall line; *c*, *c*, the footing lines. At each of these points a saw-kerf is cut, and the line is slipped into the kerf when it is put into place. Other points can be indicated on the boards, in the same way.

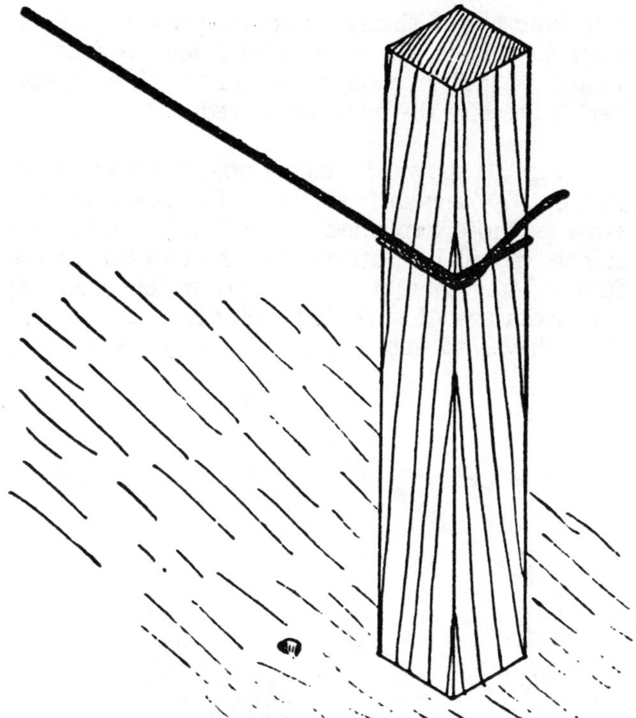
Fig. 8

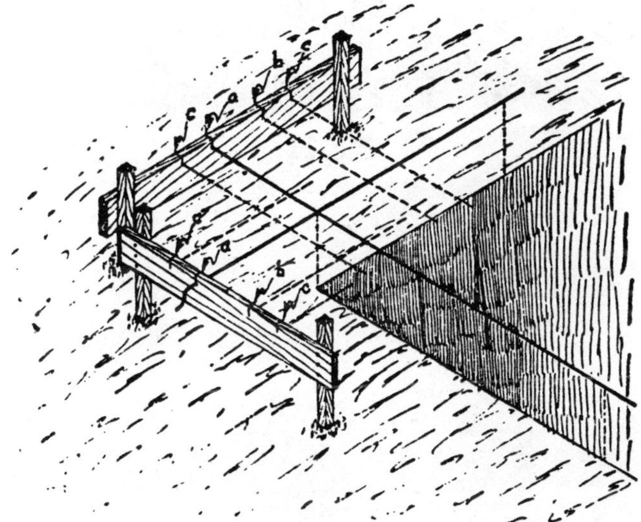
Fig. 9

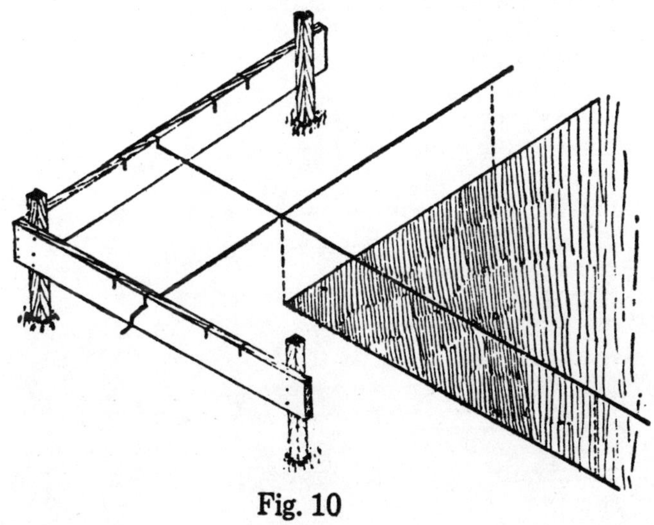
Fig. 10

For example, if there is a chimney to be built, as there usually is, the offsets for it and its footings should be indicated on the batter boards with saw-kerfs to receive the lines when needed.

Fig. 10 shows the batter boards marked *a*, in Fig. 6. The saw-kerfs shown in the boards are the same as those explained under Fig. 9. In this instance, it will be noticed that the two boards are fastened together at the corner, which produces the advantage of stiffining each other; enough so, we believe, to more than offset the advantage pointed out for the method shown by Fig. 9.

Fig. 11

The batter boards shown at *b*, Fig. 6, is shown in perspective by Fig. 11. This figure should be studied with the two locations of the boards at *b* and *b* Fig. 6.

Another method of placing batter boards, is shown by Fig. 12. This method has both advantages and disadvantages. We will leave it to the readers to find them.

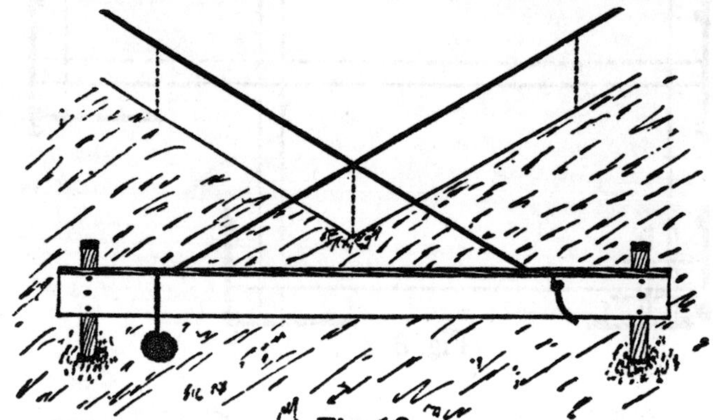

Fig. 12

Ordinarily a carpenter's chalk line gives satisfactory results for lining in staking out a building. But in cases where the distances are rather long, a soft and pliable wire is much better, for it does not stretch and is not easily broken.

2

Form Building

Plywood

In the illustrations covering form building, we are using boards for the forms, because for many years that was the material used for building forms. During that period, plywood made its advent, and soon became a leading factor in form building. At first it was used cautiously, and then by leaps and bounds. Now it is here to stay.

Plywood is easier to handle, and for that reason reduces the labor cost. When the forms are removed, it leaves a much smoother surface than what forms of boards produce. For light forms, ¼-inch plywood will answer the purpose; however, if the plywood is to be used later for rough flooring, boxing, or roof sheathing, then a thickness of plywood should be used for the forms, that will answer those purposes. As a rule, ½-inch plywood will do the job, but the decision will be up to the man in charge of the work, unless definitely specified.

While plywood is extensively used for form building, it has not eliminated the use of boards for forms, and it probably never will.

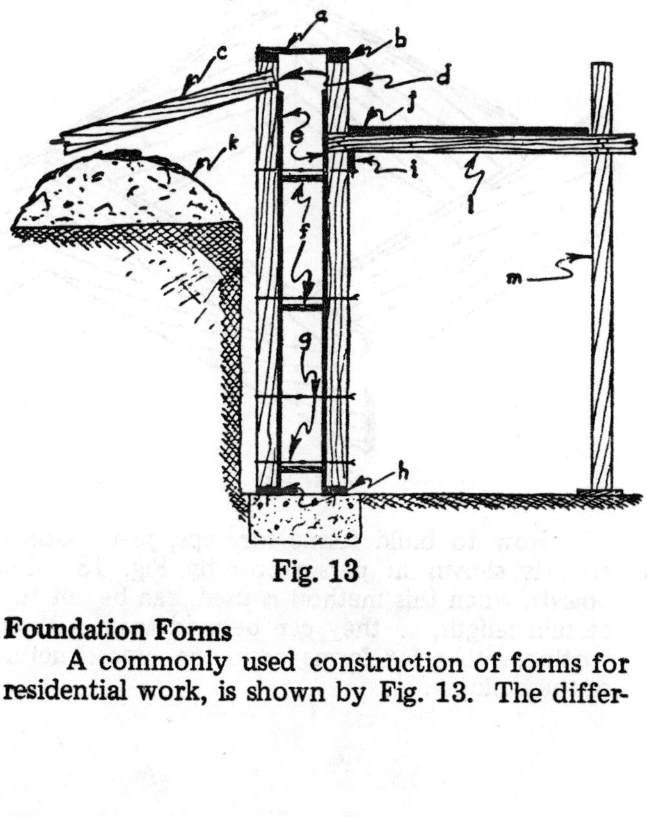

Fig. 13

Foundation Forms

A commonly used construction of forms for residential work, is shown by Fig. 13. The different parts are indicated as follows: At *a*, we show a tie cleat; at *b*, the upper plate; at *c*, the outside brace; at *d*, the studding; at *e*, form boards; at *f*, spreaders; at *g*, tie-wires; at *h*, bottom plates; at *i*, extra support for the ledgerboard; at *j*, the first plank of the runway; at *k*, loose dirt; at *l*, the ledgerboard; at *m*, the upright for the runway.

With these things in mind, let us take up the process of building the forms. The footings have been poured, and have set enough to allow working over them. Now locate the wall on the footings by means of chalk lines; the location, though, must be established by placing the lines in the batter boards, and plumbing down to the footing. From these points the chalk lines can be struck. This done, nail the bottom plates onto the footings in such a manner that when the form boards are nailed on, the face of them will come in line with the chalk lines. This will keep the bottom of the form straight. When the forms are complete, excepting bracing, start the runway, and by means of it, line up the upper part of the forms. The plank marked *j*, should be perfectly straight, then by keeping the studding tight against it, they will be in line. Wherever a joint of the runboards comes, a brace should be used as at *c*. By this method, the forms will be straight when the runway is completed. It will be noticed by referring to the drawing, that the runway is built on the basement-side of the wall, which requires less labor and less material than a similar runway built on the outside; besides it has the advantage of shortening the distance for the concrete wheelers.

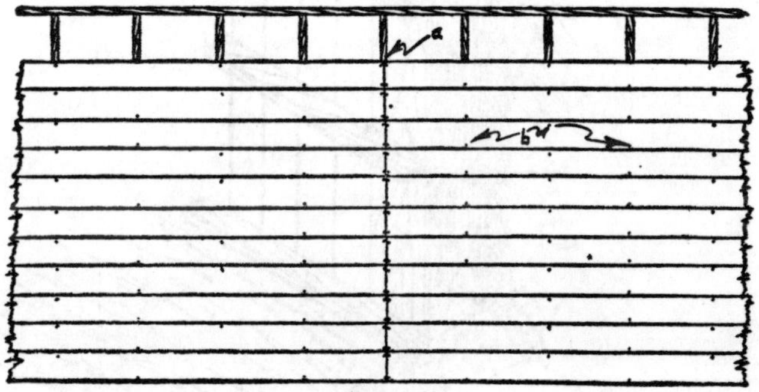

Fig. 14

How the form boards should be fastened to

the studding, is shown by Fig. 14. The joints should not be broken but should all be made on one studding, as shown at point *a*, and down to the bottom. One 6d common nail to a bearing, excepting at joints, is sufficient, and they should be staggered, as indicated by the arrows at *b;* that is to say, at one bearing the nail should be at one edge of the board, and on the next the nail should be at the other edge, and so on.

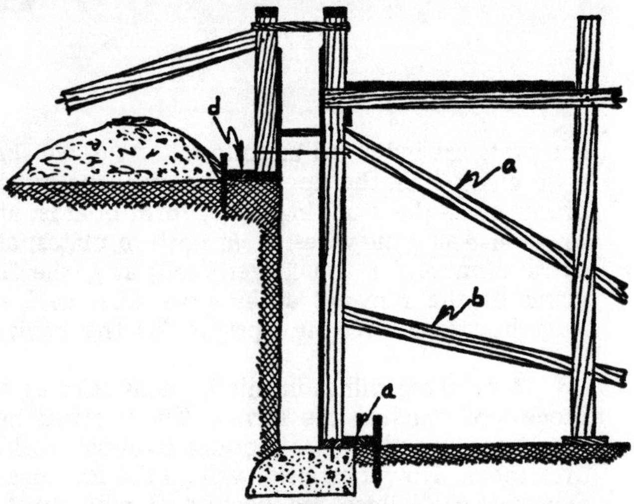

Fig.15

A method of building forms, where the dirt-bank is used for part of the outside form, is shown by Fig. 15. This method is inferior to the one shown by Fig. 13; however, some contractors, before machines were used for excavation, used it quite successfully. Little, we believe, is gained by it, though, for the extra bracing, shown at *a*, *b*, *c*, and *d*, more than offsets the saving on excavating, material and labor. The runway is exactly the same as that shown by the other method.

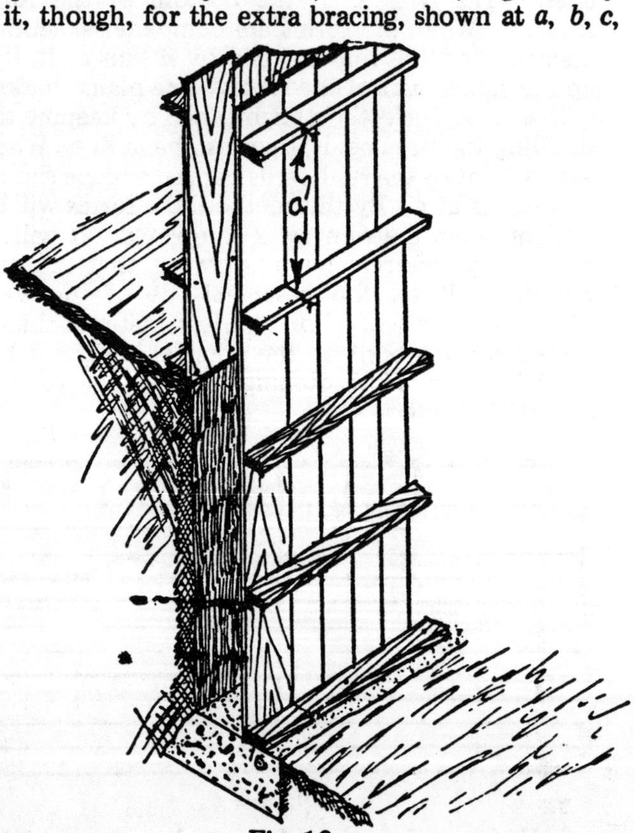

Fig. 16

Fig. 16 shows a perspective view of a method of form building where the boards are placed perpendicular. This method is more nearly suited to be used when the dirt-bank is to answer for part of the outside form. Tie-wires, as pointed out at *a*, are used at intervals wherever the forms are double; otherwise the form must be held in place by a system of bracing.

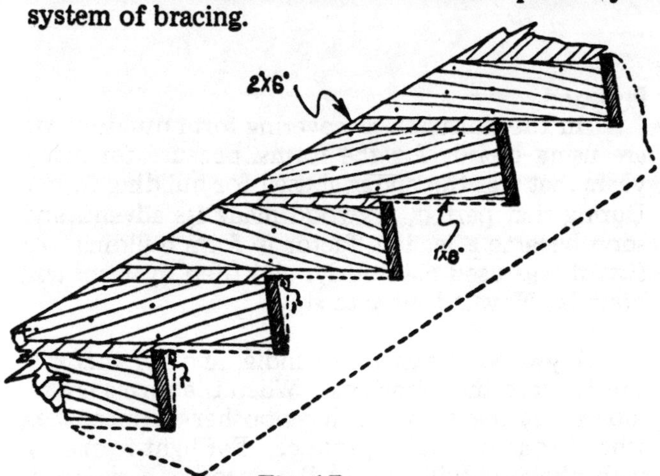

Fig. 17

A method of forming for a concrete stairway is shown by Fig. 17. The risers are shown with a 1-inch batter, which provides for more toe-room on the tread. The boards that form the risers are nailed to the end of 1 x 6's, which in turn are nailed, as shown, to a 2 x 6. The dotted lines show the outlines of the steps, in cross section.

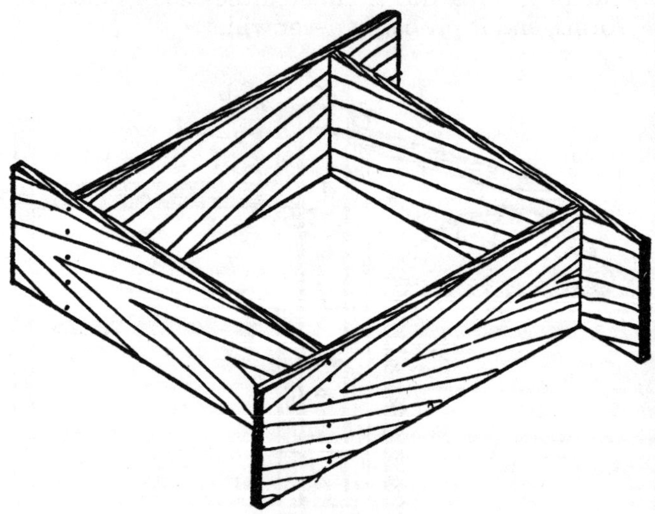

Fig. 18

How to build forms for caps, pier footings, etc., is shown in perspective by Fig. 18. The boards, when this method is used, can be cut to a certain length, so they can be used again without cutting, either for forms or in the superstructure of the building.

Forms for Slabs and Beams

Substaintial and workmanlike construction are the two important considerations in permanent structures, but form building, which is temporary construction, requires a great deal more than that. Forms must be built even more rigid than permanent work, and at the same time, provision must be made for their removal; which means that they must be constructed so they can be taken down easily.

The best of our public buildings, as well as many other buildings, are either altogether fireproof, or at least in part. The carpenter who does not know how to build forms, is, unfortunately, working at a disadvantage. Form work is not difficult, but at the same time, it requires a great deal of experience and skill to do it properly. There are several good systems of form building, two of which are presented here. Both of these systems make it possible to use the form lumber for permanent construction, either in the same building or in some other building; for nothing heavier than 2-inch material is used, and after the form lumber has been removed, the 2 x 4's and 2 x 6's can be used for studding, the planks can be used for joists, and the boards for boxing or roof sheathing.

Concrete slabs are usually too rough for footplates, and in order to overcome this unevenness, a bed of sand, as indicated in the illustration, Fig. 19, is used. The sand does not need to be clean, it can be sand that is otherwise unfit for use, or sand-screenings. The sand bed should be leveled with a straightedge to a certain point, giving it from a quarter to three-eights of an inch allowance for settling. Then the plate is put in place, on which the shores are set. The shores, which are all cut to the same length, are braced as shown.

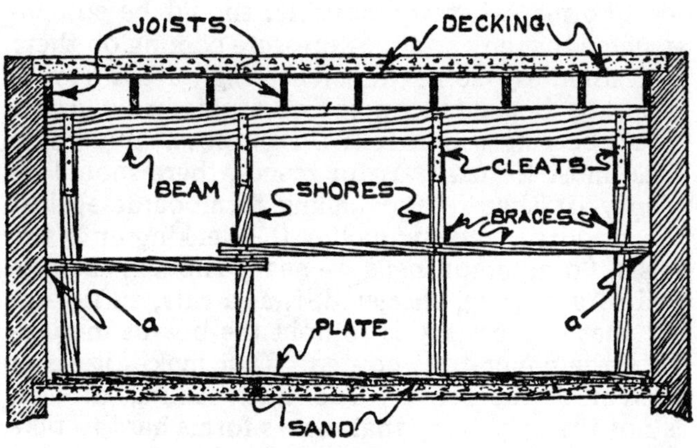

Fig. 19

The braces should butt against the side walls, as indicated at *a*, *a*, to the right and to the left of the drawing. These braces should be well nailed, to prevent the shores from buckling when the weight is applied. The beam, as can be seen, is tied to the shores by short cleats. When the beams are in place, the joists are set and nailed lightly. The decking is then laid on the joints, much like a rough floor, but nailed with 6d common nails, just enough to hold it down until the concrete is poured. The other sectional view of this form is shown by Fig. 20, which shows the sand bed, the plate, the shores, the braces, the beams, the joists, the decking, together with the forms for a concrete beam and a T-shore supporting it.

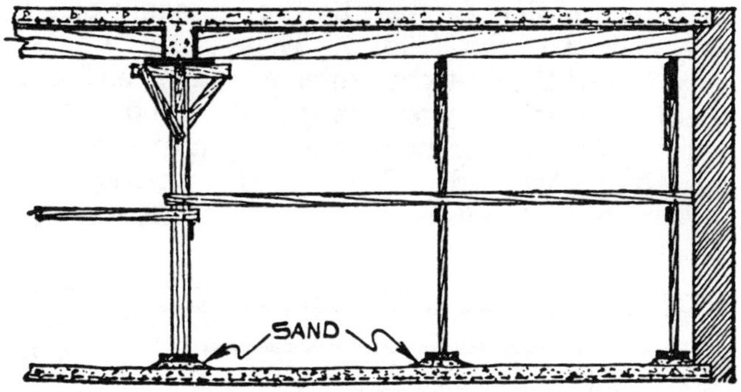

Fig. 20

A detail showing how the beam is fastened to the shores, is shown by *A* and *B* of Fig. 21. The 1 x 4 cleat should be nailed as indicated by the heavy dots. These cleats will be more substantial

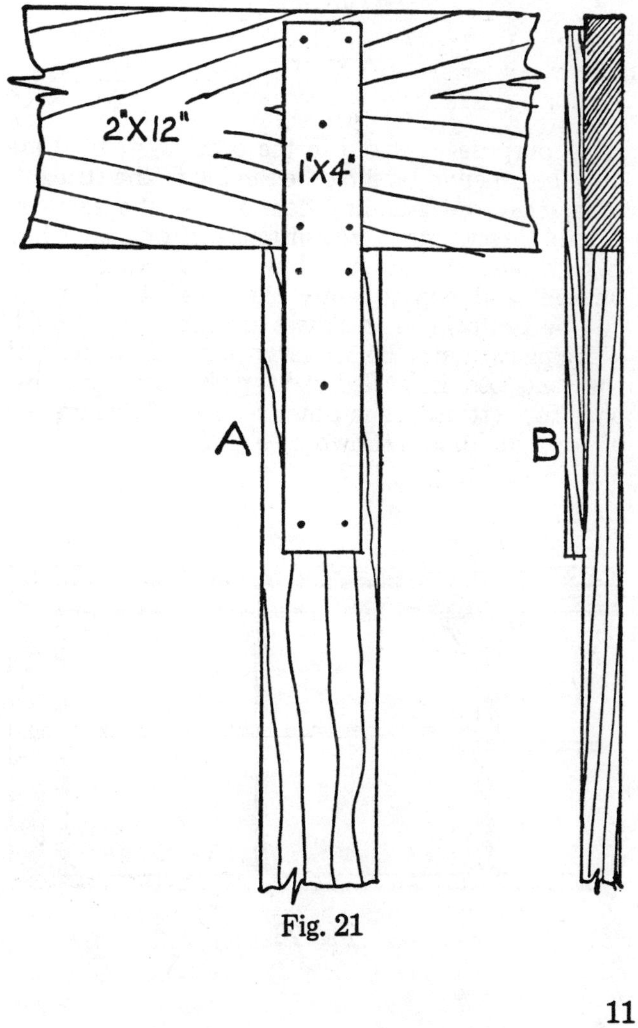

Fig. 21

if they are placed staggeringly; say, for one shore on one side of the beam, and for the next shore on the other side, and so on.

Another method of shoring, which is similar to what we have just explained, is shown in Fig. 22. Here the shores are placed practically as one would place studding for a partition. The purpose, here, is to use 2 x 4's or 2 x 6's, that later can be used as studding for partition, or in some other way utilized in the building. These shores are usually spaced 2 feet from center to center, and the plates on which they rest are spaced around 5 feet, more or less, depending on the amount of supporting strength needed. The sand bed answers the same purpose as it did in the other system. Care must be taken in leveling the sand with the straightedge, that the studding can be cut the same in length throughout. For instance, if one sand bed is high, and the other is low, the decking will be uneven, and consequently the concrete slab will not be uniform in thickness after it is completed. Wedging will not be necessary, in this system, if the sand bed is leveled right, giving proper allowance for settling. The other section of this system of form building is shown by Fig. 23.

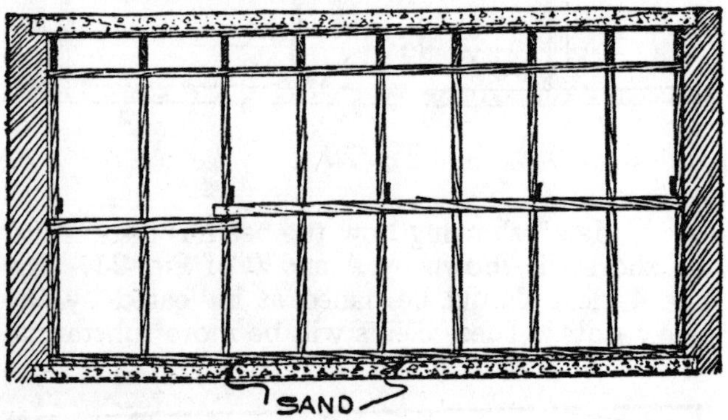

Fig. 22

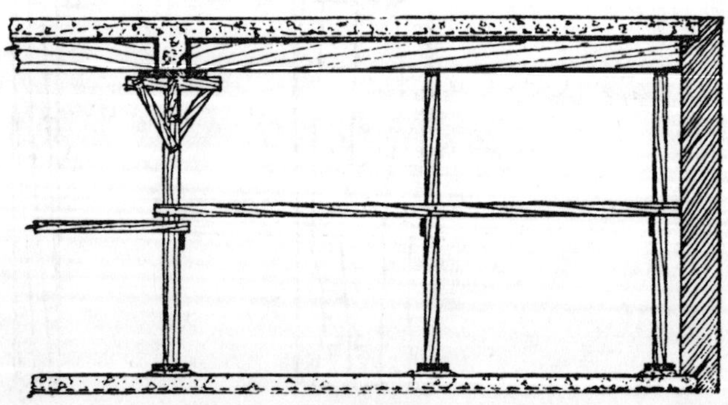

Fig. 23

Here also is shown a form for a concrete beam with a T-shore supporting it. A detail of the beam form and the T-shore, is shown by Fig. 24. The cleats and the braces should be well nailed, as the heavy dots show. We show the bottom of the beam form made of 2-inch material, but that is not altogether necessary; the bottom can be made of 7/8-inch material, but 2-inch material gives better results. The dotted lines, to the upper left, show how the beam is formed where the concrete slab on the one side of the beam is thicker than the slab on the other side.

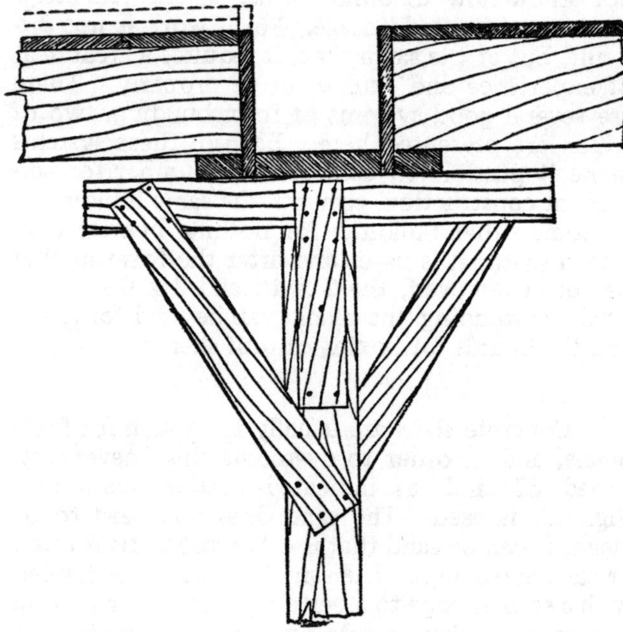

Fig. 24

Forms, wherever possible, should be so constructed that any strain or pressure coming on them is resisted by some direct bearing. Nails should never be depended on to resist pressure where a direct bearing will answer the purpose. But where nails must be used, as for braces, there should be plenty of them. For nailing formboards against the studding, and for nailing the decking onto the joists, 6d common nails are heavy enough, and one nail to a bearing, staggered, is as a rule, sufficient. All that is necessary is to hold the boards in place until the concrete is poured. That makes it easier to remove the forms. Too much nailing, though, is not the only thing that makes forms hard to take down.

Placing the form-lumber in positions so that it will be wedged in after the concrete is set, causes difficulties in removing it. Allowing ends of boards or ends of planks to become embedded in concrete always causes trouble; besides, it mars the appearance of the wall and impairs the strength in some instances. Forms should be so constructed that when they are removed, the concrete wall will have a smooth surface and a pleasing appearance. Surfaced plywood usually gives the best results.

Forming with Pans

Here we are taking up the pan system of form building. The pans are made of metal, and

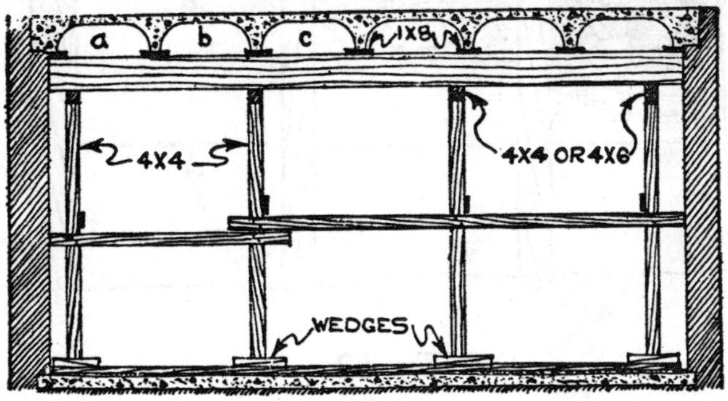

Fig. 25

bent into an arch-shape, as can be seen by referring to Fig. 25, *a*, *b*, *c*. etc. These pans are fastened to 1 x 8 boards, as indicated, which form the bottom of the small concrete beams, or joists as they are commonly called. The purpose of this construction is to reduce the weight of the floor by supporting it with a system of these reinforced concrete joists, rather than by a solid slab. The longer the span is, the deeper the pans should be, thus making the depth of the joists greater, which, in turn, increases their strength.

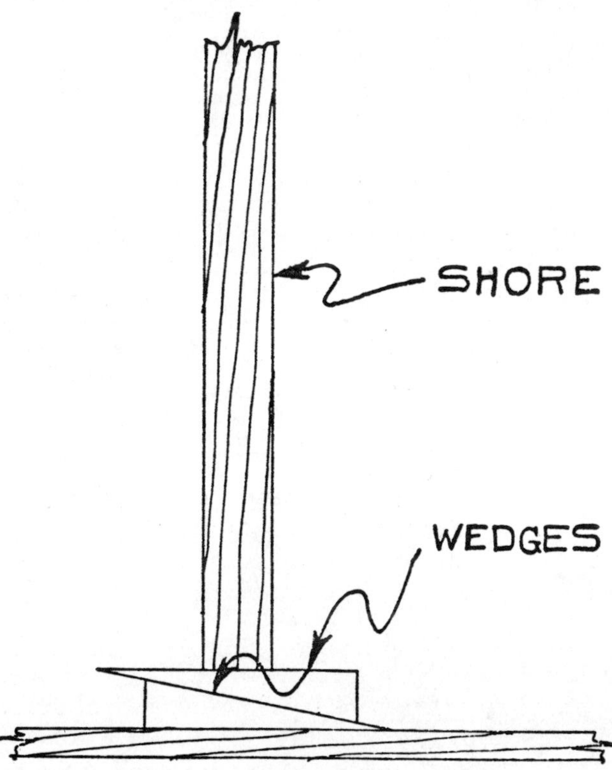

Fig. 26

Fig. 25 also shows another method of supporting forms. Here 4 x 4's are used for shores, and the joist-supporting beams can either be 4 x 4's or 4 x 6's, whichever the load might require. Wedges are used under the shores with which the forms are leveled and lined up to the proper elevation. A detail of these wedges is shown by Fig. 26, together with the lower part of a shore. Fig. 27 shows how the timber is marked for cutting the wedges. To the left is shown an end-view of the timber. This timber is presumed to be a 4 x 4, which makes good wedges where there is a great deal of adjusting necessary in leveling up the forms; however where but little adjustment is required, the wedges can be cut out of 2 x 4 stock.

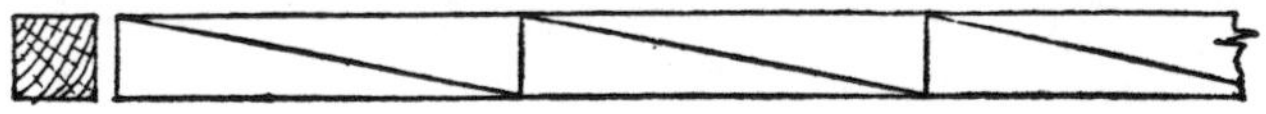

Fig. 27

A floor plan of the forms onto which the pans are fastened, is shown by Fig. 28. The dotted lines show the location of the pans when they are in place. The indicators at *A* and at *B* show how the ends are formed. Short boards long enough to reach from one joist to another are used for this.

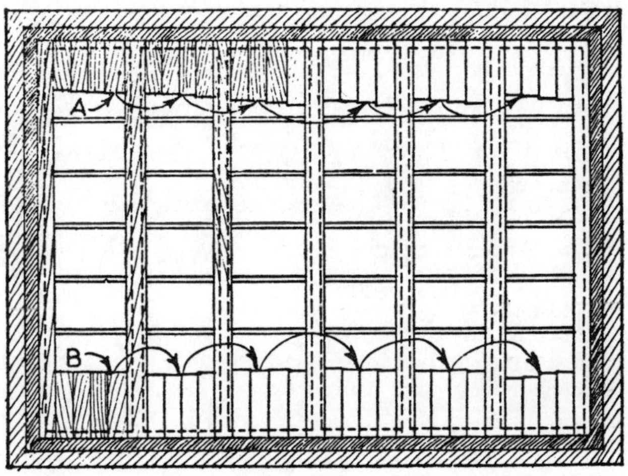

Fig. 28

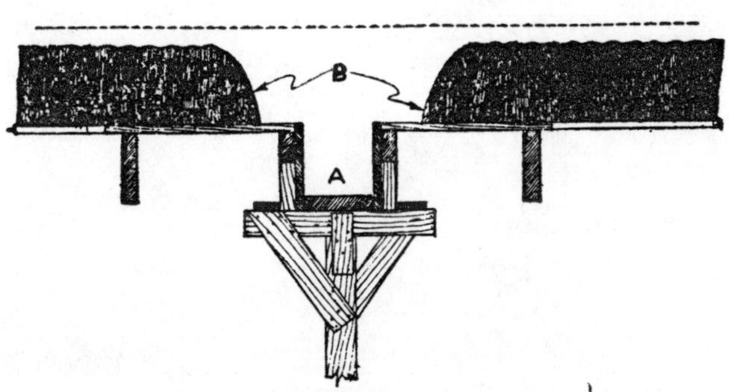

Fig. 29

A study of Fig. 25, a cross section, and Fig. 28, the floor plan, will help the student to understand this system of form building.

Fig. 29 *A*, shows a form for a large concrete beam, and *B*, shows the relative position of the pans to the beam. The dotted line above indicates the top of the concrete slab, after it has been poured. Fig. 30 is a detail of the upper right-hand corner of Fig. 28, showing the pans in place. The continuous lines *a*, *a*, *a*, etc., show the joints of the sections that make up the pans.

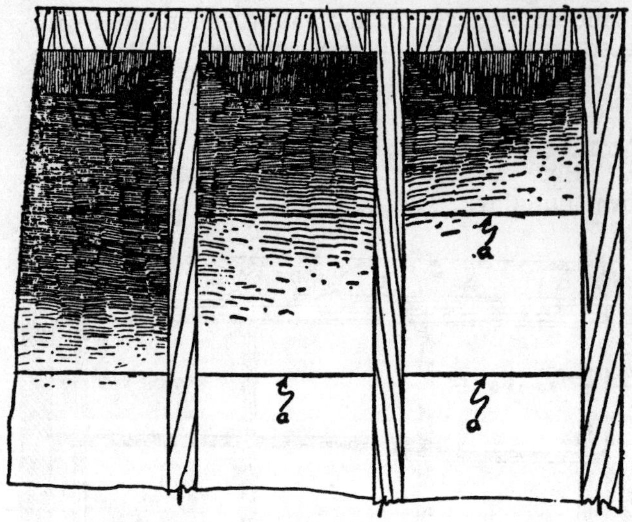

Fig. 30

3

Hollow Tile and Collapsible Pans

Treating Form Lumber

Form lumber should be carefully selected for a number of reasons. If the concrete work is to be covered with earth, or for some other reason kept out of sight, any kind of boards that will keep the mixture of concrete from escaping will do. But if the concrete, after the forms are removed, is to present the finish surface, then the form lumber should be of a much better quality. The boards should not only be free from loose knots, but they should be reasonably straight. Loose knots, even though they might be covered with heavy paper or with tin, nevertheless, leave a bad appearance on the concrete when the forms are off. Even tight knots show the grain of the wood, and sometimes become objectionable. Boards that are too crooked to make tight joints should never be used where a smooth finished surface is required. Good plywood is the answer to this proposition.

For work, such as belt-courses, window sills or imitations of stone, a very good grade of lumber should be used. And when an especially fine job is desired, it is advisable to surface the lumber and sandpaper it. In this way a very smooth surface can be obtained.

Oiling the forms makes them easier to remove, and it lessens the danger of chipping, either the edges or the surface of the concrete work. Any coating, though, that will leave a discoloration on the finished concrete should not be used when a natural cement-color effect is desired. Moreover, oiling form lumber not only prevents the lumber from absorbing the water in the concrete, but it makes possible a more uniform setting of the cement. This is particularly true in hot weather, when the concrete next to the forms often dries out, rather than sets.

Hollow Tile System

This system is somewhat similar to the pan system. We will call it the hollow-tile system. Fig. 31 shows a cross section of the forms supporting a slab of concrete, bridged with hollow tile. Between each row of tile there is a little beam of concrete, or as it is porperly called, a concrete joist.

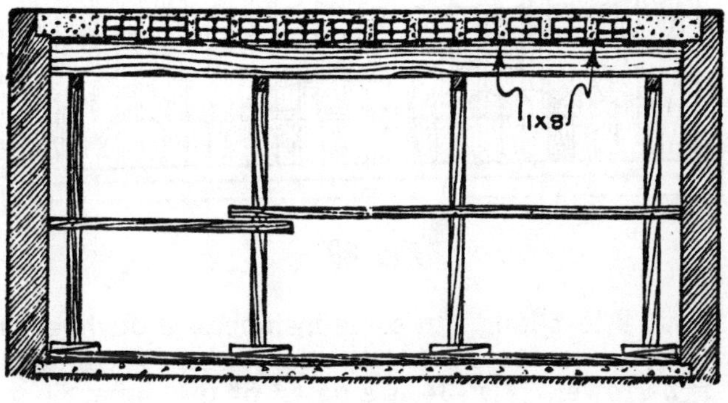

Fig. 31

Near the bottom of each joist, a reinforcing rod is shown by a heavy dot. It will also be noticed that every other form-board is missing, excepting at either side, where the form is solid for a space of about 20 inches. Fig. 32 is a detail showing how the tile are laid over the open spaces, with the edges resting on the boards. The reinforcing rods are indicated by the heavy dots.

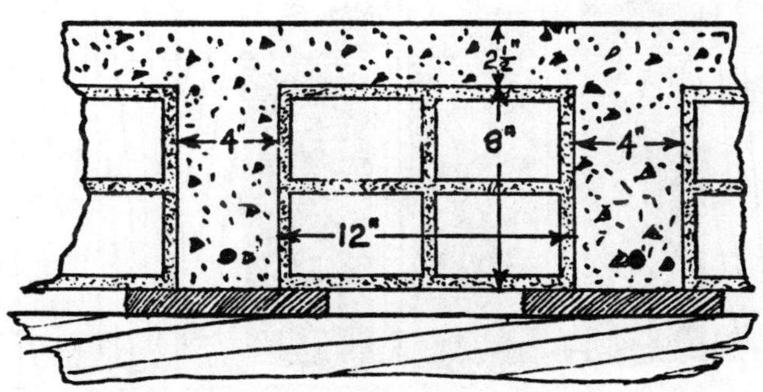

Fig. 32

A floor plan of the forms all ready for the tile to be laid, is shown by Fig. 33. The dotted lines indicate the location of the tile. It will be observed that there is a margin all around the form of about 20 inches without tile. This margin, as was seen by Fig. 31, is to be of solid concrete, reinforced with steel rods. The indicators at *A* and at *B* point out the short boards which form every other space of the margin on two sides, from the hollow tile out to the edge of the form. These boards need not be longer than to reach from the wall-joist to the next one to it, which should not be set closer to the wall than 20 inches, nor farther away than 2 feet. As a rule, these boards are cut about 2 feet long. In some instances, though, even shorter boards are used, especially where the margin is narrower. Fig. 34 is a detail of the upper right-hand corner of Fig. 33. The heavy dots shown over the location of the joists indicate how the boards should be nailed. The tile are shown shaded. The last tile, shown at *A, A, A,* of each row, is laid crosswise, in order to prevent the concrete

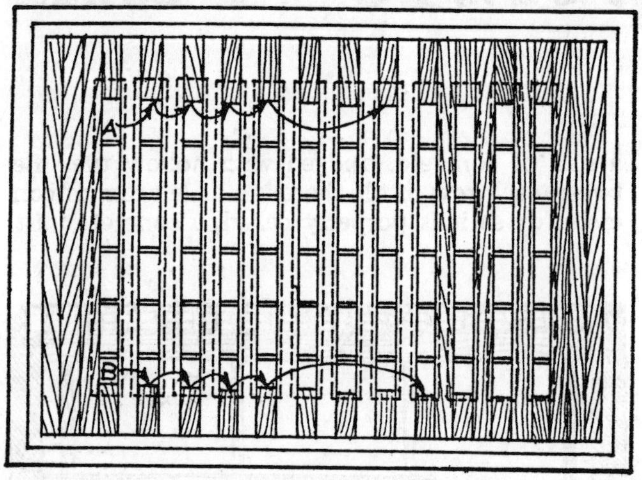

Fig. 33

Fig. 34

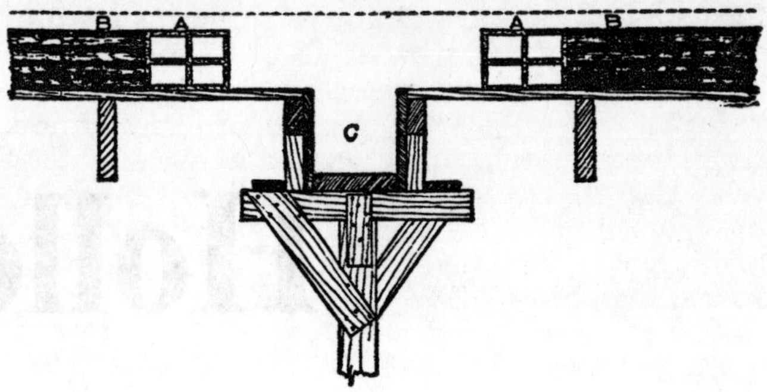

Fig. 35

from filling up the hollow of the other tile. The tile shown at *B, B, B,* are in the regular order. Fig. 35 is another detail showing a cross section of a form for a concrete beam that is to support one end of two tile-bridged floor slabs. *A* and *A* show how the end tile is turned to prevent the concrete from running into the hollows of the tile. *B* and *B* show the tile lying in the regular order. The dotted line above shows the top line of the concrete when it is in place. The detail also shows the relative position of the tile to the concrete beam, marked *C.* A method of using hollow tile bridging between reinforced concrete joists, is shown by Fig. 36. The forms are built practically the same as those shown by Fig. 33. This system of floor construction should be used cautiously, for it is suitable only for very light construction.

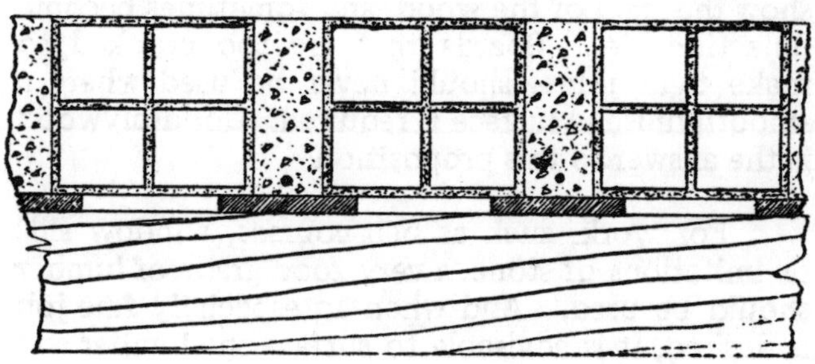

Fig. 36

In building forms for hollow-tile bridged slabs, it is important to know the thickness of the tile to be used. For instance, if the tile are only 4 inches thick, the thickness of the whole slab will be about 6 inches. A form for a 6-inch slab will not require the amount of shoring to support the load that a form for a 12 or a 14-inch slab will. These things must be taken into consideration in building forms for hollow-tile bridged slabs. We are using the term "slab," in this work, in a very broad sense, for it embraces, the whole floor construction, concrete, tile, and reinforcing.

Again, the ceiling height must be taken into account when forms for floor slabs are built, for shores and braces that will safely support the weight for an 8-foot ceiling, will not be sufficient for the same kind of a load with an 18 or 20-foot ceiling height. The additional strength is usually obtained by extra braces. There should be a full set of horizontal braces substantially nailed at regular intervals. The intervals must be governed by the size of the shores and by the strength of the braces.. As a rule, though, these braces should seldom be placed farther than 5 or 6 feet apart.

Determining Weight of Slab

Good judgment in form building is a valuable asset, but if the judgment is not based on some ascertained facts, it often becomes misleading. Determining the weight of a reinforced concrete slab, by good judgment, is merely arriving at an approximate weight, based on previous experiences. In many cases that is sufficient to answer the purpose, but there are times when that will not do, and then it becomes the part of common sense to use an established basis of facts in order to compute the weight that must be supported.

Plain gravel, sand-and-cement concrete weighs on an average 140 pounds per cubic foot, and if it is reinforced, the average weight is 150 pounds. With these figures one can easily obtain the weight of a concrete slab, provided he know what every form builder should know, the thickness of the slab. But forms, as a rule, must support more than the weight of the slab, which, as it is called, is the dead load. They must also support what is known as the live load. For example, a certain part of a concrete slab has been poured, over which concrete for another part is being wheeled, in which case the forms would have to support, in addition to the slab, the vehicle carrying the moving concrete, the concrete, the man who is doing the wheeling, plus the weight of the runway.

Besides this, there is an unknown possible load for which provision must be made. Let us suppose an instance: If for some reason a half dozen or more men come to a certain spot just at the time when the man with the load of concrete comes to the same spot, what will happen? Nothing -- if the form builder has made provision for the possible live load.

Collapsible Pans

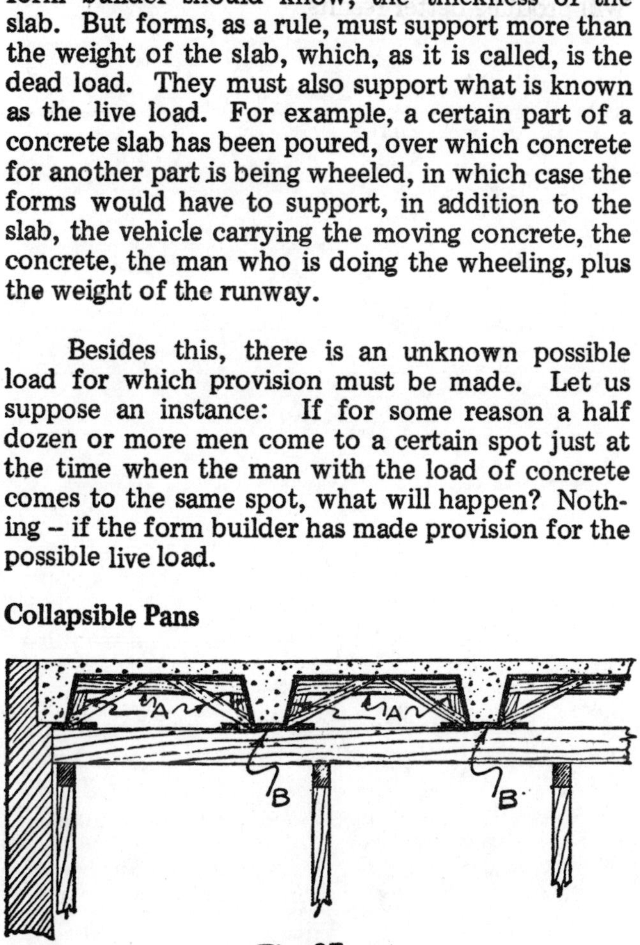

Fig. 37

A cross-section of forms for a beamed floor construction, is shown by Fig. 37. These forms can either be made up in sections and used over and over as long as they last, or they can be built in place in such a manner that they can be taken apart when the forms must be removed. It will be noticed that the forms for the sides of the beams are not perpendicular, making the concrete beam wider at the top than at the bottom. The purpose here is twofold, it makes a better beam, and at the same time, makes it easy to remove the forms. The beams must be reinforced with reinforcing rods.

The bridging, which we are indicating at *A* and *A*, are made independent of all of the rest of the formwork. These bridges are set into place, and the board beam bottoms shown at *B* and *B*, are put in afterward and fastened to the supporting joists.

When the forms are to be removed, the shores and the joists are taken down, and then the sections between the beams are removed. If they are made properly they will come out easily, especially if they were coated with oil before the concrete was poured.

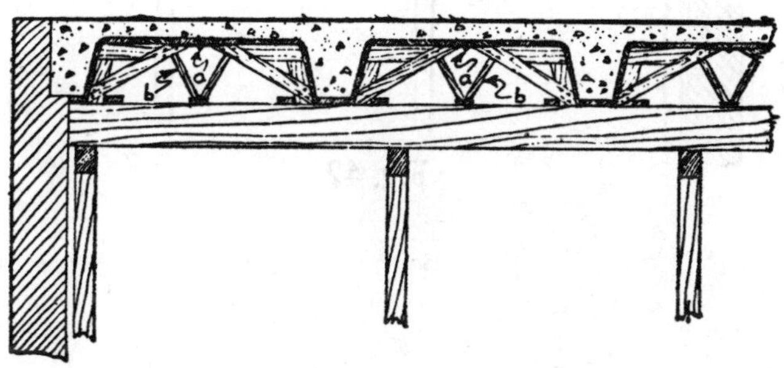

Fig. 38

The same floor construction is shown by Fig. 38, but the forms between the beams are different. They are made in sections, and are collapsible. At point a, they are held together with hinges, where they break on removing them, much as shown by Fig. 39. Center supports, as shown at *b* and *b*, Fig. 38, are necessary.

Fig. 39

Another type of collapsible form is shown by Fig. 40. Here the sides of the concrete beams are perpendicular. This form is built in sections, with 7/8-inch material, which is cut in two from *a* to *b*, as shown by the illustration. To remove this

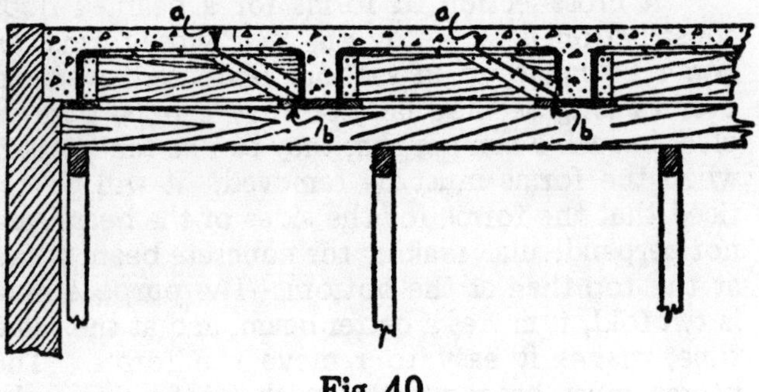

Fig. 40

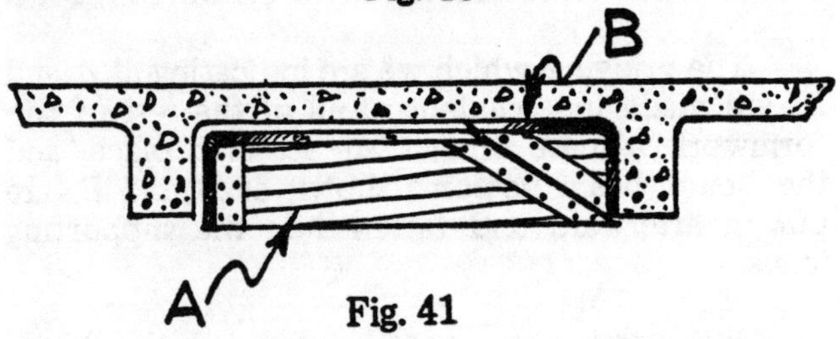

Fig. 41

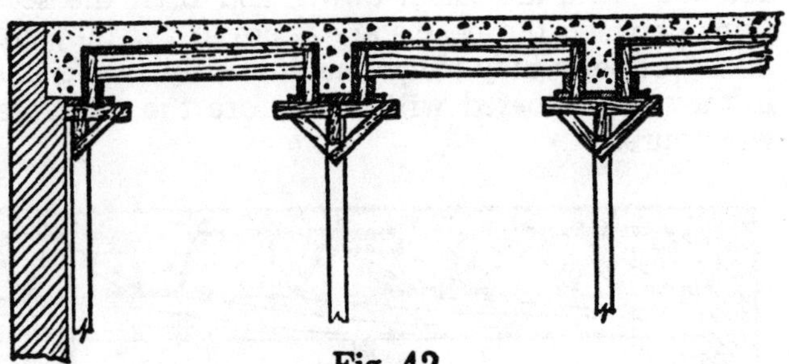

Fig. 42

type of form, the shores and joists underneath are taken down, and the part marked *A*, Fig. 41, is pried loose from the concrete. The part marked *B*, is then pried loose, and the section is ready to be used again.

In Fig. 42 we show a method of supporting a system of non-collapsible forms for a floor construction similar to what we have been dealing with. Here we have the whole form supported by T-shores. This method is suitable, where the concrete beams are spaced from 5 to 7 feet on center.

The reader should bear in mind that all of the types of form construction shown, if they are to be used in actual practice, must be modified to suit the conditions under which they are to be used. This is true with all of our illustrations. Moreover, there are other types of collapsible forms that we have not shown. Every form that is made in sections, so that the sections can be used again and again, is a collapsible form.

The forms shown in this chapter are made with boards, but the use of plywood and other manufactured material, that is suitable for forms, can and should be used whenever such materials will produce better results.

4

Shores and Beams

Factor of Safety

Circumstances and conditions must be taken into consideration, in order to establish factors of safety for form building, and the form builder himself is the man who would do that.

The architect does not specify what kind of materials are to be used for forms; he is concerned only with the permanent structure. As a rule, though, a safety factor of 6 will give a wide enough margin to protect the form builder in most cases. That is to say, if he will build his forms strong enough to support 6 times as much as the weight of the concrete slab, he will not have to worry, unless some unusually heavy live load comes onto the forms. Let us assume, for example, that we are building a form which must support a 6-inch slab. If a cubic foot of reinforced concrete weighs 150 pounds, as it does, then every square foot of that slab would represent 75 pounds of dead weight. By multiplying 75, the dead weight, by 6, the factor of safety, we have 450, or the amount of weight, in pounds, for every square foot of slab that would have to be provided for.

This done, we would proceed to build the forms, regulating the size of the shores and spacing them in such a way that they would support a maximum load equal to 450 pounds for every square foot of slab . . . But how are we going to know what size timbers, or how many of them to use so as to support this maximum load? We will give the solution at the close of this chapter.

Shores and Beams

A system of shoring up metal forms by means of screw-jacks, is shown by Fig. 43. The jacks can also be used for supporting forms made of wood. The screw at the bottom of the jacks makes it easy to level or line up the forms, as well as to take them down when they have to be removed. The jacks are shown without braces, which in this case are not necessary; but when the load is liable to cause buckling, then the jacks should be reinforced with braces.

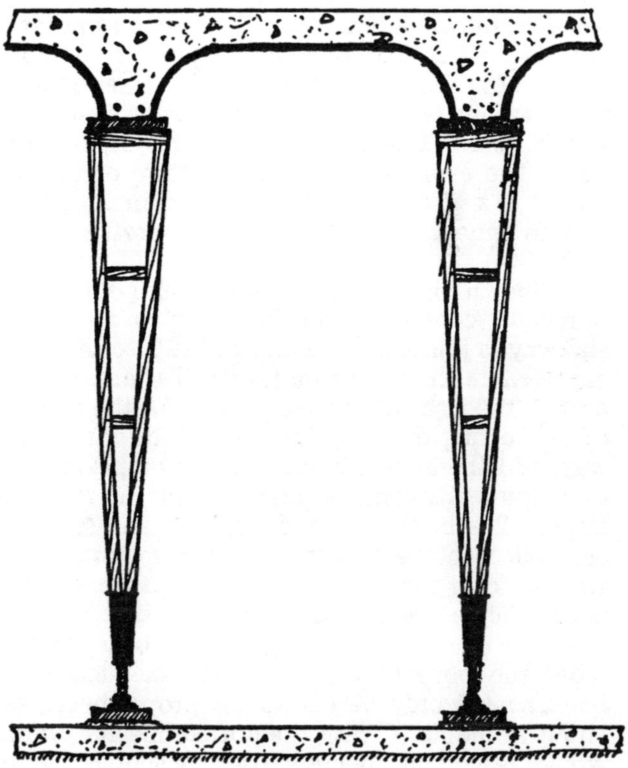

Fig. 43

A cross section of a reinforced concrete cantilever and the form for it, is shown by Fig. 44. The slab part of the cantilever is braced by tapering concrete beam-like brackets under it. The dotted line represents the bottom line of these brackets,

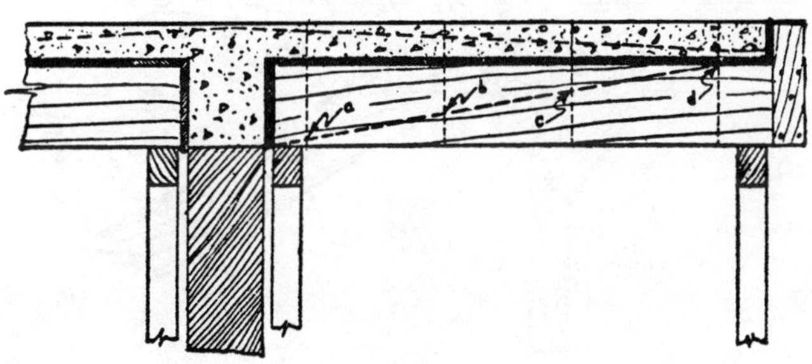

Fig. 44

running through points *a*, *b*, *c* and *d*. Fig. 45 shows the longitudinal section. Here can be seen the shapes of the beam-like brackets at various points.

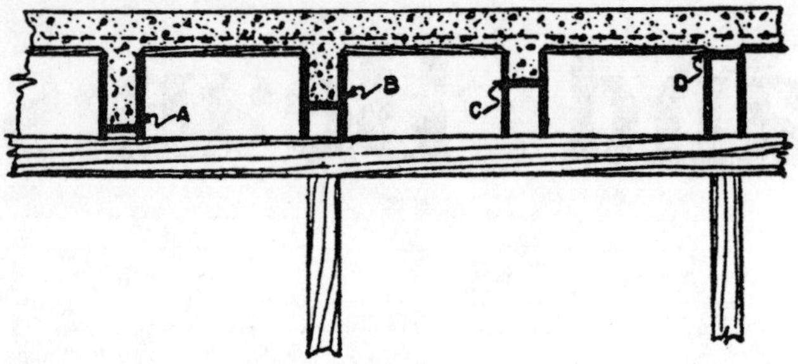

Fig. 45

At *A* is shown a bracket as it appears at point *a*, Fig. 44, and *B* shows one cut at point *b; C* shows one at point *c*, and *D*, one at *d*. The sides of these forms are made of straight boards, held together by 4-inch boards that form the bottoms of the brackets. The form boards for the bottom of the slab are cut at a 45-degree angle, edgewise, making them easy to remove when the concrete has set.

Shown by Fig. 46, is a special T-shore supporting a concrete beam, onto which a concrete stairway is jointed. The arm of the T to the left is exactly like the arms of a regular T-shore; but the arm to the right has been cut short for the purpose of supporting the form for the bottom of the stairway. To the upper right is shown a simple method of holding the forms in place for the risers of the steps. The boards marked *A*, *A*, *A*, and so forth, are nailed to the timber marked *B*, making a substantial form for the steps. The steps shown have perpendicular risers, but frequently it is required that the risers slant to the front in order to give more toe-room to the steps. In such cases the boards need only be cut to the proper bevel, say about as indicated with the dotted line at *a*, which would leave the concrete riser as shown by the dotted line at *b*.

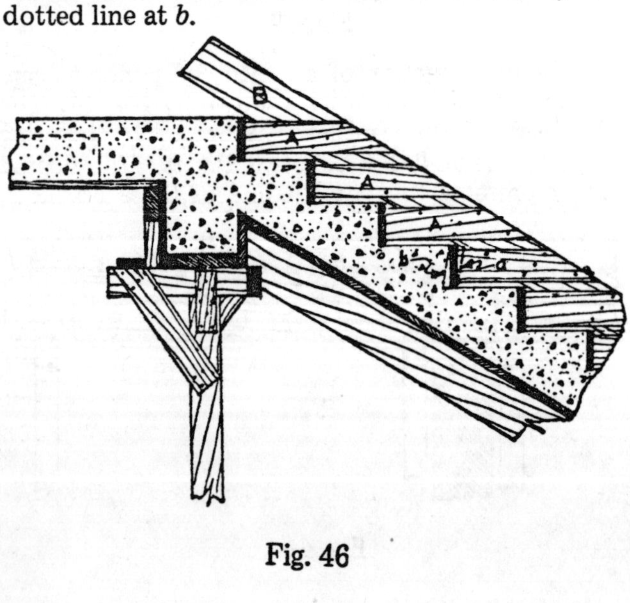

Fig. 46

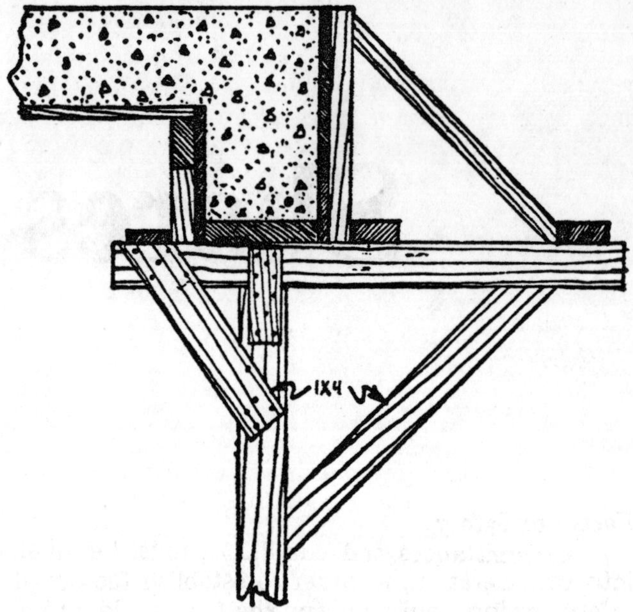

Fig. 47

Another special T-shore is shown by Fig. 47. This shore supports an outside beam, and the long arm to the right provides a toe-hold for the braces of the outside form. A similar T-shore is shown by Fig. 48. Here the long arm is braced at a 15-degree

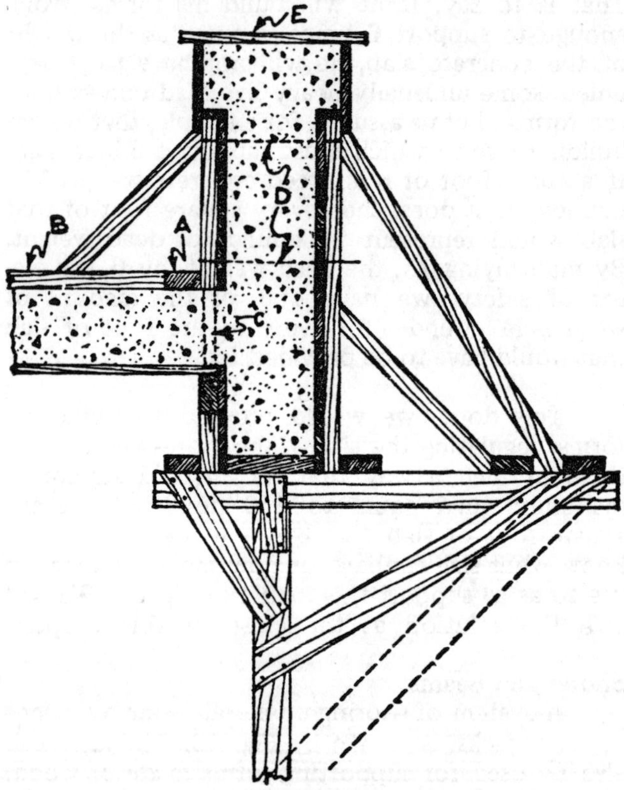

Fig. 48

angle, however, the dotted lines show the safer method of bracing T-arms, which is at a 45-degree angle. This long arm provides toe-holds for the braces of the outside form for the beam and the firewall. Also shown here is a method of forming and bracing the inside form for the firewall. The timbers pointed out at *A* and *B*, if the concrete is still somewhat green, can be spiked to the concrete. In case the concrete is completely set, these timbers should be weighted down enough to hold them in place. Sometimes it is required that the slab and the firewall be poured with one continuous pouring. In such cases the inside form should be supported with a prop something like that shown by the dotted lines at C, and wired as indicated at D. The top is tied together with cleats nailed at such intervals as will keep the side forms of the coping from spreading and in line.

Forms for beams, copings, belt-courses and other exposed concrete work, should not only be straight when completed, but should be braced in such a way that they will remain straight until the concrete is thoroughly set. Concrete slabs, even though they are eventually completely covered, should come to their proper elevation, and should be as nearly level as possible. Carelessness in these things often causes grief and extra expense later on. Concrete, after it is set, cannot be chipped off without more or less impairing its strength. This is especially true of slabs, beams, and of columns. If the forms are not set right, the concrete work cannot be right and the form builder will be to blame.

Formula

And now we will give a solution to the problem of ascertaining the number of shores necessary to support the weight of a concrete slab plus the weight of a possible live load equal to five times the weight of the slab, or, in other words, using a safety factor of 6.

The first thing we want to know is the compressive strength, with the grain, of the various kinds of wood that might be used for shores. According to good authority, the average extreme load per square inch, with the grain, for shores, not over 15 times as long as the narrowest side, is as follows:

White oak, 4,500 pounds; white pine, 3,500 pounds; Georgian yellow pine, southern long-leaf pine, Canadian white and red pine and chestnut, 5,000 pounds; Douglas, Oregon, and yellow fir, 6,000 pounds; short-leaf yellow pine, red pine, Norway pine, eastern fir, hemlock, cypress, 4,000 pounds.

The formula for obtaining the extreme load per square inch of sectional area for shores longer than 15 times the width of the narrowest side of the shore, is:

$$S = U - \left[\frac{U \times L}{100\,D}\right]$$

In this formula

S represents extreme load per square inch of the sectional area of the shore;

U represents compressive strength of the wood per square inch with the grain;

L represents length, in inches, of the shore;

D represents, width of narrowest side of the shore in inches.

Here is an example:

What safe load will a 4 x 4 short-leaf yellow pine shore, 10-feet long, carry, using a safety factor of 6?

In the list given above, we find that short-leaf yellow pine will support an extreme load per square inch of 4,000 pounds, or the value of U. The shore is 10 feet long, which reduced to inches gives us the value of L, or 120 inches. The narrowest side of a 4 x 4 is 4-inches, giving us the value of D. Substituting these values for the various terms in the formula, our problem reads:

$$S = 4{,}000 - \left[\frac{4{,}000 \times 120}{100 \times 4}\right] = 2{,}800 \text{ pounds}$$

or the extreme load per square inch of sectional area of the shore. By dividing 2,800 by 6, the factor of safety, we will have 466, or the safe load in pounds per square inch of sectional area. The sectional area of a 4 x 4 is 16 square inches. So by multiplying 466 by 16 we will have the safe load in pounds for a 4 x 4 shore 10 feet long, or 7,456 pounds.

Previously, in this chapter, we found that the dead weight, plus the possible live weight, that would have to be taken into consideration, for every square foot of 6-inch reinforced concrete slab, was 450 pounds. By dividing 7,456, the safe load for a 4 x 4 shore, by 450, the combined live and dead load per square foot of slab, we get 16, plus, or the number of square feet of 6-inch reinforced concrete slab that a 4 x 4 shore will safely support. Now 16, plus, will go into 100, six times, which means that it will require six 4 x 4 shores for every 100 square feet of slab.

Spacing Shores

Fig. 49 shows a plan, 20 feet by 30 feet, in which the shores are so spaced that 6 shores will support 100 square feet of slab, without horizontal bracing. The locations of the shores are indicated by the heavily shaded squares, and the dotted lines show the direction of the beams that rest on the shores and support the joists. It will be noticed that the lines of shores are spaced 5 feet from center to center, and the shores in the lines

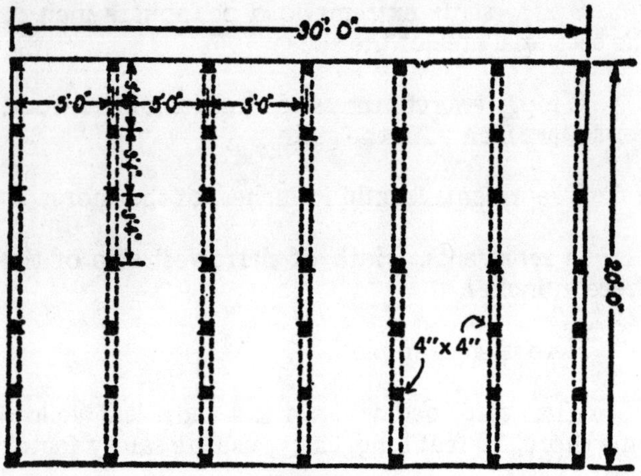

Fig. 49

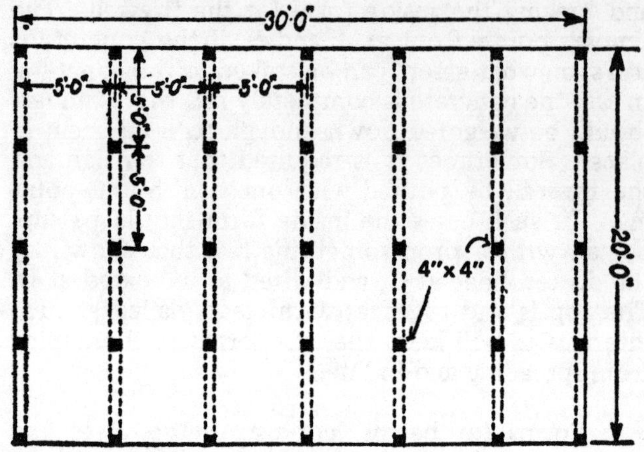

Fig. 50

are spaced 3 feet, 4 inches. While we show these shores spaced 3 feet 4 inches, in practice, they would probably be spaced 3 feet, or 3 feet, 6 inches. It is a matter of common sense to make the spacing as convenient as possible. Shifting the spacing a few inches one way or the other will affect the margin of safety so little that it is justifiable. Of course, if the margin of safety is narrow, then the shifting should be made in its favor, but when there is a wide margin of safety, the shifting can safely be made against it.

If the shores are substantially braced horizontally, and the braces are spaced not over 15 times the width of the narrowest side of the shore, which in a 4 x 4 would be 60 inches, or 5 feet, then the solution of the problem is simple. For example, we would divide the average extreme load per square inch of sectional area of the shore, which in this case is 4,000 pounds, by 6, the factor of safety, and we would have 666, or the safe load in pounds for every square inch of sectional area. The shore being a 4 x 4, we would have a sectional area of 16 square inches. Multiplying 666 by 16, would give us 10,656, or the safe load in pounds for a short-leaf yellow pine shore. By dividing 10,656 by 450, the combined live and dead load, in pounds, of a square foot of 6-inch concrete slab, we would have 24, or the number of square feet of concrete slab a 4 x 4 shore will safely support, if braced substantially every 5 feet. Now it will require as many shores to support 100 square feet of slab, as 24 is contained in 100, or 4 shores.

One way of spacing the shores so that 4 shores will support 100 square feet of slab, is shown by Fig. 50. Here both the lines of shores and the shores in the line are spaced 5 feet from center to center. This spacing is objectionable, unless the beams that rest on the shores and support the joists are heavy enough to carry the weight safely. Nothing less than 4 x 6's should be used for beams when the shores are spaced in this way. It should be remembered that this spacing requires that the shores be braced horizontally every 5 feet and both ways.

Perhaps a better way of spacing the shores so that 4 shores will support 100 square feet of slab, is shown by Fig. 51. In this method, the lines of shores are spaced 7 feet, 6 inches, which is just right for 8-foot or 16-foot joists. The shores in the lines are spaced 3 feet, 4 inches, making the spacing close enough to use 4 x 4 beams for supporting the joists. At *A*, are pointed out two braces running lengthwise, and at *B*, are shown braces running crosswise. Fig. 52 shows how the braces are fastened to the shores. At *A*, we show a plan of the joint, and at *B*, an elevation.

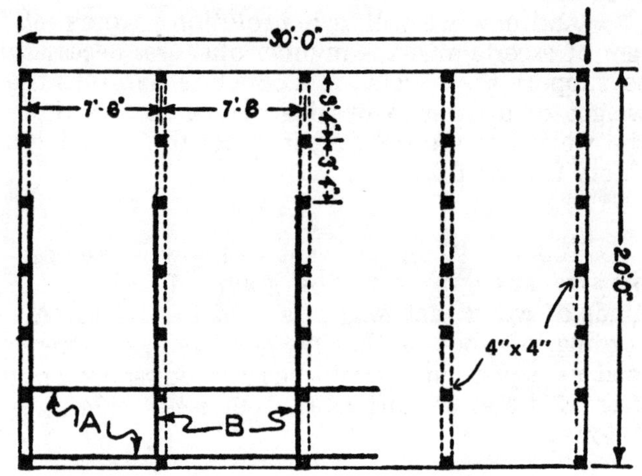

Fig. 51

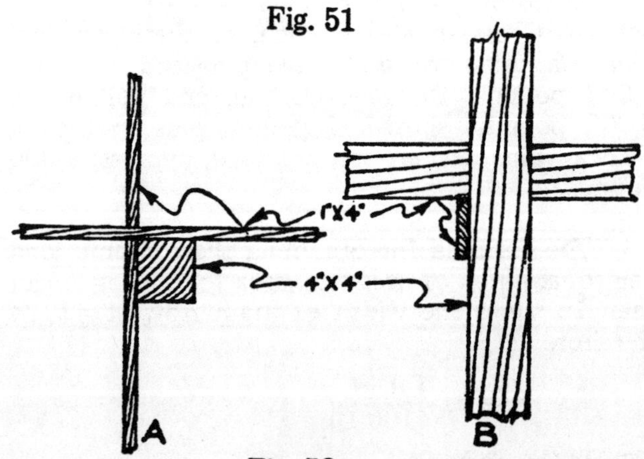

Fig. 52

5

Simplified Engineering Formulas

Problems Worked

In our explanation of Fig. 50, we said that nothing less than a 4 x 6 beam should be used, because the shores were spaced 5 feet from center to center. We felt that the margin of safety was rather narrow, but that was only a matter of judgment. In this lesson we are going to determine whether or not the beam suggested there will safely carry the load.

In explaining Fig. 51, where the shores were spaced 3 feet, 4 inches, we mentioned a 4 x 4 beam, which was also just a matter of judgment, so we will compute the safe load for this beam also. While the load in these cases is uniformly distributed above the joists, it comes on the beams at intervals of 16 inches or 2 feet, depending on the spacing of the joists; nevertheless, we are going to use the formula for determining the extreme uniformly distributed load for beams, which is:

$$B = \frac{2\,K\,S}{3\,L}$$

In this formula

B represent breaking load of beam;

K represents sectional modulus of the beam;

S represents modulus of rupture of the wood;

L represents length of beam, in feet.

In order to proceed with the problem, we must have the value of K, S and L. The value of S, the modulus of rupture of the wood, is the only one that is not given in the problem. Here is a list of woods, giving the modulus of rupture in pounds per square inch:

White oak and northern short-leaf yellow pine, 6,000 pounds; white pine, Norway pine, eastern fir and spruce, 4,000; Georgian yellow pine, 7,000 pounds; Douglas fir, Oregon fir and yellow fir, 6,500 pounds. Washington fir, red pine, cypress, cedar, chestnut and California spruce, 5,000 pounds; hemlock, 3,500 pounds.

Now let us determine whether the 4 x 6 beam we suggested by Fig. 50 in the previous chapter, is strong enough to carry the load.

Here is the problem:

What uniformly distributed load will break a 4 x 6 Georgian yellow pine beam, the span in this case being from one shore to the next, or 5 feet? What will be the safe load, using a safety factor of 6?

The problem gives us the value of L, Fig. 53, or 5 feet. The modulus of rupture in pounds per square inch for Georgian yellow pine is 7,000, the value of S, and the size of the beam, 4 inches by 6 inches, gives us the basis for obtaining the sectional modulus of the beam, or the value of K. The problem for determining the sectional modulus of the beam can be stated thus:

$$\frac{b\,d^2}{6} = \frac{4 \text{x} 6^2}{6} = 24, \text{ or the value of K}$$

Now, the value of B is what we want to know;

the value of K is 24;

the value of S, as given in the list, is 7,000;

the value of L, as given in the problem, is 5.

Proceeding further with the problem, we have:

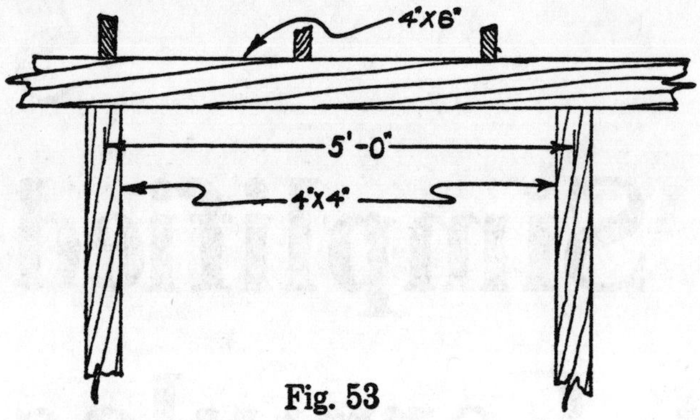

Fig. 53

$$B = \frac{2\ K\ S}{3\ L} = \frac{2x24x7{,}000}{3x5} = 22{,}400, \text{ or the}$$

breaking load in pounds of the beam. By dividing 22,400 by 6, the factor of safety, we have 3,733, or the safe load for the beam, in pounds. That means that a 4 x 6 Georgian yellow pine beam, over a span of 5 feet, will safely carry 3,733 pounds. The beam in question must carry a uniformly distributed load equal to 25 square feet of 6-inch reinforced concrete. Reinforced concrete weighs 150 pounds per cubic foot, so every square foot of 6-foot slab would represent one-half of 150 pounds, or 75 pounds. Multiplying this by 25 gives us 1,875, or the dead load in pounds. Then if we subtract 1,875 from 3,733, we find that, according to these figures, the beam is strong enough to safely carry the load of a 12-inch reinforced slab of concrete, giving us a wide margin of safety.

The second problem deals with the beam suggested by Fig. 51 and shown in Fig. 54. Here is the problem:

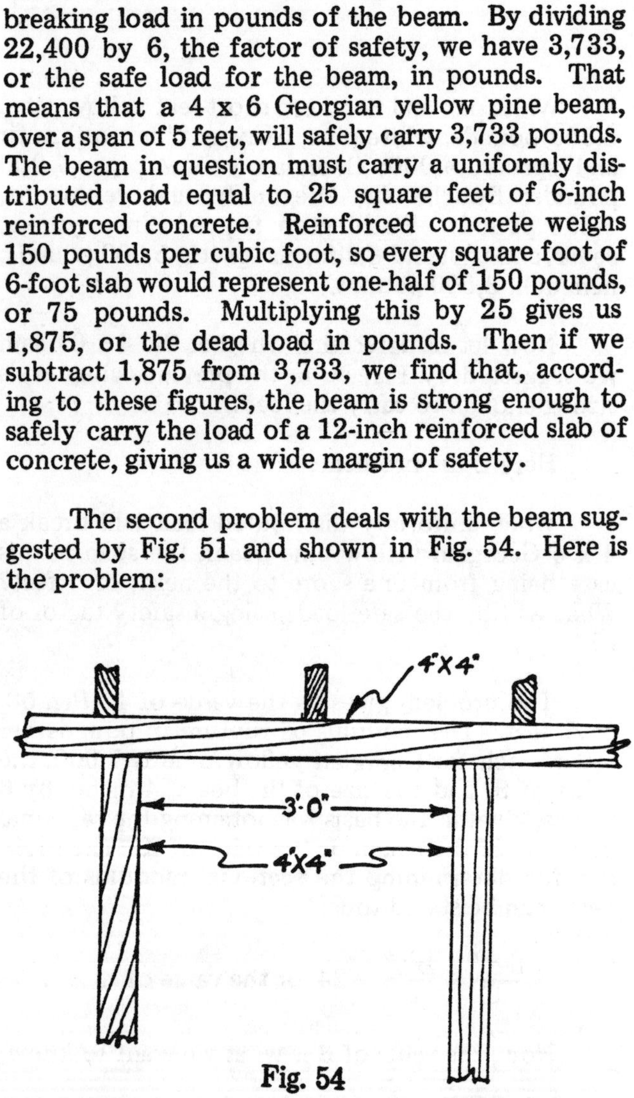

Fig. 54

What uniformly distributed load will break a 4 x 4 Georgian yellow pine beam, the span in this case being from one shore to the next, or 3 feet? What will be the safe load, using a safety factor of 6?

The first thing we want to know is the sectional modulus, which we obtain by multiplying the length of the beam by the depth of the beam, squared, and dividing by 6, or the problem could be expressed:

$$\frac{b\ d^2}{6} = \frac{4x4^2}{6} = 11, \text{ the sectional modulus.}$$

The value of B is what we want to know;

the value of K is 11;

the value of S is given in the list, or 7,000;

the value of L is given in the problem, or 3.

Proceeding just as we did before, we have:

$$B = \frac{2\ K\ S}{3\ L} = \frac{2x11x7{,}000}{3x3} = 17{,}111, \text{ or the}$$

breaking load in pounds of the beam.

Dividing 17,111 by 6, the factor of safety, will give us 2,852, or the safe load in pounds for the beam. The uniformly distributed dead load the beam must carry is 1,875 pounds, which leaves 977 pounds, or a wide margin of safety.

Explaining the formulas:

What does $\frac{2\ K\ S}{3\ L}$ means?

It means the same as 2 K S divided by 3 L. If this is true in the formula, then it must be equally true in the problem after it has been written out in figures. Take the first problem of this chapter: $\frac{2x24x7{,}000}{3x5} = 22{,}400$, means the same as 2x24x 7,000 divided by 3x5. By using cancellation, 3 will go into 24, 8 times, canceling 3 and reducing 24 to 8. And 5 will go into 7,000, 1,400 times, canceling 5 and reducing 7,000 to 1,400. Now the problem stands: $\frac{2x8x1{,}400}{1x1}$, which would be the same as 2x8x1,400, or 22,400.

Thus far we have shown how to compute the strength of shores and beams, but we have said nothing of joists. Joists, in fact, are small beams, and so the same rule for determining their strength will apply.

An example:

What safe uniformly distributed load will a 2x8 Georgian yellow pine joist carry over a span of 8 feet, using a safety factor of 6?

Let us determine the sectional modulus:

$\frac{b\,d^2}{6} = \frac{2x8^2}{6} = 21.$

In this problem:

The value of B, is what we want to know; the value of K is 21; the value of S is 7,000, the modulus of rupture of Georgian yellow pine, according to the list given in the forepart of this chapter.

The value of L, as given in the problem, is 8.

Using the formula we used before:

$$B = \frac{2\ K\ S}{3\ L} = \frac{2x21x7{,}000}{3x8} = 12{,}250,$$ or the

breaking load. Dividing 12,250 by 6, the factor of safety, we have 2,042, or the safe uniformly distributed load a 2x8 Georgian yellow pine joist will carry. If the joists are spaced 2 feet on center, then the joist would carry the weight of a slab 2 feet wide and 8 feet long, or the equivalent of 16 square feet of slab. If the slab is 6 inches thick, it would represent 75 pounds for every square foot, or 1,200 from 2,042, we have 842 pounds left, a wide margin of safety, and unless an extremely heavy live load comes onto this construction, the form builder will not have to worry.

We have been dealing with problems relative to computing the strength of beams; now we will reverse the order, and determine what size beam is necessary to safely carry a known load. Take this example:

How wide, in inches, will a Georgian yellow pine beam have to be, assuming its depth as being 12 inches, in order to support a uniformly distributed load equal to 600 pounds per lineal foot, the span being 15 feet, and using a safety factor of 6?

In this problem we have assumed the depth of the beam, and it remains for us to determine the breadth. It is necessary in solving this problem that we find the maximum bending moment in inch-pounds. This formula will give us foot-pounds:

$M = \frac{W\ L}{8}$, in which,

M represents maximum bending moment in foot-pounds;

W represents load of beams;

L represents length of beam, in feet.

The load per lineal foot, or 600 times 15, the number of feet in the span, equals 9,000, or the total load.

Now,

$M = \frac{W\ L}{8} = \frac{9{,}000x15}{8} = 16{,}875$, or the maximum bending moment in foot-pounds. Multiplying 16,875 by 12, gives us 202,500, or the maximum bending moment in inch-pounds.

The depth of the beam, as we assumed in the problem, is 12 inches. By determining the sectional modulus of 1 inch by 12 inches, and multiplying it by the modulus of rupture, 7,000, divided by a safety factor of 6, we would have the maximum bending moment in inch-pounds for 1 inch of the width of the beam. Or the problem could be stated: $\frac{b\,d^2}{6} = \frac{1x12^2}{6} x \frac{7{,}000}{6} = 28{,}000.$

There is a maximum bending moment of 202,500 inch-pounds to be provided for. The maximum bending moment for 1 inch of the breadth of the beam in question, is 28,000; so by dividing 202,500 by 28,000, we will have 7, plus, or the breadth of the beam, in inches. Therefore, an 8-inch by 12-inch Georgian yellow pine beam will safely carry 600 pounds per lineal foot.

What we have been doing is illustrated by Fig. 55. We assumed the depth of the beam, 12 inches, which is shown in figures to the left of the diagram. The two indefinite dotted lines, one at the top and one at the bottom, from a to *b*, indicate an unestablished idea of a beam. We found that the maximum bending moment was 202,500 inch-pounds, that is what the whole beam must support.

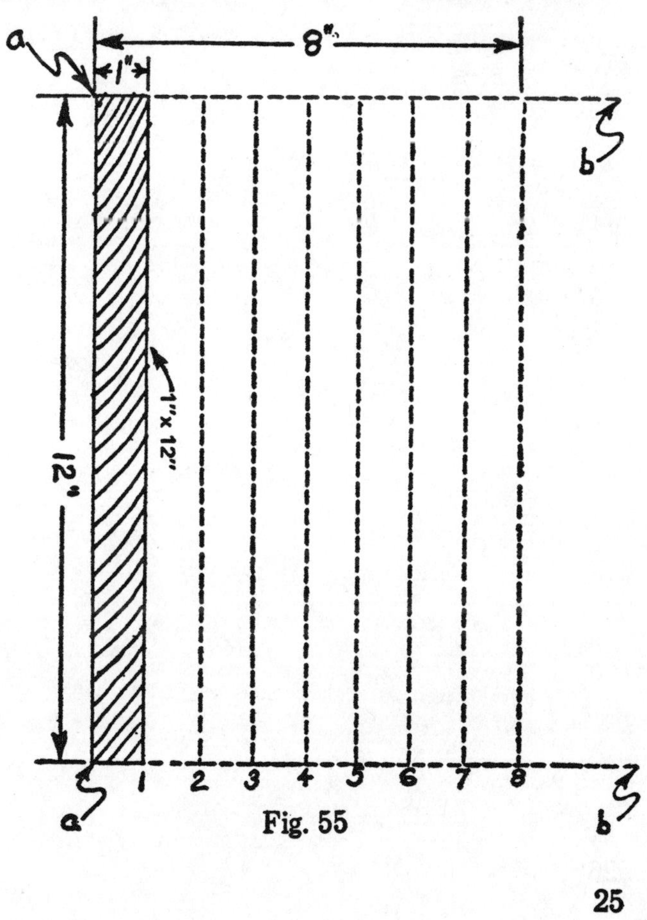

Fig. 55

Then we found how many inch-pounds the 1-inch by 12-inches, shown to the left on the diagram and shaded, would support, which is 28,000 inch-pounds.

Now if the total load in inch-pounds is 202,500, and a part of the beam equal to 1-inch by 12-inches will support 28,000 inch-pounds, then there will have to be as many inches in the breadth of the beams, as 28,000 is contained in 202,500, or 7, plus; let's say, to be safe, 8 inches. Now we have a full-sized beam of 8-inches by 12-inches, which will safely carry a uniformly distributed load equal to 600 pounds per lineal foot.

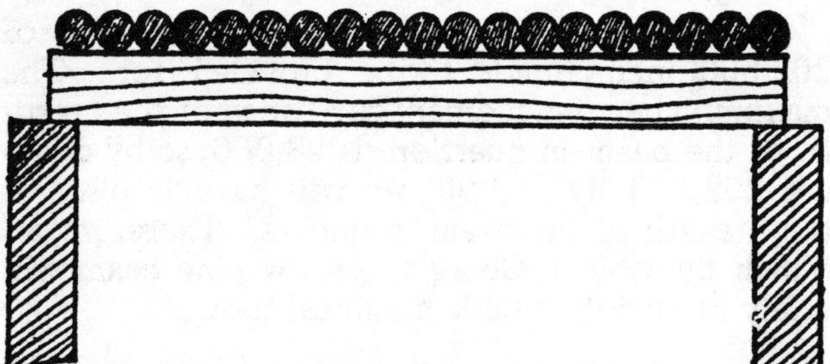

Fig. 56

What is meant by a uniformly distributed load, is shown by Fig. 56. The shaded disks represent weight, and the combined weight of all of the disks constitutes the load of the beam. This kind of load is what the form builder usually deals with. Fig. 57 shows a beam loaded with a non-uniform load. Such loads the form builder must provide for occasionally, but not as a general rule. Reinforcing the beams, either with extra shores, or by increasing the size of the beams, or by placing the beams closer together, will take care of such loads. The large disks could easily represent piles of sand or cement, and the smaller ones could mean other building material of lighter weight.

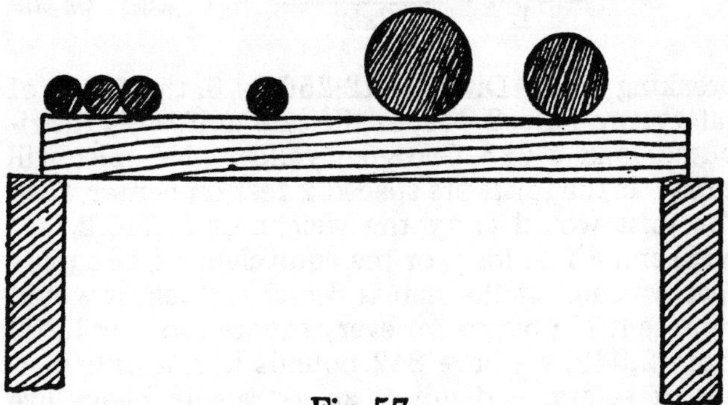

Fig. 57

6

Column Forms and Other Forms

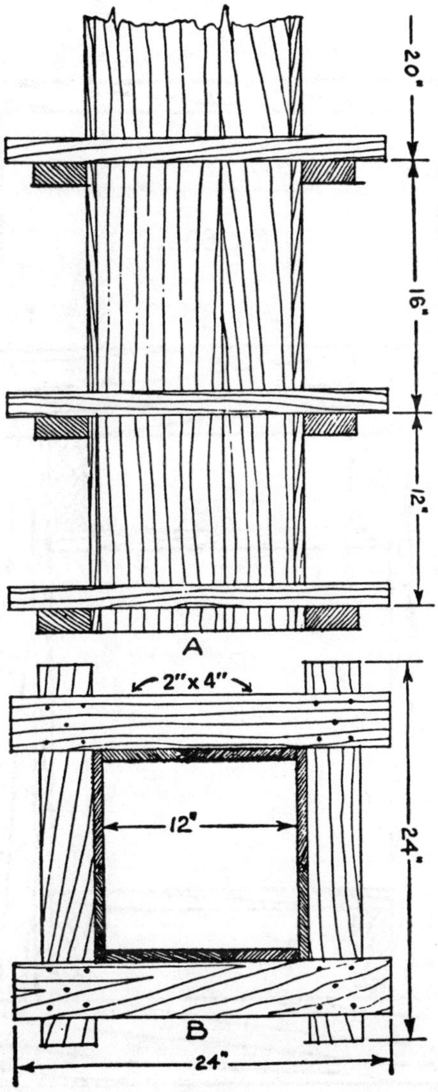

Fig. 58

Fluctuating Pressure

It is a demonstrable fact, that a perpendicular sheet of water one inch thick has the same outward pressure per square inch as a sheet of water much thicker. This, at first, seems unreasonable, but it is true.

It does not take any more force to hold a thick sheet of water in place than it does to hold a thin sheet of water in place. That is provided the water is perfectly still, but whenever the water is disturbed in any way, the pressure fluctuates. The thicker the sheet of water, the greater will be the pressure fluctuation. The outward pressure, though, is always greatest at the bottom, and gradually decreases all the way to the top.

What is true of water is more or less true of unset concrete. If concrete could be placed in the forms without disturbance, the forms for a 2-foot wall would not have to be braced or tied any more than the forms for a 4-inch wall. The difference lies in the fact that a wheelbarrow load of concrete poured into the forms for a 2-foot wall at once, would cause a much greater outward pressure per inch, than dropping the same amount of concrete into a 4-inch wall. Also, green concrete falling 8 to 10 feet will produce a greater outward pressure than if it fell only 8 or 10 inches. Thus we can clearly see that the pressure is increased at the bottom, where it is already the greatest, and gradually decreases all the way to the top.

This being true, it must follow that it will require a greater amount of bracing and wiring at the bottom than it will at the top, in order to resist the outward pressure of the concrete. For instance, the bottom part of a form for a column is shown by Fig. 58, *A*, and it will be noticed that the bands around the column are spaced, respecttively, 12-inches, 16-inches and 20-inches. Each space is increased 4-inches, but the increasing of the spaces should stop when they reach 24-inches.

What the form builder should strive for is to have his forms uniformly substantial, rather than have the bands or braces uniformly spaced. The latter leaves the forms too weak at the bottom, and overly braced or tied at the top.

Column Forms

At *B*, Fig. 58, we show a plan of the column form shown in elevation at *A*. This method of building forms for columns is substantial. The 2x4's used for the bands should be cut to a length that will hold the salvage value at a maximum. The pieces are shown cut to a length of 24 inches, but there are other lengths that will salvage well. In fact, circumstances under which the pieces are to be reused will have to be taken into consideration.

The nailing at the angles, indicated by the heavy dots, should be substantially done, but the nails should not be clinched, for then they can be driven back and pulled with a wrecking bar. The form builder should cultivate a habit of holding the salvage value of the material at a maximum.

Another method of holding the bands of column forms together, is shown by Fig. 59. Here

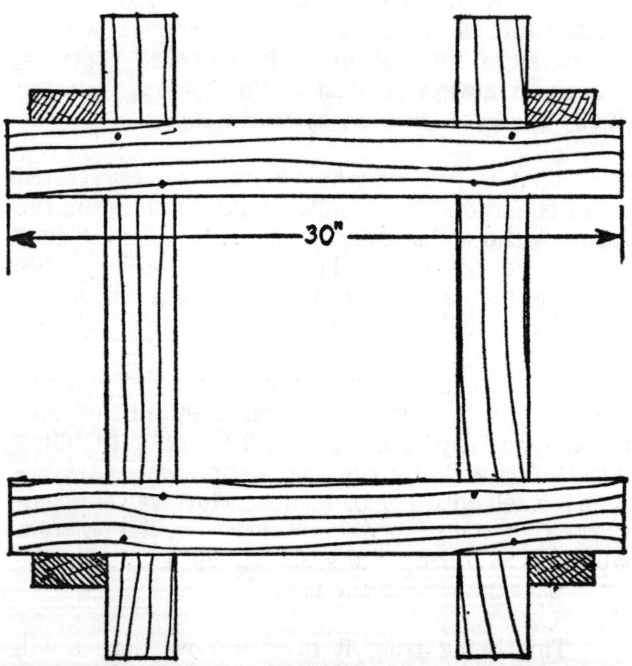

Fig. 59

the nailing at the angles is light, but at each corner there is a 2 x 4 nailed into the angle of the lugs, which keeps the joint from spreading. The pieces,

Fig. 60

Fig. 61

in this case, are cut long enough so that the lugs will receive the 2 x 4 in the angle and provide nailing for it.

A method of building column forms with the boards running horizontally is shown by Fig. 60, which is a floor plan of the form. Fig. 61 shows how this form can be reinforced with bands. These bands should be spaced to meet the requirements of the case in hand. A long column form, quickly filled, must resist more pressure than a short one. Again, a form that is filled slowly does not have to resist as much pressure as one that is filled quick-

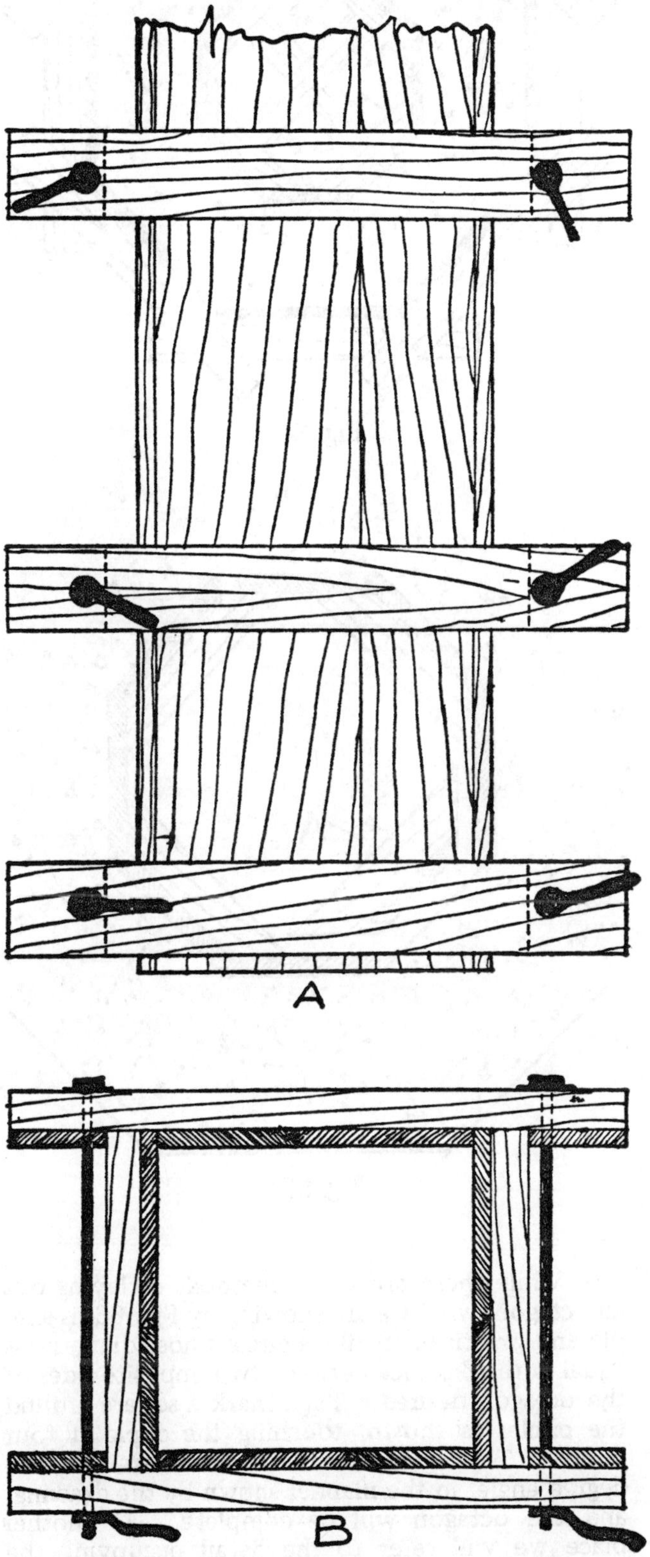

Fig. 62

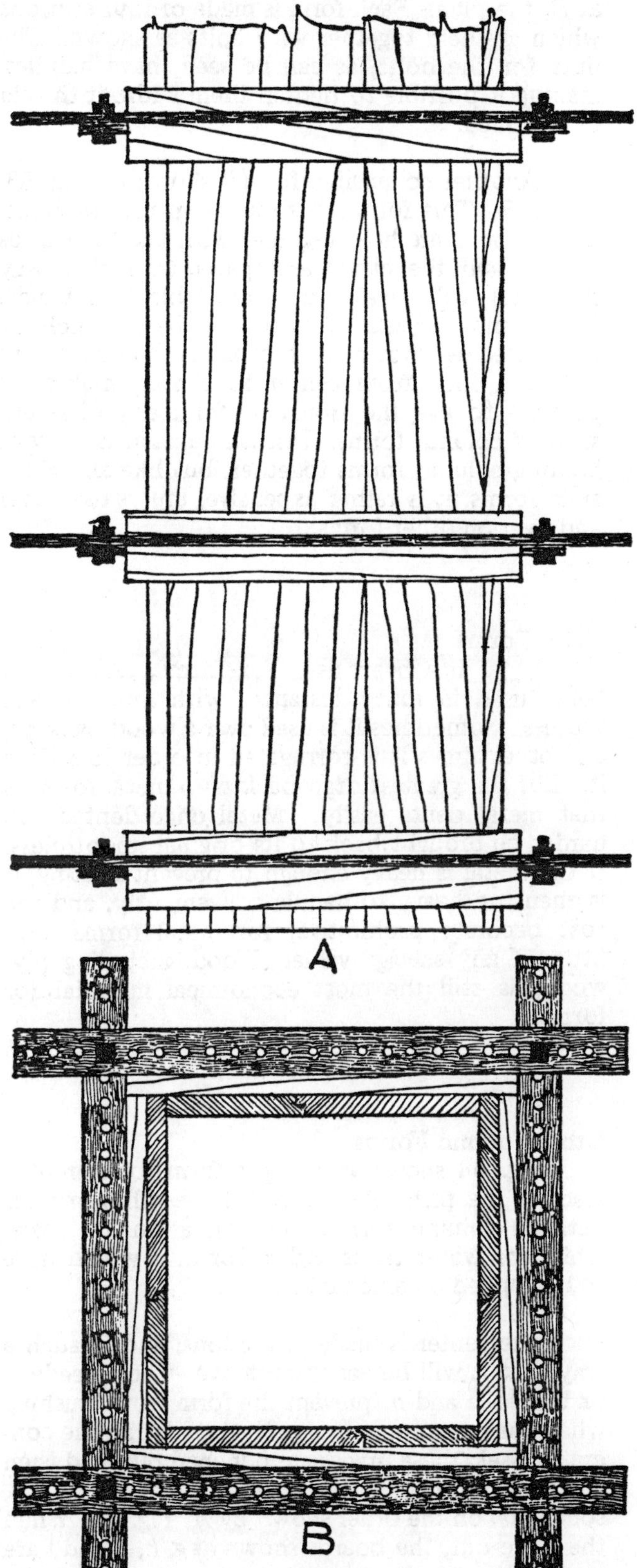

Fig. 63

ly, for the concrete sets almost as fast as the pouring is done. These things cannot be governed by fixed rules, but they must be determined by the form builder's judgment.

A substantial collapsible form for columns, is shown by Fig. 62. At *A* is shown the elevation, and at *B*, the plan. Each form is made of four sections which are held together with bolts as shown. The nuts for the bolts, as can be seen, have handles, making it possible to tighten them without the aid of a wrench.

Another collapsible form is shown by Fig. 63, *A* and *B*. This form is also made in four sections. The sections are held together with 1 x 4 cleats, as shown, and the cleats are spaced in such a way that they will receive the adjustable iron bands. Each band consists of four iron bars which are perforated as shown by the plan. The holes are spaced 1 inch from center to center, making it possible to use the iron bars for many different sizes of column forms. This is a substantial way of holding column forms together, but like all collapsible forms, it is rather expensive, unless used over and over on other forms.

Metal Forms

Metal has been used for forms, and is still being used in many instances with more or less success. Sometimes it is used over a wood backing, and other times it is corrugated in order to stiffen it. But the greatest draw-back with metal forms is that metal dents easily. Metal once dented can hardly be brought back to its original smoothness. If the metal is heavy enough to prevent denting, it is then too heavy to handle conveniently, and the cost becomes prohibitive, for metal forms have little, if any salvage value. Wood, including plywood, is still the most economical material for forms.

Other Column Forms

Fig. 64 shows an octagon form for a smokestack. The principle involved here will work for octagon column forms, only on a smaller scale, while for water tanks and so forth, it would have to be applied on a larger scale.

The center is made in sections, and in such a way that it will be easy to remove. The spreaders, or braces, *a* and *b*, prevent the form from crushing when the pressure comes against it. When the concrete is set, these braces are punched out, and then the four sections, *c*, *d*, *e*, and *f*, are taken out, somewhat on the order shown by *A*, Fig. 65. When these are out, the boards shown at *g*, *h*, *i*, and *j* are removed.

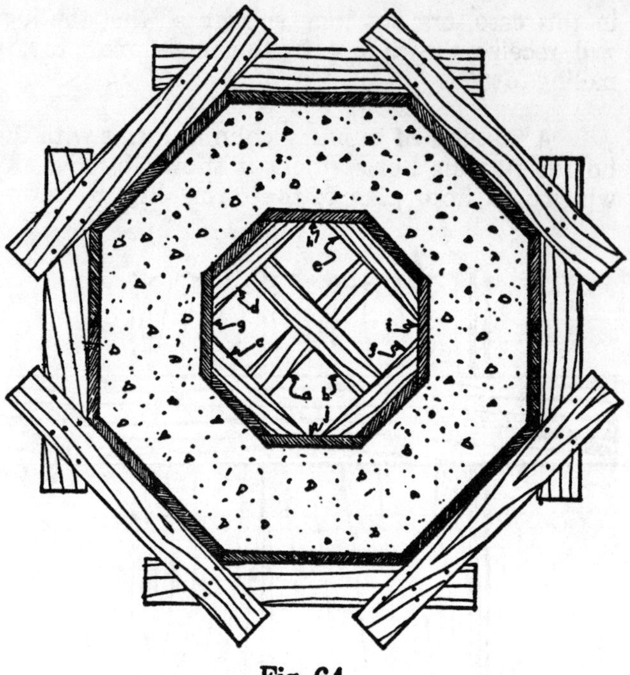

Fig. 64

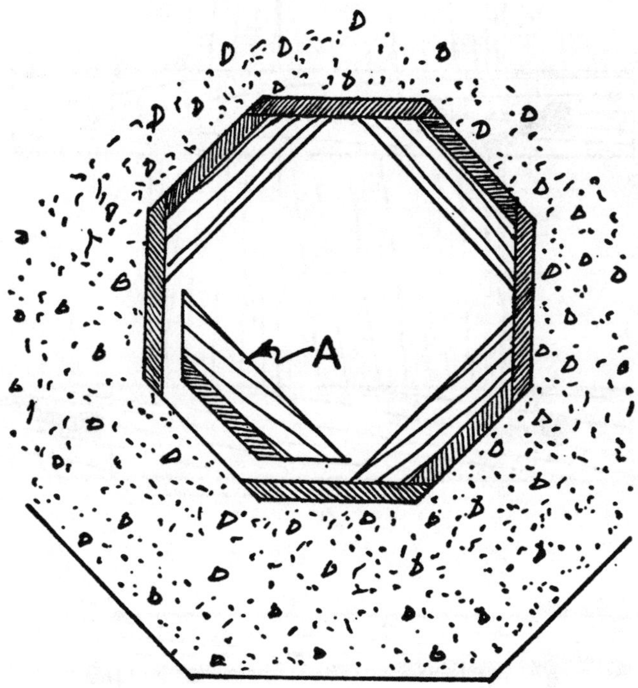

Fig. 65

While there are many methods of laying out an octagon, what we are showing by Fig. 66, is simple and practical. Strike a circle whose diameter is equal to the distance between two opposite sides of the octagon desired. Then mark a square around the circle, as shown, touching the circle at four points. This done, cut off the corners at a 45-degree angle, in the manner shown by the drawing, and the octagon will be complete. At another place we will refer to the detail occupying the space within the circle.

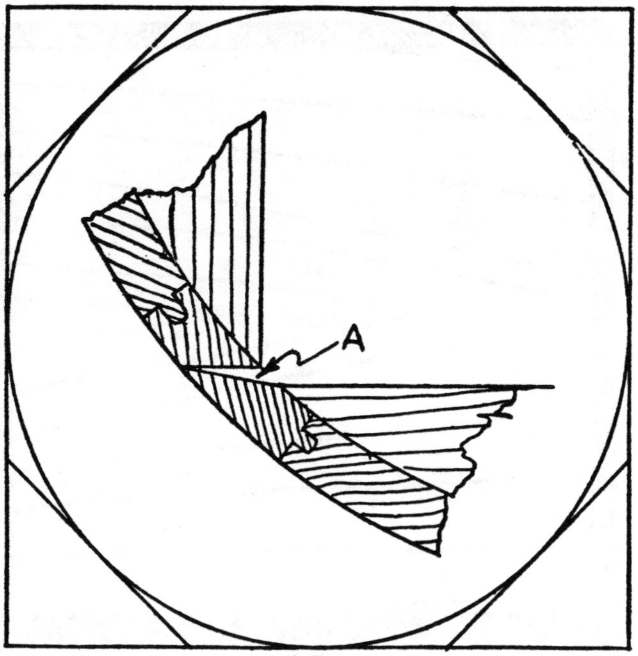

Fig. 66

Fig. 67

The construction of a form for a circular smokestack, is shown by Fig. 67. In this, as in the case of the octagon form, the principle will apply to columns, but on a smaller scale, and to water tanks, etc.

The form for the flue is made in four sections, held together by the pieces pointed out at *a, a, a, a* and braced by the spreaders marked *b* and *b*. The detail shown within the circle of Fig. 66, shows at *A* the construction of the joints marked *c*. After the concrete has set, and the form is to be taken out, the braces marked *b*, and the pieces marked *a* are punched out first, and then the section shown to the left of the drawing will come out somewhat on the order indicated by the dotted part-circles. Fig. 68 shows this form with the braces all out and the right and left sections partway removed. The other two sections, it can readily be seen, will come out easily.

We are showing by Fig. 69 *A*, what is perhaps the best method of forming for flues, which is an elevation. At *B* the plan is shown. Here can be seen how lath are nailed to a spreader board. The laths should be kept apart around 1-half of an inch, and the spreader boards should be spaced close enough to keep the laths from bending. When the laths are on, the skeleton is covered with a good grade of waterproof building paper, and the form is ready to be set. The paper is shown partly removed, on the elevation. The dotted lines show the location of the spreaders. The plan shows a square flue, but it will work just as well on polygons and circles.

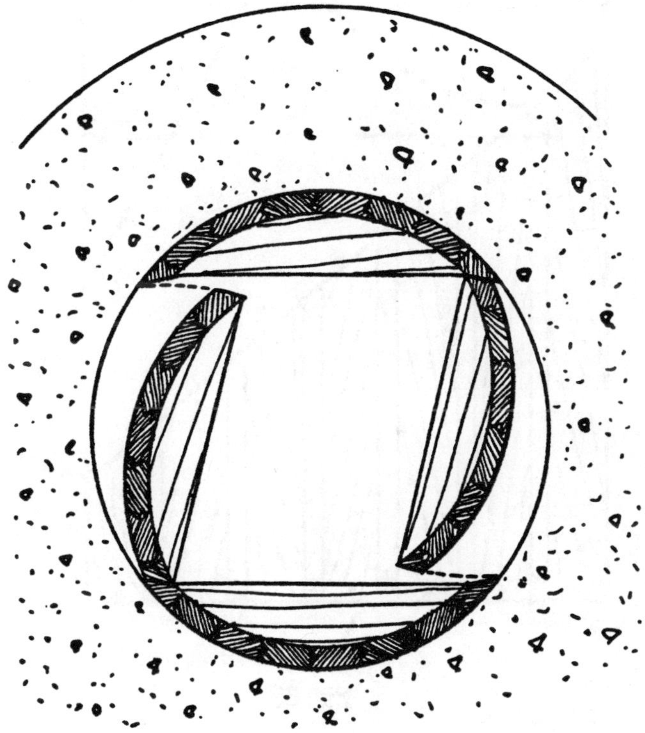

Fig. 68

Miscellaneous Forms

Two views are shown of what we call permanent forms by Fig. 70. They are forms that are not removed when the concrete is set. They are used in places such as entrances to stores, bathroom floors, hearths for fireplaces, or wherever a concrete slab is permanently supported by wood joists.

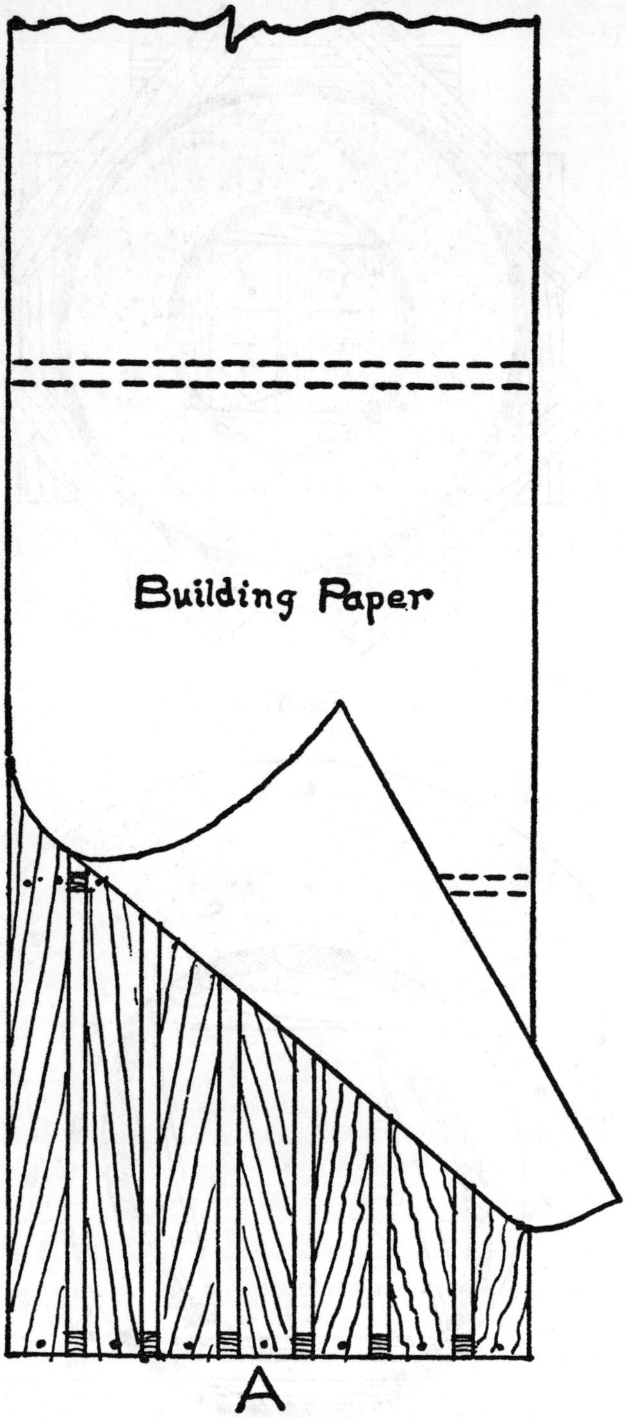

Fig. 69

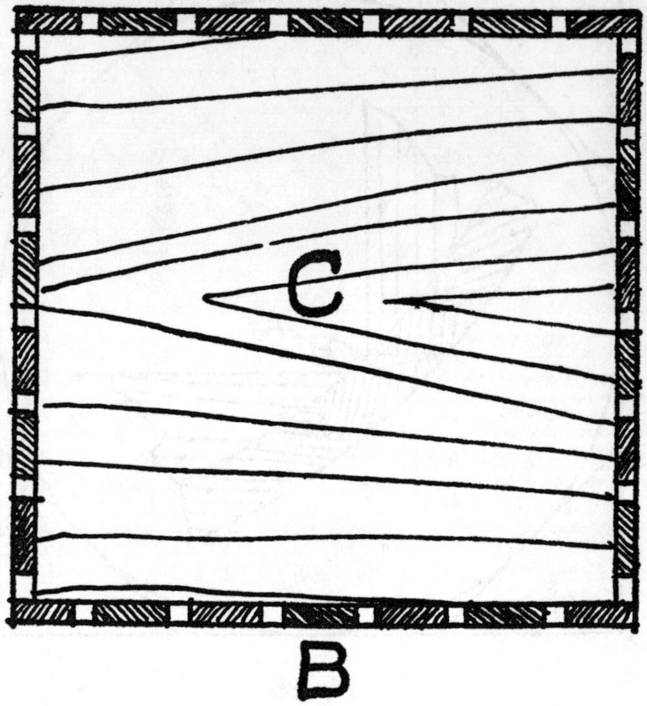

Fig. 69

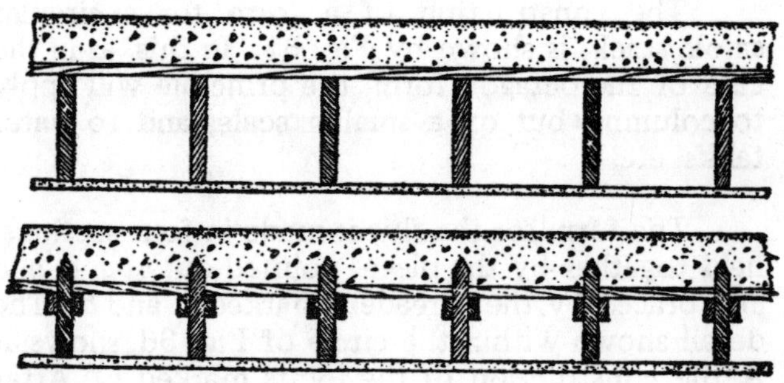

Fig. 70

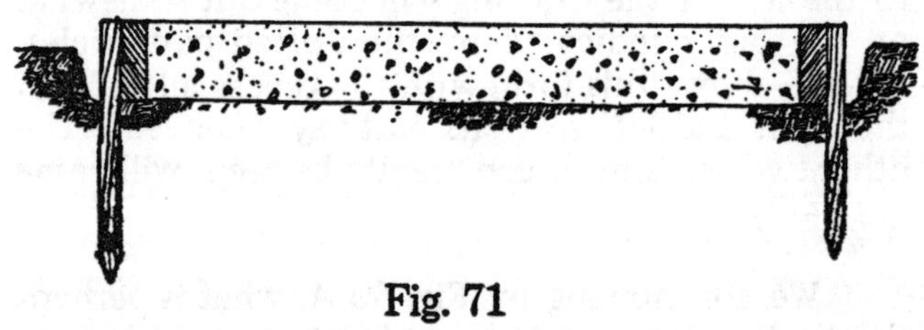

Fig. 71

The method shown by the upper drawing is the best, for it leaves the concrete slab as well as the joists unimpaired. But this method requires more room than is often available, and when that is the case the joists are chamfered on the upper edge somewhat as shown by the bottom drawing, and the forms are placed between the joists, resting on cleats nailed to the joists, as indicated by the illustration.

Sidewalk Forms

A cross section of forms for a sidewalk is shown by Fig. 71. Such forms are usually built either with 2 x 4's or 2 x 6's, and the stakes can be 2 x 2's, 1 x 3's, or 1 x 4's. Fig. 72 shows a plan of a sidewalk, in which the forms have been built in such a way that every other block of the walk is poured and finished, and when the concrete is set, the end forms are removed and the omitted blocks of walk are poured and finished. This method prevents the walk from cracking, excepting at the joints. Of course, provision must be made for ex-

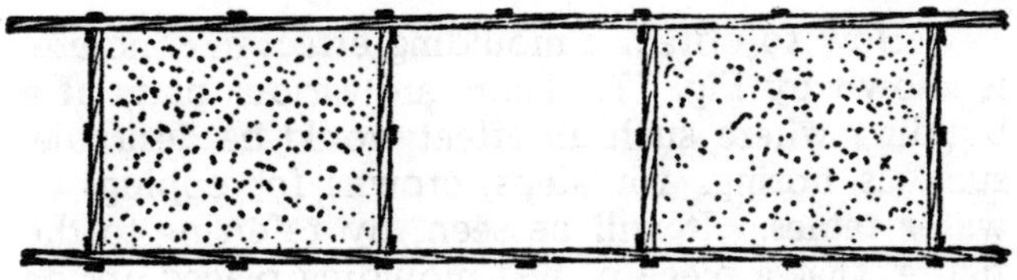

Fig. 72

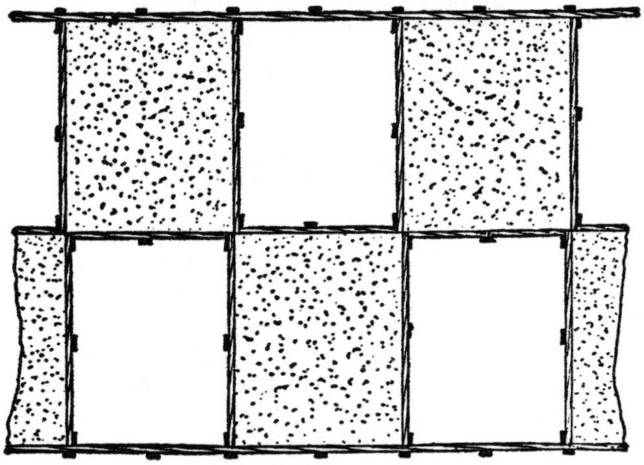

Fig. 73

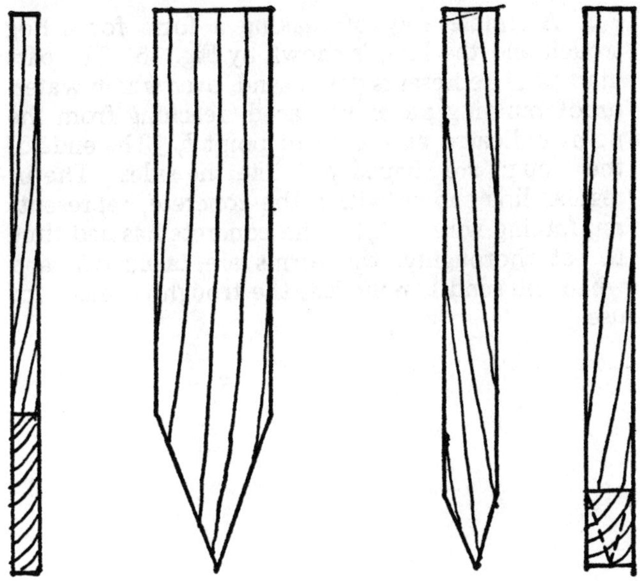

Fig. 74

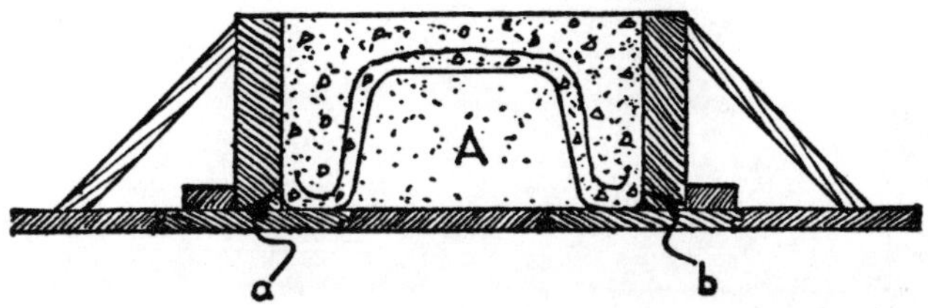

Fig. 75

pansion and contraction. Usually the cross forms are not used, and the whole walk is poured, as it were, at once. When this is done, the joints are cut into the concrete while it is being finished.

How to place the forms for a sidewalk wide enough to require a double row of blocks, is shown by Fig. 73. The dotted sections represent blocks that have been poured and finished, while the clear sections show the blocks that must be poured after the inside forms have been taken out.

In building sidewalk forms it is important to have straight, smooth and unwarped material, and the forms must be set to a grade in order to conform with the lay of the ground, as well as to provide drainage for the walk. Ordinarily, a stake every 4 or 5 feet is sufficient to hold sidewalk forms in place. The ends of the stakes should not project above the upper edge of the forms, so that in spreading the concrete the forms will answer for screeds. The length of the stakes depends on the condition of the soil. If the soil is wet and soft, the stakes should be longer than in cases where the soil is firm.

Details of two kinds of stakes, are shown by Fig. 74. At the left are shown two views of a 1 x 4 stake, and how it should be sharpened. To the right are shown two views of a 2 x 2 stake, which can either have a chisel point, as shown by the continuous lines, or a pick point, as indicated by the dotted lines on the detail to the right.

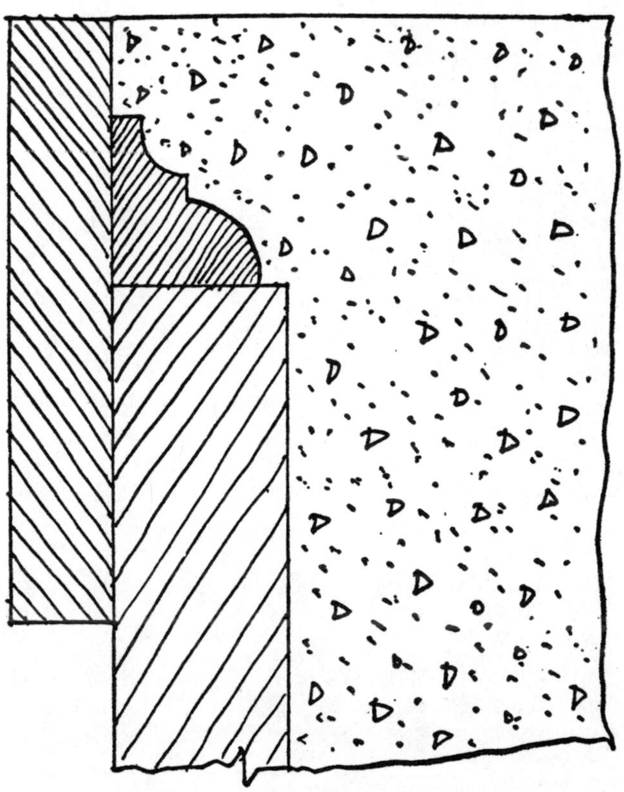

Fig. 76

A simple way of making a form for a hog trough and the like, is shown by Fig. 75. The part marked *A* represents damp sand, over which waterproof building paper is placed, reaching from the point indicated at *a*, over to point *b*. The ends of the trough are shaped just like the sides. The irregular line shown within the concrete, represents reinforcing wires. After the concrete has had time to set thoroughly, the forms are taken off, and when the sand is removed, the trough is ready for use.

How to obtain a moulding effect in concrete, is shown by Fig. 77. There are various parts of a building where such an effect would be desirable, such as nosings for steps, crowns for copings or water tables. It will be seen, by referring to the figure, that a piece of bed moulding placed upside down on a piece of 2-inch material, and a piece of 7/8-inch board, nailed to the outside, projecting above the moulding, produces the effect. In this way, the form builder, by carefully selecting the moulding, can obtain many and pleasing effects.

7

Sills and Fireproof Construction

Mouseproof Job

One of the things that every builder should keep in mind from the very beginning of a job, and especially a residential job, is to make the building mouseproof. A mouseproof job does not cost any more than a job that is not, excepting that it takes a little more care and attention in choosing the form of construction. When the work is done by contract, and there are plans and specifications, then the builder has little choice in the matter, for the architect has established the form of construction to be used, both by his drawings and by his specifications.

Sills

Here we take up problems that pertain more nearly to residential work. Of course we are not going to confine ourselves strictly to that, for some of the projects, in order to be thorough, will make it necessary for us to go outside of this branch of carpentry, but in the main, we will deal with residential problems.

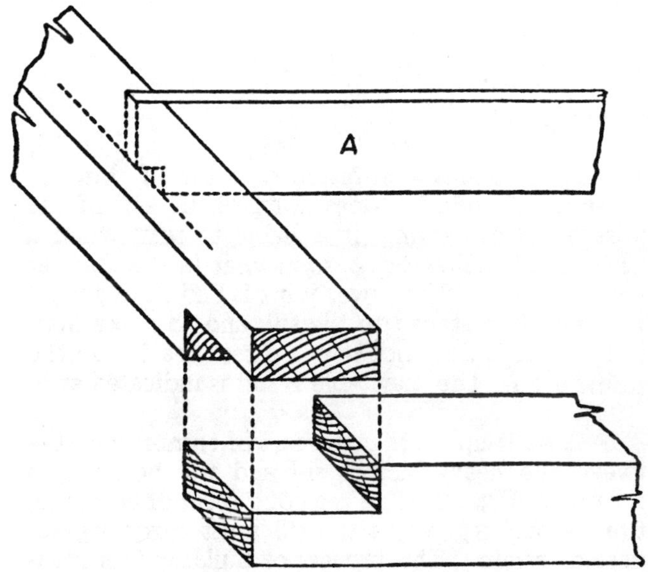

Fig. 77

The first thing above the foundation that the carpenter has anything to do with, is the sill, and we shall show in this chapter a number of sill constructions. Fig. 77 shows a perspective view of a corner of a heavy-timber sill. The joint shown is a half-lap joint, but in many instances mortise-and-tenon joints are used. At *A* is a joist and to the left, by dotted lines, is shown how the joist is fastened to the sill. Mortise-and-tenon joints are also used in fastening joists.

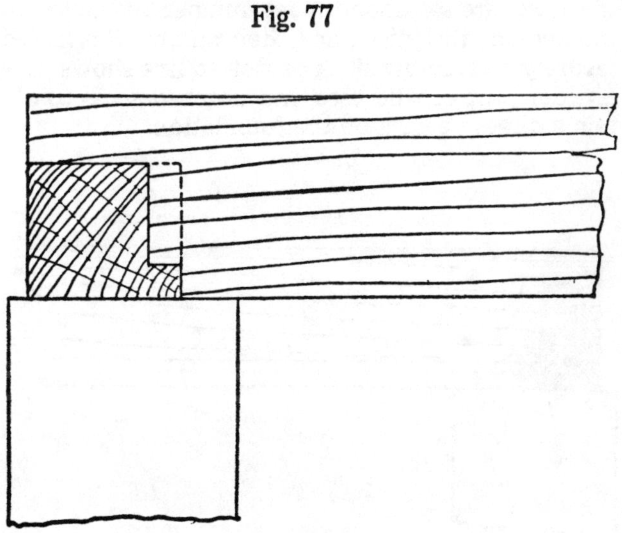

Fig. 78

Another heavy-timber sill is shown in cross section by Fig. 78. This joint between the sill and the joist, is substantial. It will be noticed that the joist is well fastened to the sill, but besides that, the joist has a direct bearing on the foundation, as can be seen by referring to the drawing. The heavy timber sill, though, is rarely, if ever used in our time on residential work.

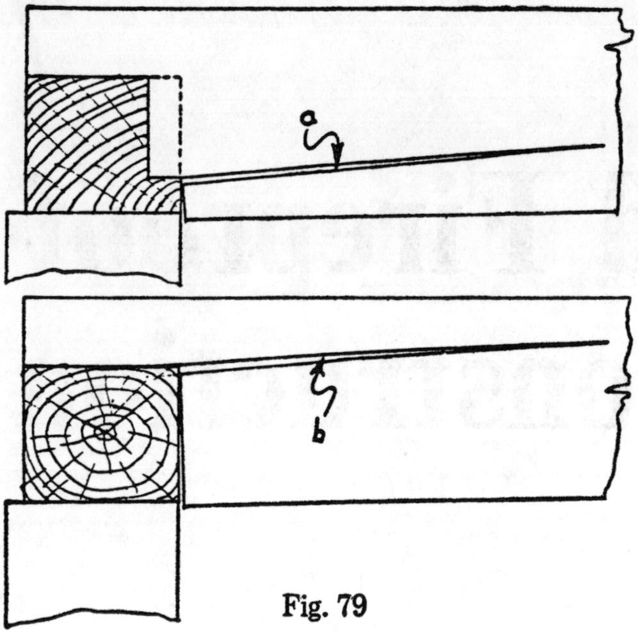

Fig. 79

Two bad constructions are shown by Fig. 79. The upper one is well fastened to the sill, but because the foundation does not give the end of the joist a direct bearing, it is liable to split when a heavy load comes on it, somewhat in the manner indicated at *a*. The lower joint is bad throughout. It is poorly fastened to the sill, and to make matters worse, it does not have a direct bearing on the foundation. The inevitable result is indicated at *b*.

A built-up sill that is a sort of compromise between the heavy timber sill and the box sill, is shown by Fig. 80. This is a good form of construction, especially where the sill must carry a great deal of strain. The process of building this sill is this: After the timber pointed out at *1* is in place, the joists are set, then timber number *2* is nailed to the ends of the joists, and then number *3* is nailed securely to number *2*. The dotted line shows how a wider joist can be used and notched so as to obtain a direct bearing on the foundation.

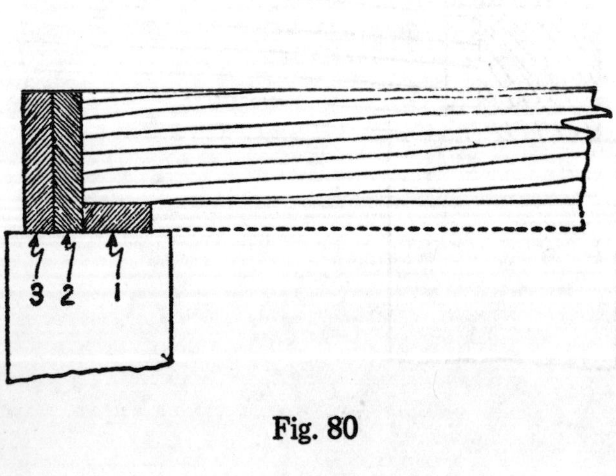

Fig. 80

In cases where there is danger of the foundation settling unevenly, heavy sills are necessary to keep the superstructure from losing its shape, but when foundations are made of reinforced concrete, heavy sills are not altogether necessary. Simple sills, such as shown by Fig. 81, are sufficient where the foundations do not settle. The upper sill, as can be seen, is merely a plank laid on the foundation, onto which the joists are fastened. The lower one, is a simple box sill. Both of these constructions can be called mouseproof. However, the upper one has a weak point at *a*, where, if the joint should be somewhat open, mice will gnaw holes through into the rooms above.

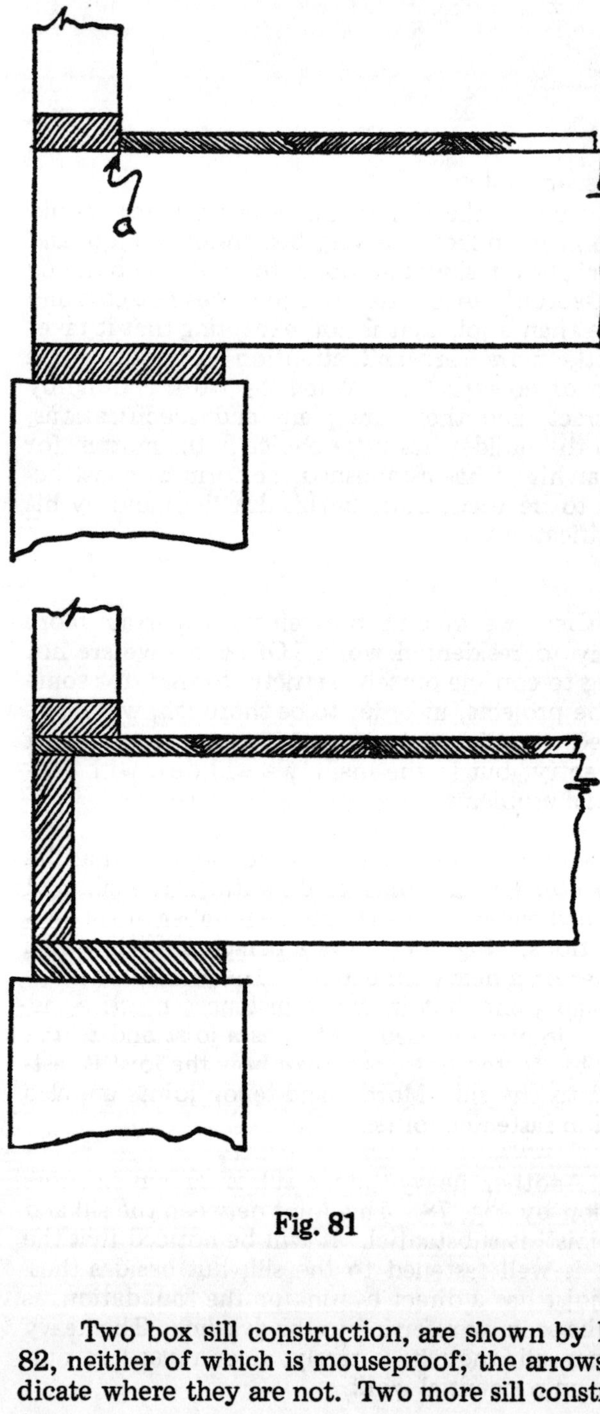

Fig. 81

Two box sill construction, are shown by Fig. 82, neither of which is mouseproof; the arrows indicate where they are not. Two more sill construc-

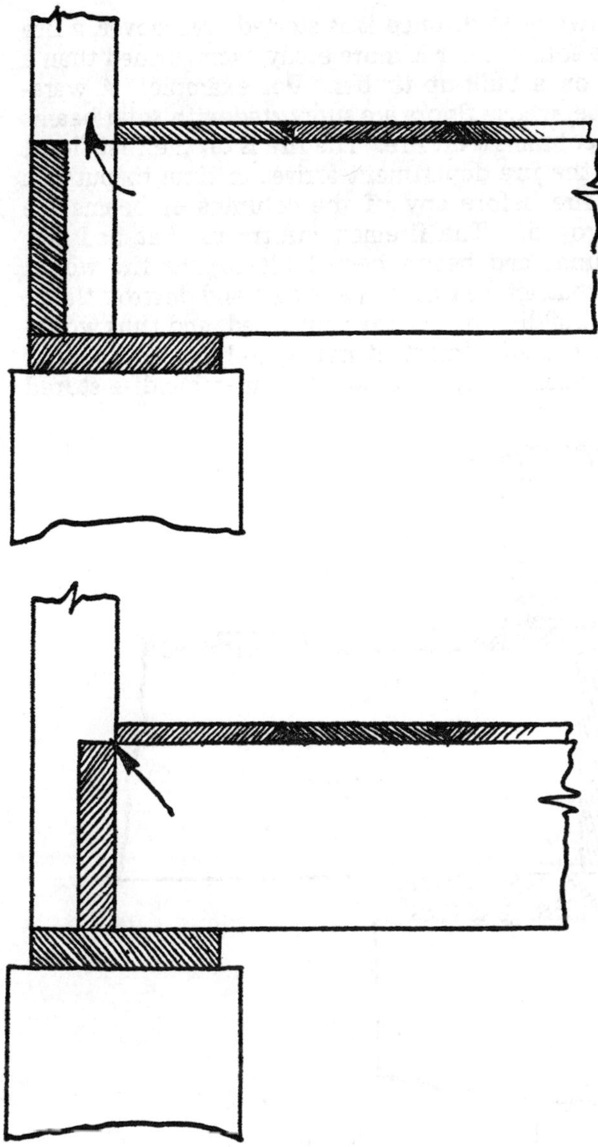

Fig. 82

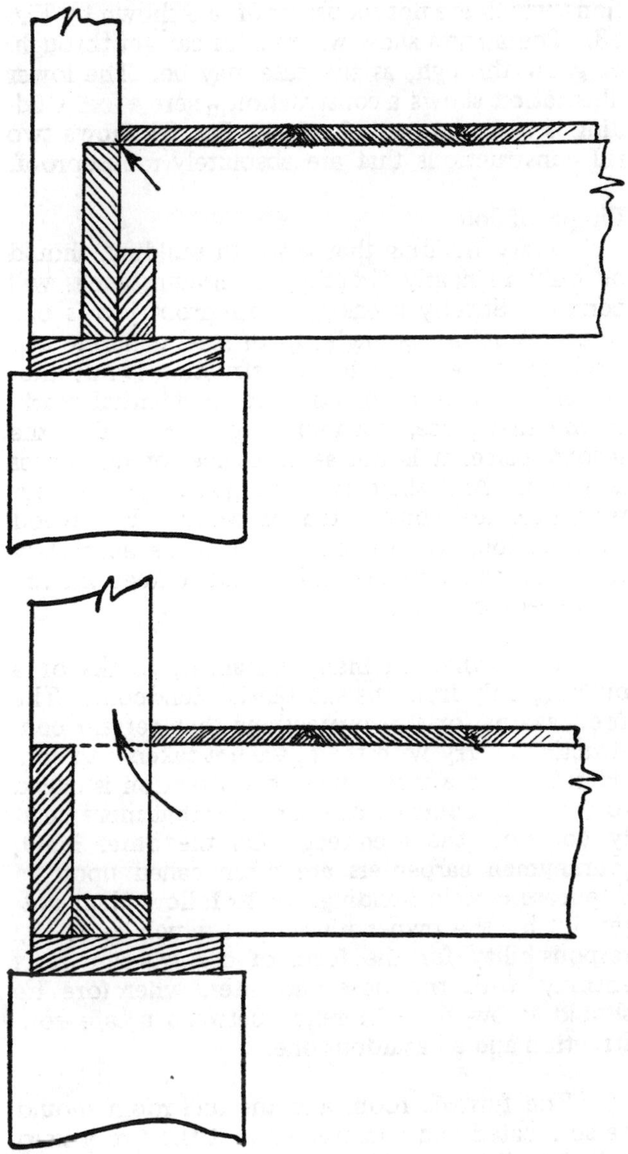

Fig. 83

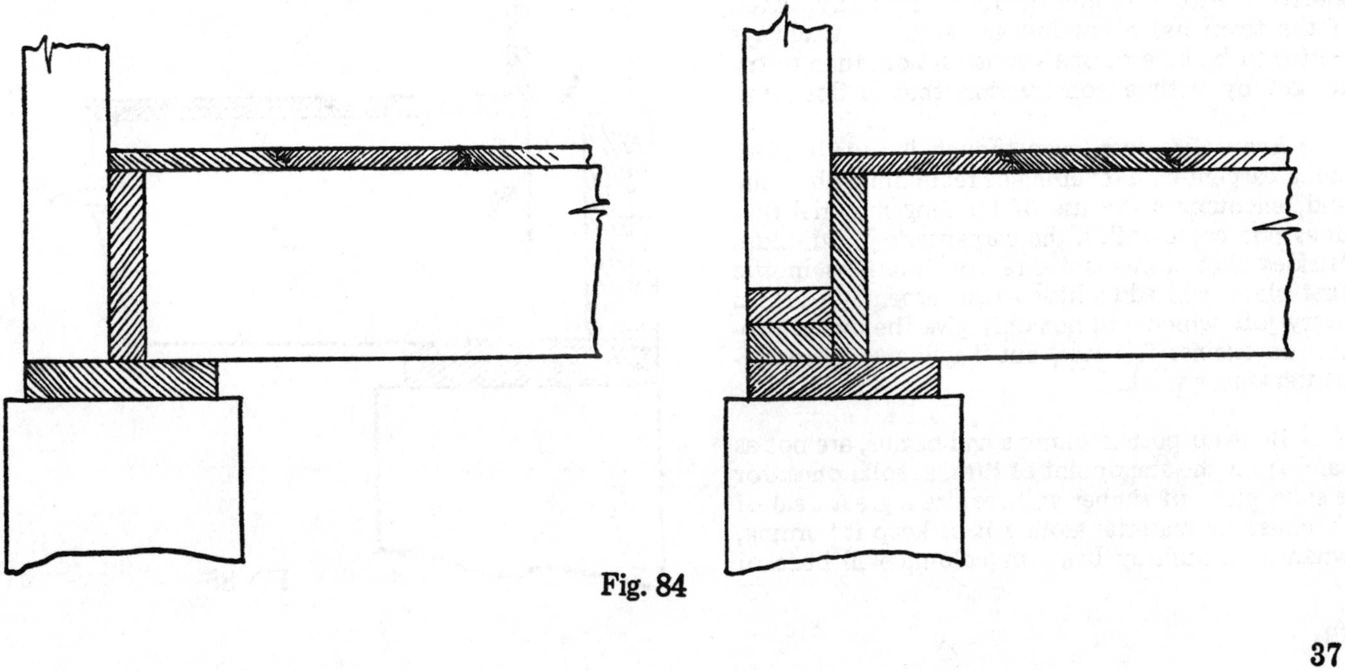

Fig. 84

tions, which are not mouseproof, are shown by Fig. 83. The arrows show where mice can get through, or gnaw through, as the case may be. The lower illustration shows a construction where a 2x6 studding is used, instead of a 2x4. Fig. 84 shows two sill constructions that are absolutely mouseproof.

Fireproof Job

Every building that is worth building, should be built as nearly fireproof as circumstances will permit. Strictly speaking, a fireproof job is one whose structure is made up of steel concrete and brick or stone. But this construction is not altogether desirable or practical for residential work. In the first place, it is more expensive, and in the second place, it is not as adaptable for residences as wood. And therefore, the greater part of our residences are constructed of wood. But wood construction, while it cannot be made absolutely fireproof, can nevertheless be made to keep the fire hazard at a minimum.

The owner, in many instances, thinks of a building only from the substantial standpoint. The fire hazards, or fire protections that certain constructions carry with them, are not taken into consideration, as a rule, unless his attention is called to them. Of course, this is a matter that more nearly concerns the architect. On the other hand, journeymen carpenters are often called upon to duplicate certain buildings, or to follow plans gotten up by the owner himself, in which cases the responsibility for the form of construction rests entirely with the boss carpenter, wherefore he should know the difference between a safe construction and a hazardous one.

The furnace room and the fuel room should be so located and constructed that the fire hazard will, for all practical purposes, be eliminated. The constructions around chimneys should always be anti-fire constructions, and carpentry work around electrical wiring, should be done cautiously. Even if the town has a building inspector, it is always better to be sure of one's construction, than to try to get by with a construction that is doubtful.

Every fire loss, even though it usually gives some carpenters a repairing or rebuilding job in the end, encourages the use of building material that does not come within the carpenter's jurisdiction. Besides that, using anti-fire constructions in the first place, will add a little extra carpentry work to every job, which will not only give the owner protection against fire loss, but it will give more carpenters more work.

Built-up posts, columns and beams, are not as safe, from the standpoint of fires, as solid ones; for a solid piece of timber will require a great deal of combustible material around it to keep it burning, whereas, a built-up beam or column will burn of its own accord, once it is started. Moreover, a fire on a solid timber is more easily extinguished than a fire on a built-up timber. For example: A warehouse, whose floors are supported with solid beams and columns is on fire. The fire is on the first floor, but the fire department arrives in time to put out the fire before any of the columns or beams are destroyed. The firemen inform us that had the columns and beams been built-up, the fire would have had ample time to penetrate and destroy them. The building would have collapsed, and thus would have caused, almost, if not entirely, a total loss of the building and the stock of merchandise stored in it.

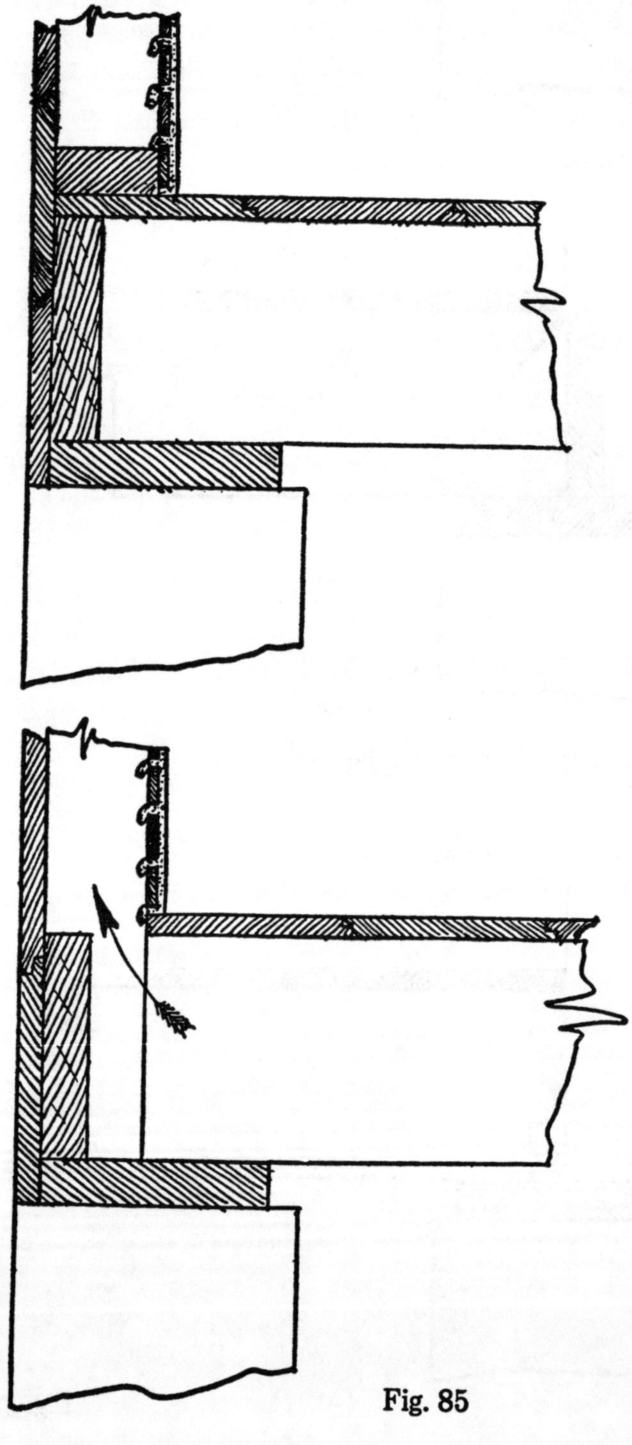

Fig. 85

In keeping with what we have just said, we shall show a number of anti-fire constructions, and some that are not. With these illustrations we want to show how some otherwise substantial constructions, judging from the fire-hazard viewpoint, are, nevertheless, faulty.

With reference to the two sill constructions in Fig. 85, the upper one is what we call an anti-fire construction, while the lower one is faulty. The arrow shows how, in case of fire, the draft between the studding of a wall will suck the fire up into it, making it difficult to put out. In fact, if there are no fire-stops anywhere in the wall, a few minutes of fire in the basement would be enough to set the whole building on fire.

Fig. 86 shows two second-floor constructions, where the joists are joined to the outside wall. The upper one of these constructions, it can easily be seen, is an anti-fire construction. The lower one, as indicated by the arrows, will give a fire right-of-

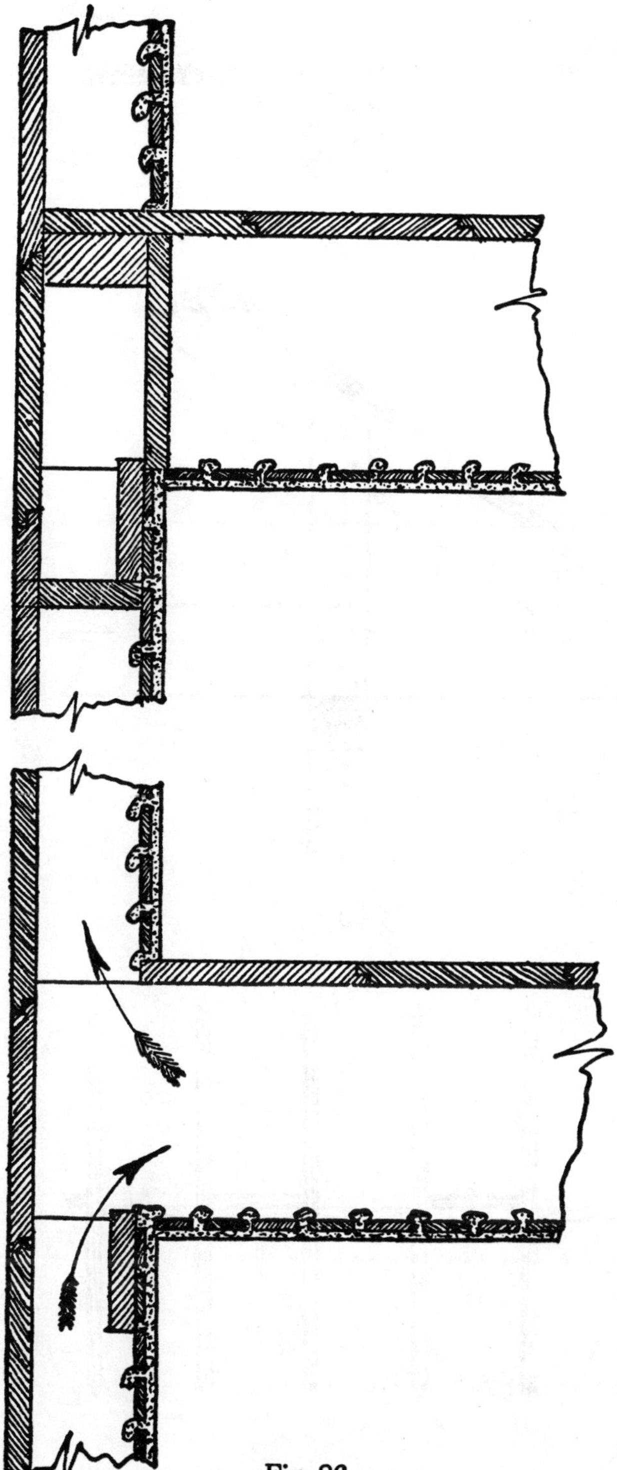

Fig. 86

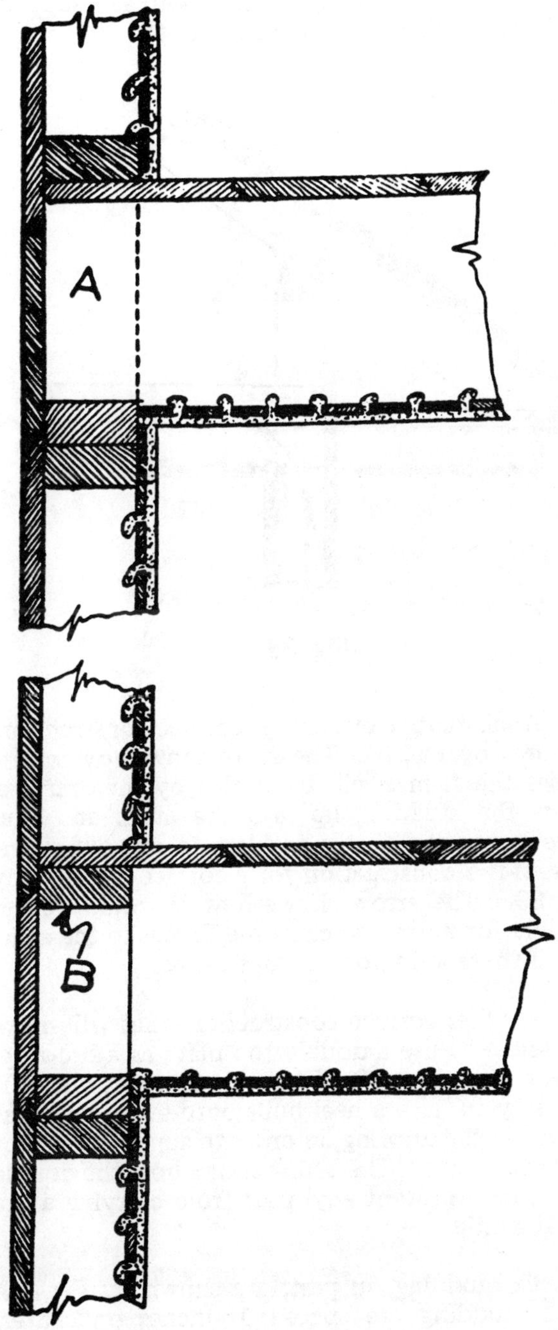

Fig. 87

way to almost every part of the building. Fig. 87 shows two more anti-fire constructions of the second floor, where it joins the outside wall. These constructions are brought up story-high with a double plate, onto which the joists of the second floor are set, as shown by the drawings.

In the upper construction we know the outside wall of the second story, setting on the joists, and by dotted lines, shown at *A*, we are indicating where a shrinkage block should be nailed in order to carry the weight of the upper story. The construction shown at the bottom is the same, excepting that the studding of the outside wall extends down to the top plate of the first story. A block is placed as shown at *B*, to serve both as a support for the rough flooring, and as a fire-stop.

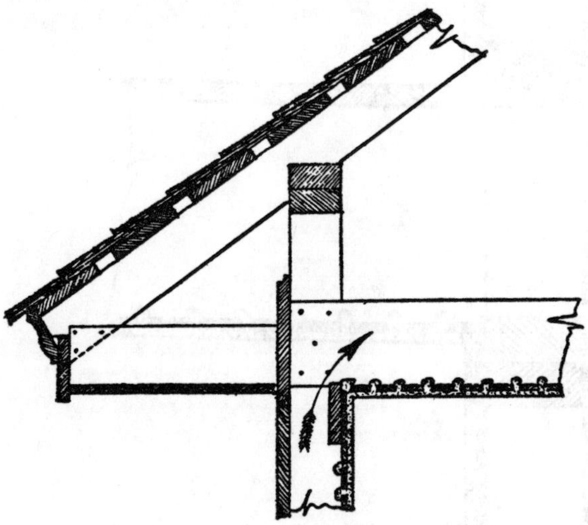

Fig. 88

A substantial, but faulty, cornice construction, is shown by Fig. 88. The arrow shows how in case of fire the flames will be carried by the draft between the studding up into the attic, and thus make it a very difficult problem to put out a fire. An anti-fire construction for a cornice is shown by Fig. 89. The arrow shows how it would be impossible for a draft to carry the flames to the attic, even if there were no fire-stops below.

Another cornice construction, that will make it possible to use a double top plate as a fire-stop, is shown by Fig. 90. This construction makes it necessary to have a heel built onto every rafter, as shown by the drawing, in order to support the roof above the plate. The arrow shows how the double plate would prevent any draft from carrying a fire into the attic.

Six studding, in part, are shown by Fig. 91. These studding are spaced 16-inches on center. The spaces between the studding have been stopped with blocks nailed between them, as shown.

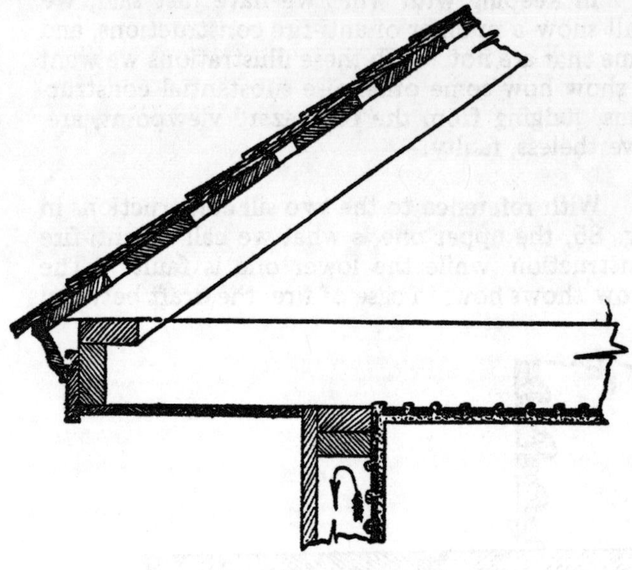

Fig. 89

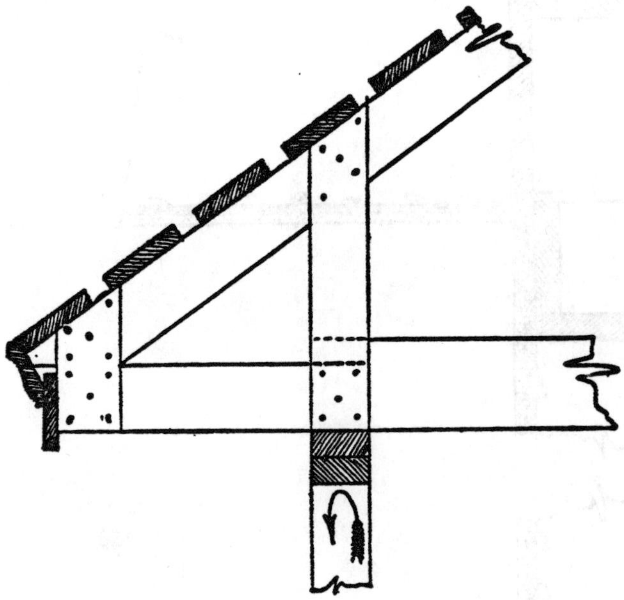

Fig. 90

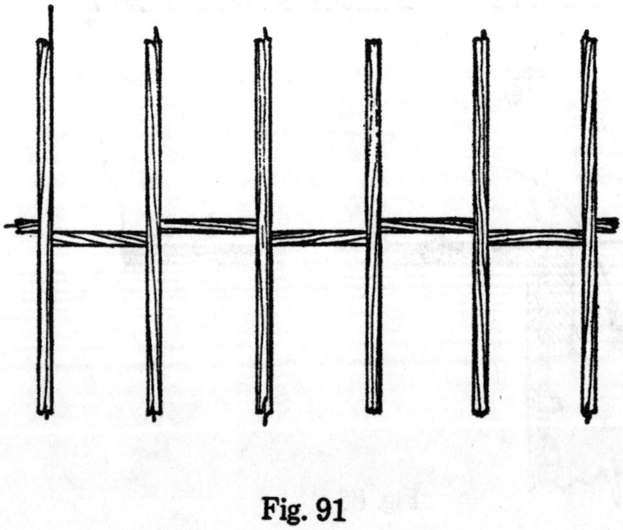

Fig. 91

The purpose of these blocks is two-fold. First, they help to hold the studding in place and stiffens them, and from that viewpoint they are sometimes called bridging. Second, they answer the purpose of fire-stops. Such fire-stops should be spaced not wider than five feet apart.

We are not advancing hard-and-fast construction rules, but we are dealing with craft problems from the principle viewpoint. Slavishly carrying out all instructions of this work to the letter, would be as foolish as to obstinately disregard them all. Modifications of rules or of drawings, in order to fit circumstances, are always in order.

8 Corners

Gypsum Wallboard

The framing we are dealing with here, was done on the basis of lath and plaster, but the advent of sheetrock is practically making lath and plaster obsolete. However, the frame construction for lath and plaster is exactly the same as the construction needed to accommodate sheetrock, gypsum and wallboard.

Anti-fire Material

We spoke of steel as being a fireproof material, which it is, but only in the sense of being non-combustible. Steel, if used for columns or for beams unprotected, is more nearly an anti-fire material than a fireproof material. An extremely hot fire around a steel beam or a steel column would soon bring the metal to a degree of heat where it would soften and bend, consequently, everything supported by it would come down. In such cases, a heavy timber beam or heavy timber column would hold up longer than steel. For it would take longer for heavy timbers to be destroyed by fire than it would take to soften steel. But steel, protected by a substantial covering of concrete, becomes one of the most fireproof materials known to the building world.

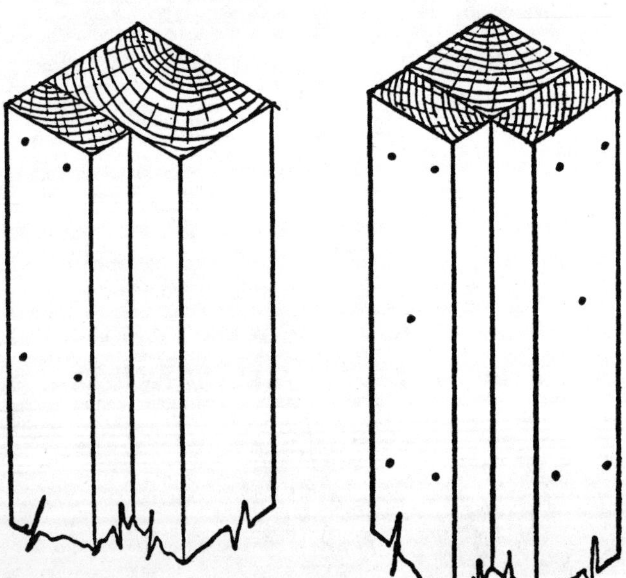

Fig. 92

Substantial Construction

In rough carpentry, if a construction is firm, strong, and will give service, the unwritten law of substantial construction would apply. In finishing, it is much harder to apply this law for in it, harmony and a pleasing effect are more important than the strength of the construction.

In the absence of specifications, a substantial, construction, will, in most cases meet the requirements. At any rate, it gives the builder a good alibi. However, the instances are rare, where a builder is justified in using a substantial construction in place of the specified one, without first taking it up with the owner or the architect. Yet, there are a great many items on every job, too numerous to mention in the specifications, that must be governed by the law of substantial construction. On the other hand, there are many unmentioned things that come up on a job, for which a mere substantial construction is not enough - they must be done in keeping and in harmony with the rest of the work.

Corners

The corners of a building are almost as important as the sill or the foundation -- and when the foundation is good, the corners are even more important than the sill. When the building is plastered on the inside. There isn't anything that will show up its defects as quickly as the corners. Whenever the plastering cracks in the angles of a room, it is almost always an indication that the corners were poorly constructed. It is true though, that seasoning of the material often causes a small crack at the corners, but when that is the case, the crack does not, later on, become an ugly and unsightly one. But cracks caused by the seasoning of the material can be prevented by using metal lath in all the angles.

A two-piece solid corner, which consists of a 4x6 and a 2x4, is shown to the left in Fig. 92. The latter is nailed to the former, as indicated by the drawing. A three-piece solid corner is shown to the right. Here two 2x4's are nailed to a 4x4 in the manner shown. Both of these corners are substantially constructed corners, but they are not used as much today as they used to be.

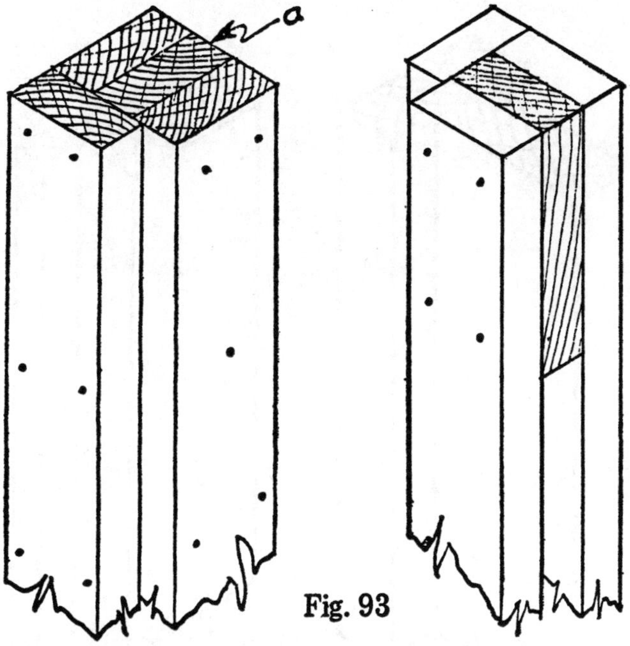

Fig. 93

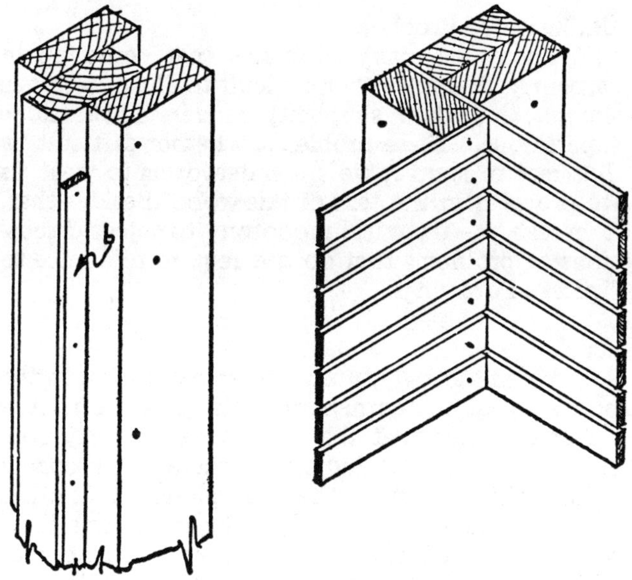

Fig. 94

Perhaps the most commonly used solid corner of today, speaking of residential work, is the one shown in Fig. 93. This is a built-up corner, made either of four 2x4's as shown to the left, or of three 2x4's with the part pointed out at *a*, in the corner to the left, made up of blocks placed at intervals of about four feet. The shaded part of the drawing to the right shows the end block in place.

Another good construction for a corner is shown to the left in Fig. 94. At *b* a strip which is necessary in cases where the 2x4's are, as they now are 1½ inches by 3½ inches, in order to bring the inside angle to the proper place. When the 2x4's are full-size the stripping is not necessary. The construction omitting the stripping, is good for joining partitions with the studding set "flat" to full-width walls. The construction shown to the right also is good but is not convenient because the lathing must be completed on one side before the third studding of the corner can be nailed in place.

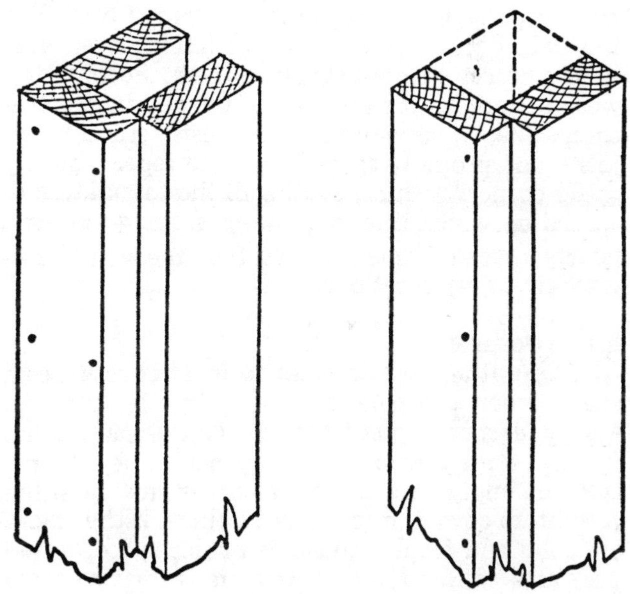

Fig. 95

The corner construction, shown by Fig. 95, to the left, is not a bad one, but it cannot be said to be foolproof. To the right is another non-foolproof corner construction. The dotted lines show how the boxing would form the outside corner. This is not a bad construction, if the boxing for one side is sawed off in such a manner that the boxing for the other side can be nailed to it. But it should not be used, excepting in cases where it is necessary to hold down the expense.

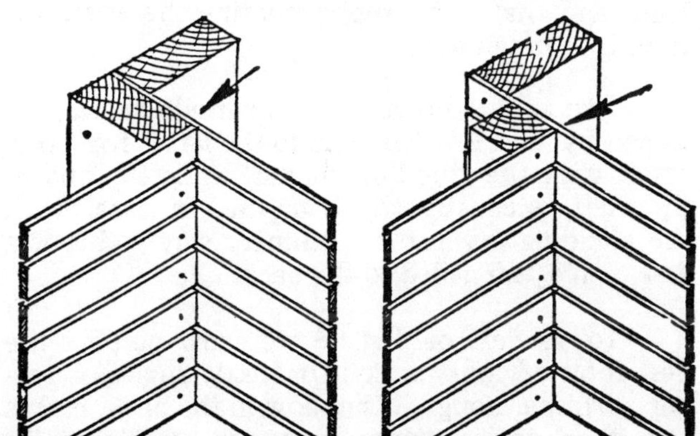

Fig. 96

The constructions shown by Fig. 96 are not to be recommended, excepting for cheap work. But even these constructions can be made to be in a way, substantial, by nailing the lath to the back of the 2x4, shown to the left, and the 2x2, shown to the right. Where the extra nailing should be done is pointed out by the arrows.

Unillustrated Problems

There are many problems that come up in carpentry that are either difficult to illustrate, or so simple, that their simplicity renders illustrations superfluous. These problems, whether difficult to illustrate or simple, we are endeavoring to treat, as we proceed, in the text of this work. Besides that, it breaks the technical monotony to inject discussions of problems that do not require reference to figures or drawings.

Cracks

Whenever ugly, unsightly cracks appear in the plastering at the corners or angles of a room, it is either evidence that the corners were poorly constructed, or else it indicates that a poor corner construction was used. Cracks in plastering at other places, besides the corners, often are as unsightly as those in the corners. There are a number of things that might cause these cracks. We will point out some of them:

First, poor workmanship in lathing, such as, not properly breaking joints, not properly nailing the lath, poorly spaced laths, or lathing over wide solid surfaces without stripping them. Second, too weak material in the skeleton of the building, something like, weak studding or joists, studding or joists not properly spaced, and, we repeat, poorly made corners. Third, settling of the foundation of a building, which is to say, uneven settling. Fourth, poorly braced framework, so that the wind pressure will cause vibrations.

Inside Corners

Disputes often arise as to what constitutes a solid corner, and what does not. Strictly speaking, a solid corner is just what the word means, solid. But the term as applied in carpentry, does not necessarily imply that the whole corner must be solid. A built-up corner, more or less hollow, if the angles are solid and firm, with ample nailing space, would pass as a solid corner. There are so many corner constructions that could be called solid corners, that there is only one way to certify a particular kind, and that is by supplementing the specifications by a detail drawing.

Two solid constructions for inside angles, are shown by Fig. 97. The one to the left is for a partition with the studding set flat, while the one to the right, is constructed to be used in a wall with the studding set in the customary way -- that is a 2x4 with a 1x6 nailed to the back.

To the left of Fig. 98 we show perhaps the best form of construction for inside angles or corners. This, although it is hollow in the back, makes two good angles, where a regular partition joins another regular partition. Of course, blocks, which are not shown by the drawing, should be placed in the back of proper intervals. A poor construction for joining a partition with the studding set flat, to another wall is shown to the right of Fig. 98. In this case the laths have been slipped back of the end studding, and unless they are nailed to the back of this studding, as they seldom are, the plastering will crack at the angles. Metal lath reinforcing will prevent cracking in such an angle.

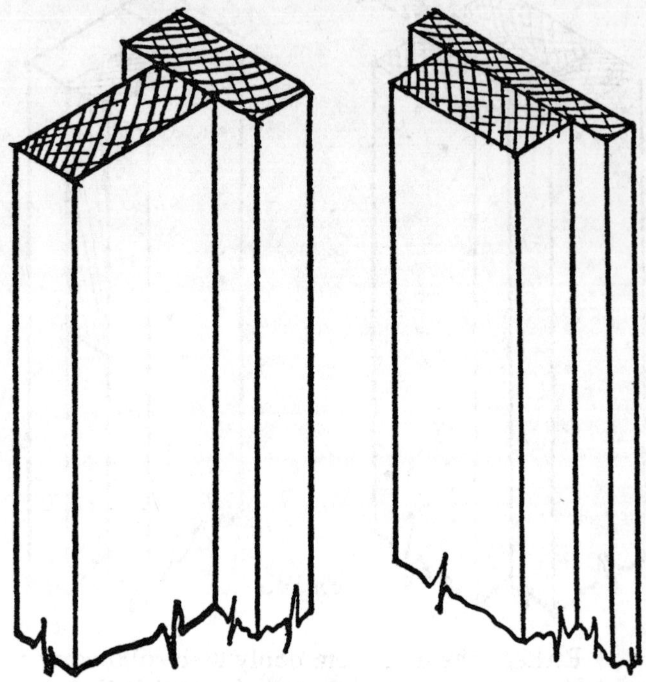

Fig. 97

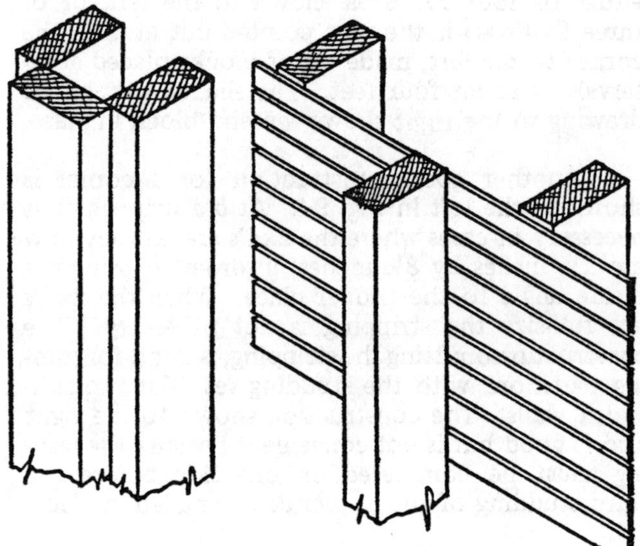

Fig. 98

A detail of a good solid inside angle construction, is shown by Fig. 99, *A*. At *B* is shown a construction where the laths are slipped behind the end studding, or if a ceiling, over the top plate. This is not a bad construction for cheap work, for the studding is nailed to the cross pieces placed back of the lath for that purpose. At *C* is shown

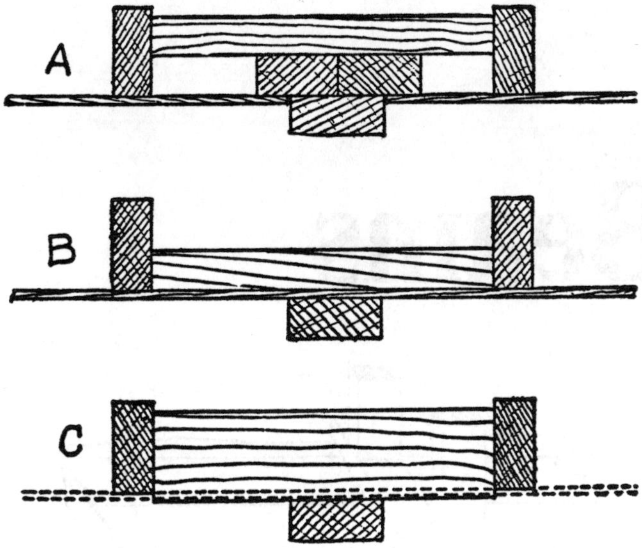

Fig. 99

a similar construction, but the cross piece is placed edgewise, and projects a trifle more than the thickness of a lath beyond the inside edge of the wall studding. This makes it possible to nail the end studding to the cross pieces before the lathing is done. The dotted lines show how the laths are slipped back of the end studding. This, also, is not a bad construction for cheap work.

How base-nailing blocks should be placed at the corners of rooms is shown by Fig. 100. The blocks are made conspicuous by the shading. Such nailing blocks should be placed at either side of the rough jambs of door openings, and at other places, where for some reason, there is no nailing for the base, back of the lath and plastering. Fig. 101

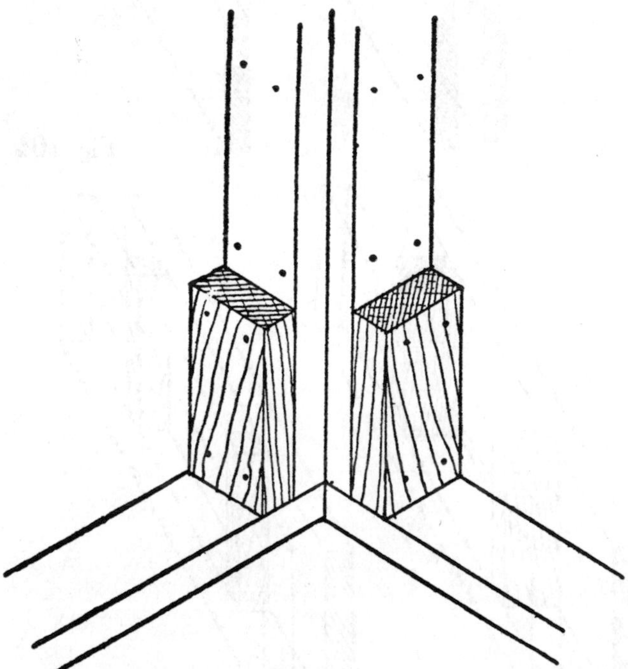

Fig. 100

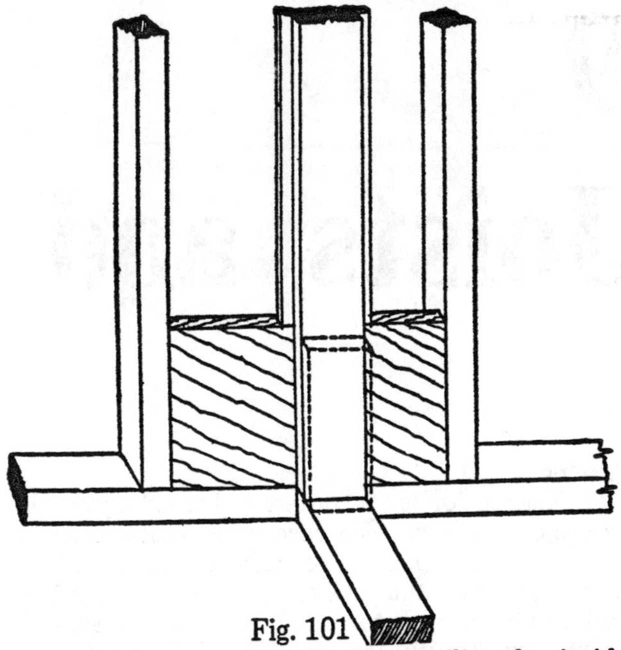

Fig. 101

shows how to provide for base-nailing for inside corners that are made by nailing a backing piece onto the end studding, as shown by the shading and the dotted lines.

Three good constructions for ceiling angles, especially at the ends of a building, are shown by Fig. 102. Number 1 shows the most commonly used construction. Here is used a double 2x4 plate, with a 2x4 nailed to the studding above to provide nailing for the ceiling laths or sheetrock. The double plate provides good nailing for picture moulding, besides the strength it gives to the wall. Number 2 is, in some ways, even a more substantial construction than number 1. The 2x6 that projects over the 2x4 plate, onto which the laths are nailed, can never pull away from the outside wall, as can the 2x4 in number 1. Number 3 is a good construction for cheap work, and especially for small build ings where the plate reaches from corner to corner, for joints in a single plate are weak. To the extreme right is shown the first ceiling joist.

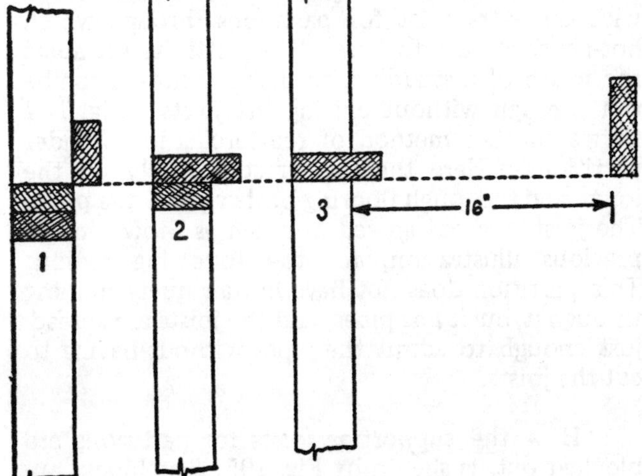

Fig. 102

9

Joists and Beams

Spacing

Spacing for joists, rafters and studding should be accurately done. To become careless and allow the square to slip, so as to make the spaces either shorter or longer than the requirement, is a practice that should not be tolerated. It is just as easy to do this work with accuracy as it is to do it carelessly, if the habit for accuracy is once established. Of course, there are other things that must be remembered as the work proceeds, but through them all accuracy should play an important part.

The spacing for the studding should govern the spacing for the joist. For instance, the joists should come directly under the studding, whenever possible. And the studding should be spaced so that when lath or sheetrock are put on, they will not have to be cut. For both lath and sheetrock require a distance of 4 feet, and when they are butted against the angle, the other edge should come about to the center of the studding. To accomplish this, the first space will have to be one-half the thickness of a studding shorter than 16-inches, or 15-3/16-inches.

Reinforced Joists

When sills are laid off for the joists, the partitions must be located in order to reinforce the joists where the partition will be set. Fig. 103 shows, in part, how the joists should be reinforced with an extra joist for partitions through which hot-air ducts are to pass. The joists are set apart the width of a studding, so that the hot-air can be run through without cutting the joists. Fig. 104 shows another method of reinforcing joists under partitions. Here the plate rests directly on the joists, and the rough flooring butts against the plate. The joists are not spread as much as shown in the previous illustration, for the following reason: This partition does not have hot-air ducts running through it, but it has pipes, and the joists are spread just enough to admit the pipes without having to cut the joists.

How the supporting joists for partitions are blocked out, is shown by Fig. 105. The blocks are shown shaded, and fastened to one of the joists.

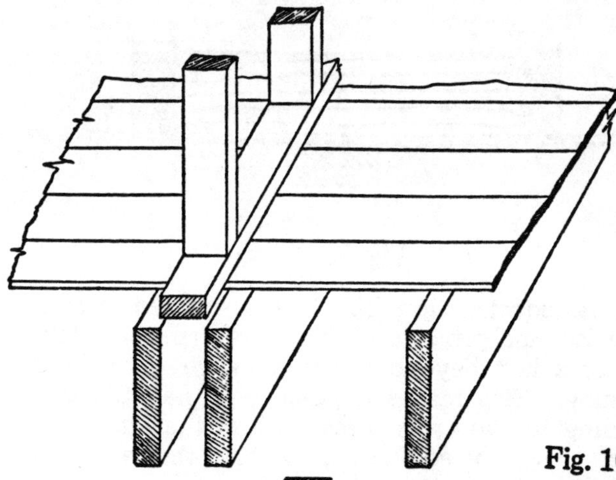

Fig. 103

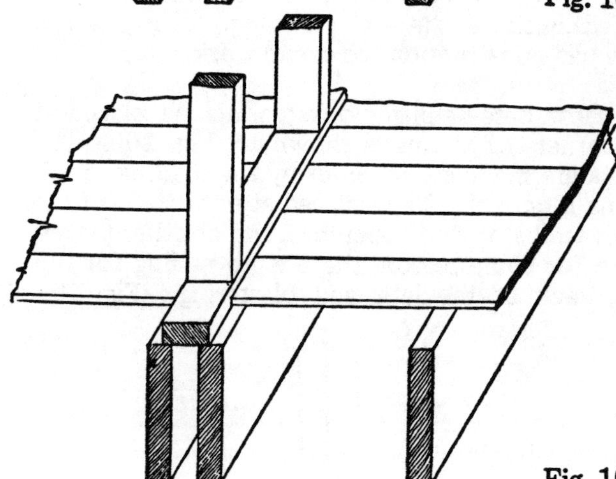

Fig. 104

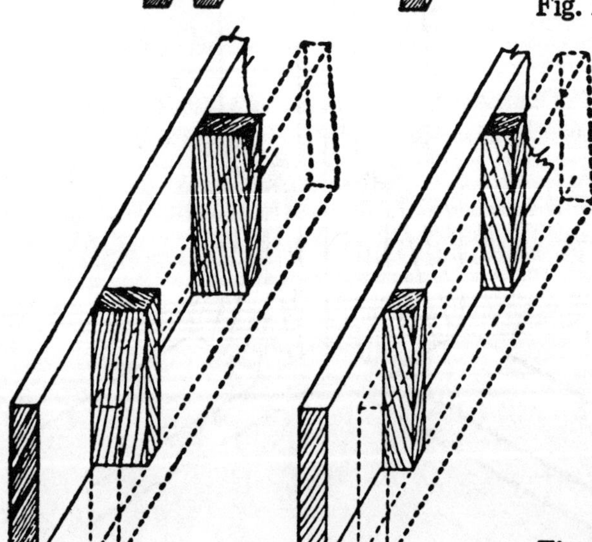

Fig. 105

The other joist is shown by dotted lines. This is true in both drawings. The one to the left pertains to the method which is shown in Fig. 103, and the one to the right shows the blocking for the second method, or that which is shown in Fig. 104. The spacing of the blocks, or spreaders, depends on circumstances and the judgment of the builder. If the joists are straight, and there is to be a rough floor, few spreaders if any, will be necessary, but if the rough floor is omitted, then there should be enough spreaders to hold the joists in line.

Where neither pipes nor hot-air ducts pass through the partition, and there is a rough floor, the reinforcing can be done in the way shown by Fig. 106. But in the absence of the rough floor, the joists should be spread enough to give support for the flooring, on the order shown by Fig. 104.

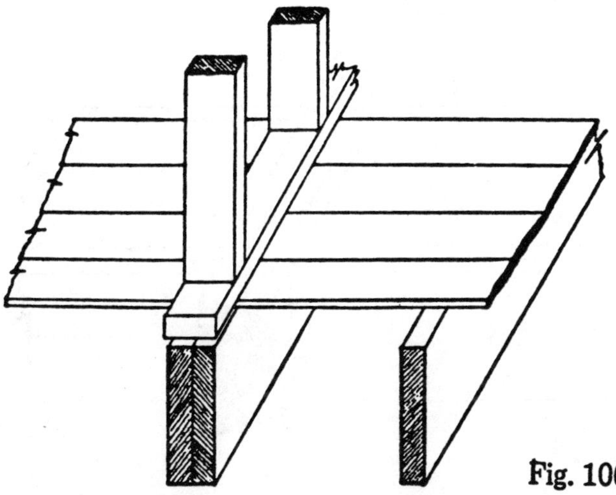

Fig. 106

A partition running at a right-angle to the joists, is shown by Fig. 107. In this case the plate rests directly on the joists. It should be noticed that the studding rest on the plate right over the joists. Fig. 108 shows the same layout, excepting that the partition is set on the rough flooring. The rule for the studding is the same, they should come directly over the joist.

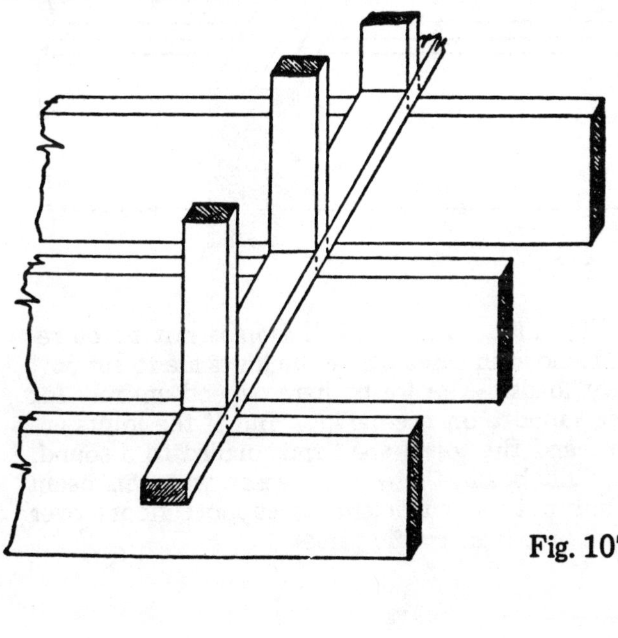

Fig. 107

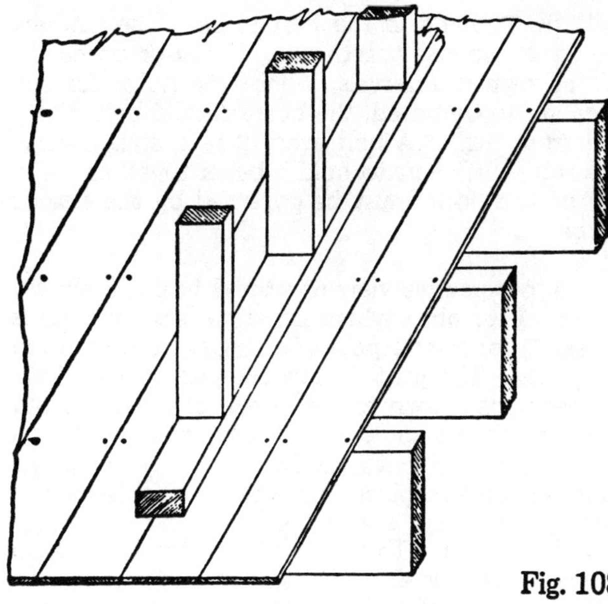

Fig. 108

A construction that is often used on cheap work is shown by Fig. 109. Here the studding are set on the rough flooring without a foot-plate. The dotted line represents a chalk-line, from which the laying-off is done, and up to which the studding are set. This makes a saving of the bottom plate, but the saving in labor is little.

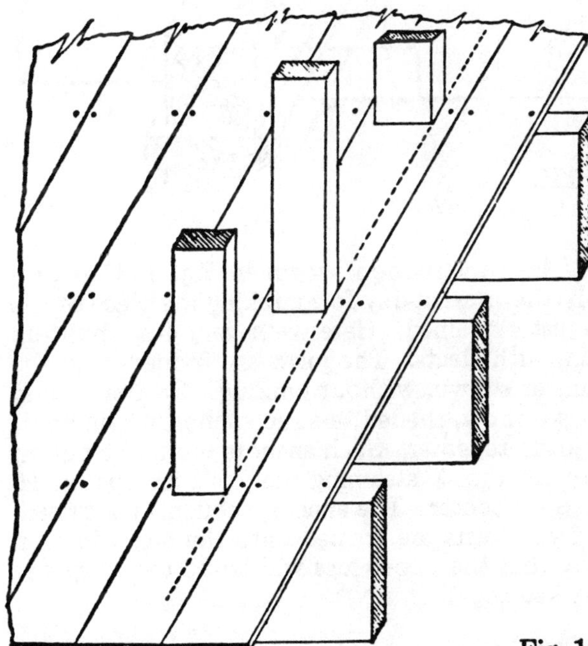

Fig. 109

Beams

A beam should be started straight, and kept straight until it is completed. Allowing a beam to lie in an uneven position while it is built will invariably produce a crooked beam. Such a beam, even though it might be straightened, in a way, afterward, is never as strong as a beam that was built up straight in the first place. The timbers should be spiked together as the work proceeds, but at regular intervals, where the holes for the bolts are to be bored, the beam should be entirely free from nails. A bolt every 2 feet, staggered, is as a rule, sufficient to hold a beam together. The size of the bolts must be governed by the size of the beam.

A perspective view of a solid beam, is shown by Fig. 110, onto which on either side, a cleat is fastened, for the purpose of giving extra support to the joists. The joist, in this case, are housed into the beam, as shown to the upper left. The dotted lines and spears show how the joist is to fit into the housing and onto the beam. The joist to the right is shown in place, but when both joists are in place the lugs lap above the beam and are firmly spiked together. This style of framing joists onto beams is rarely used in our day. It is a form of construction that belongs to the period in which heavy timber played a more important part in carpentry than it does now.

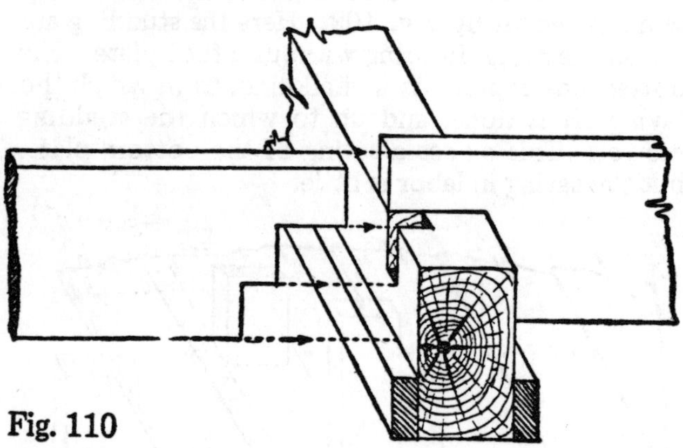

Fig. 110

The construction shown in Fig. 111, is probably the present-day form, taking the place of the one just explained. Here we have a 4-ply built-up beam, with cleats. The joists are framed onto the beam, as shown, without housing. On the second joist we show, shaded, a strap anchor, which holds the joists together. Such anchors should be put on every fifth joist, assuming the joists are spaced 16 inches on center. The anchor, though, is not needed if the joists are framed onto the beam in such a way that the upper lugs will lap so that they can be spiked together.

The same kind of beam is shown in Fig. 112, but the joists are framed onto it differently. Here the joists are notched over the cleats, and the upper edge of the joists come flush with the top of the beam, eliminating the lugs entirely. This method of framing joists onto beams is widely used in our time, and for most purposes is a good construction.

A simple 4-ply beam without cleats, onto which the joists are butted and toe-nailed, is shown

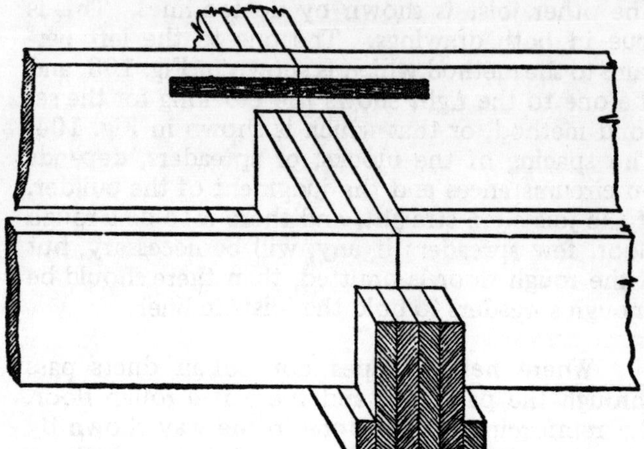

Fig. 111

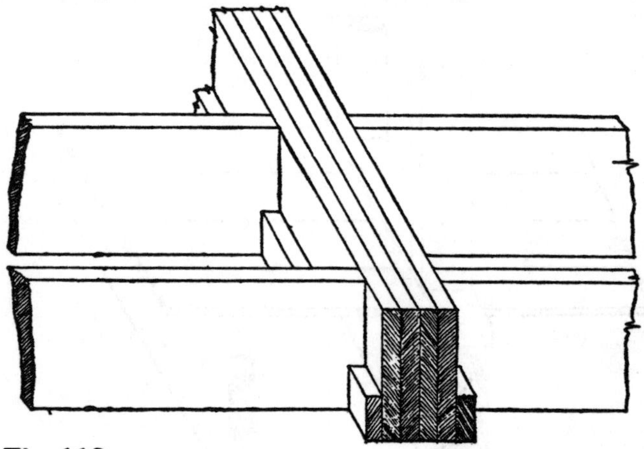

Fig. 112

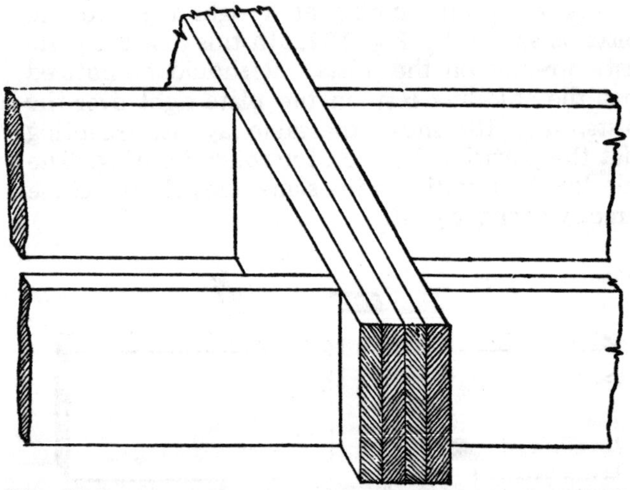

Fig. 113

by Fig. 113. This construction is not to be recommended in cases where the joists are to support heavy loads. The joists, here depend entirely for their support on toe-nailing. But if the joints are tight, and the joists are firmly nailed to a sound, well-built beam, there is no reason why this beam should not be sufficient to support floors over which only light traffic passes.

Another simple construction, that is substantial, is shown in Fig. 114. In this, the joists rest on

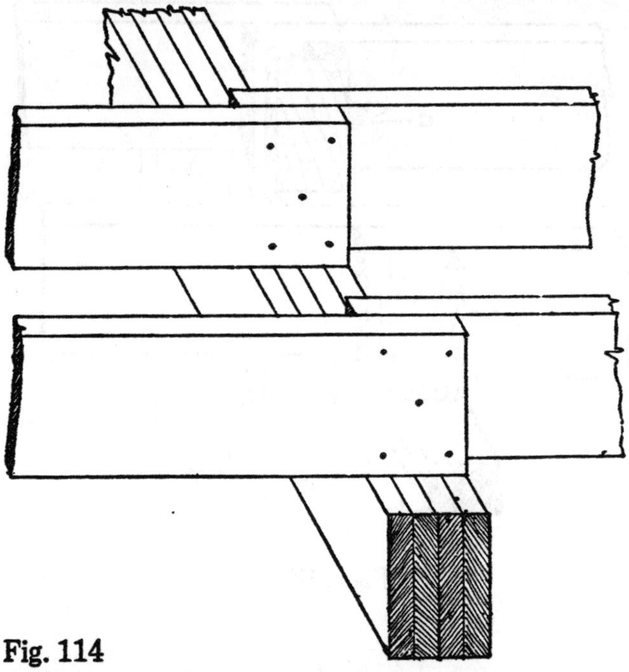

Fig. 114

top of the beam, and are lapped and spiked together. The objection, though, to this beam is that it is not concealed, and for this reason is seldom used, excepting in basements or for other rough work.

Fig. 115 is a perspective view of the ends of three joists, showing how they are framed when they are to be built into a wall. The ends are cut on a batter, and for ordinary purposes, have about a 4-inch bearing on the wall. Every fifth joist spaced 16 inches on center, should be provided with an anchor, somewhat on the order of the one shown by the illustration. If a strap anchor is used at the center, as shown in Fig. 111, the wall-anchor should come on the same joist.

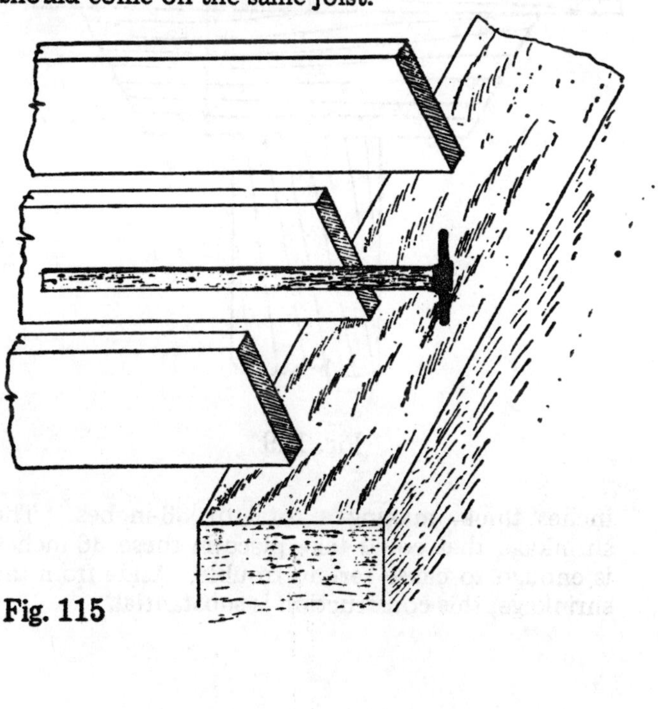

Fig. 115

How the anchors are fastened to the joists, where they run parallel to the wall, is shown by Fig. 116. These anchors should reach over at least three joists, and should be placed at intervals of not over 5 feet.

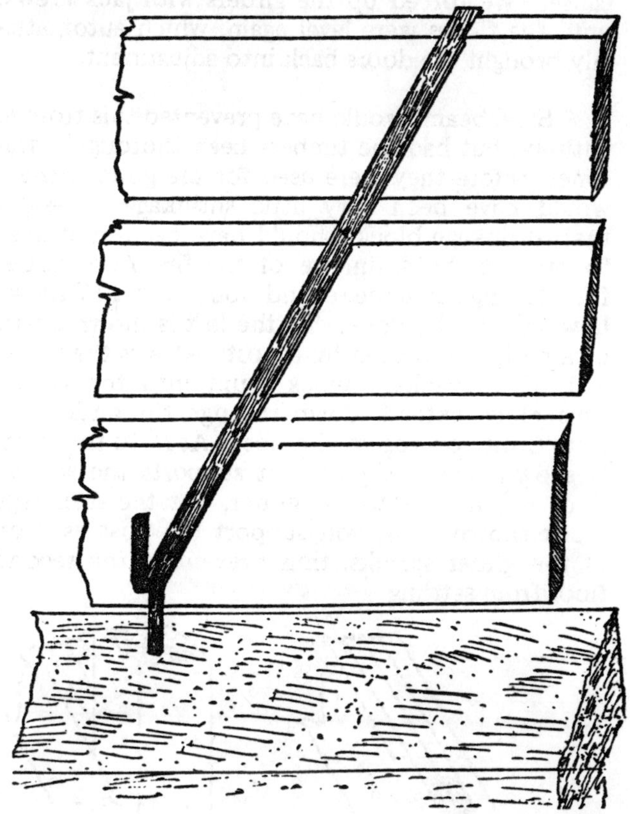

Fig. 116

Shrinkage

A business building, after it had been completed about a year, began to pop and crack and repeat, until, for fear the building would collapse, some of the tenants were preparing to move out. The owner called up the contractor who erected the building, and he in turn, called on us, and together we went through the building, to find the trouble.

We inspected the footings on which the posts in the basement rested, but they seemed all right. Going to the first floor, we found that the center was down ¾ of an inch. The second floor, though, was sunk in the middle about 1½-inches. All of the doors on that floor were out of adjustment, while some of the doors on the first floor still worked, but with difficulties. The difference between the settlement of the first and second floors convinced us that the trouble was not in the footings, but that it was purely a matter growning out of the shrinkage of the material. We examined the girders through the center, and found that they were built up of 2x14-inch stock – and no doubt, rather green.

These girders, during the process of seasoning, had shrunk about ¾ of an inch, letting the first floor down that much and the second floor twice as much, or an inch and a half. The solution of the problem was easy after we had discovered the cause. We forced up the girders with jackscrews, until the floors were level again, which automatically brought the doors back into adjustment.

Steel beams would have prevented this trouble entirely, but had the timbers been thoroughly seasoned before they were used for the girders, there would have been very little shrinkage. Besides that, shrinkage blocks should have been used so as to prevent the shrinkage of the first-floor girder from letting down the second floor. Fig. 117 shows how this can be done. To the left is shown a part of a girder supported by a post. At *A* is the upper end of a shrinkage block, built into the girder. This block extends down through the girder and rests firmly on top of the post. At *B*, to the right, is shown how the post that supports the second floor is joined onto the girder, but the shrinkage block shown at *A*, will support the post as soon as the girder shrinks, thus preventing the second floor from settling.

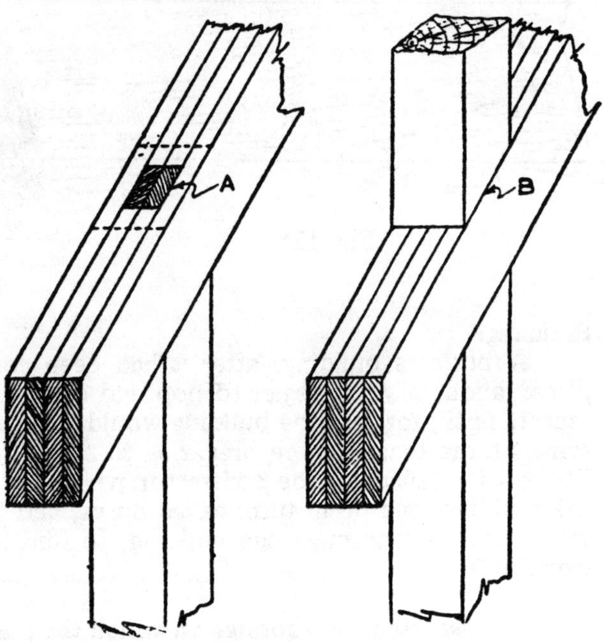

Fig. 117

We show in Fig. 118, *a, a,* how shrinkage blocks should be placed directly under joist-supporting partitions. Such blocks will keep the partition from settling when the joists shrink. At *a, a, a,* Fig. 119, dotted lines show where similar shrinkage blocks should go. However in this case, they would help the situation but little, because there is 36 inches of construction that is liable to shrink. Here we have resting on the post a block 10-inches wide, on it a girder 12-inches wide. Then the joists, 12-inches wide and a rough floor and footplates, 2-inches thick, making a total of 36-inches. The shrinkage that could take place in these 36-inches is enough to cause serious trouble. Aside from the shrinkage, this construction is substantial.

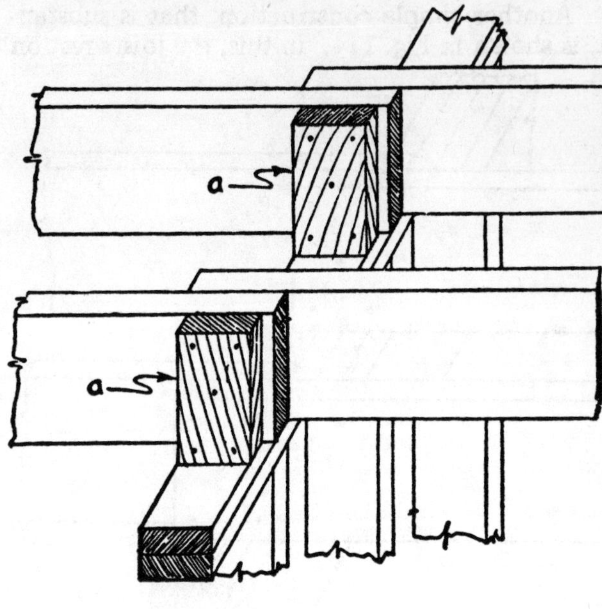

Fig. 118

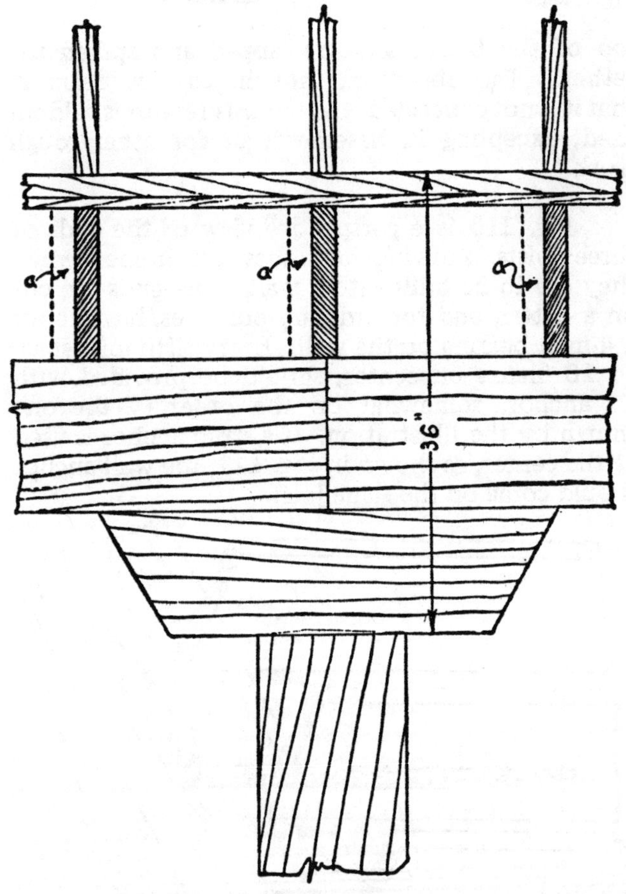

Fig. 119

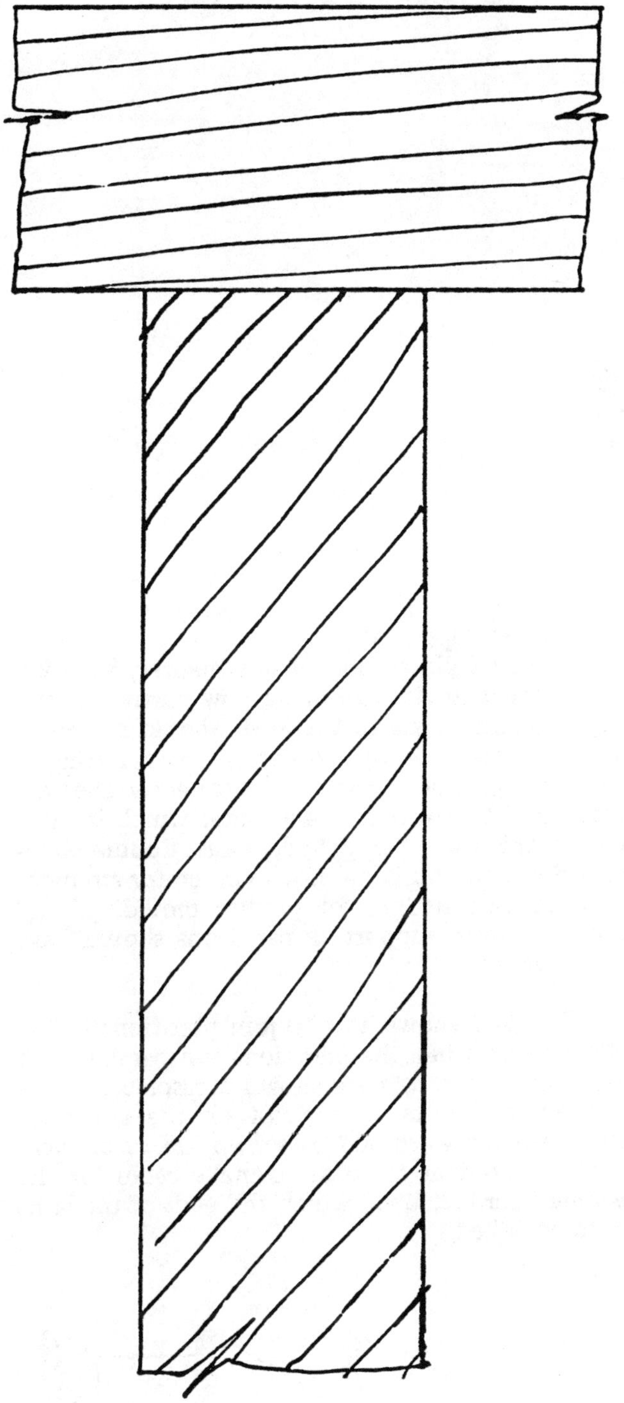

Fig. 120

It is a well-known fact that wood shrinks but little with the grain, but should a cross-grained timber be used for a post, the shrinkage would be noticeable. Fig. 120 shows a cross-grained post, in part. Such a post should never be used to support a load. Besides the danger of shrinkage, there is the liability of the post snapping in two when the load is placed on it.

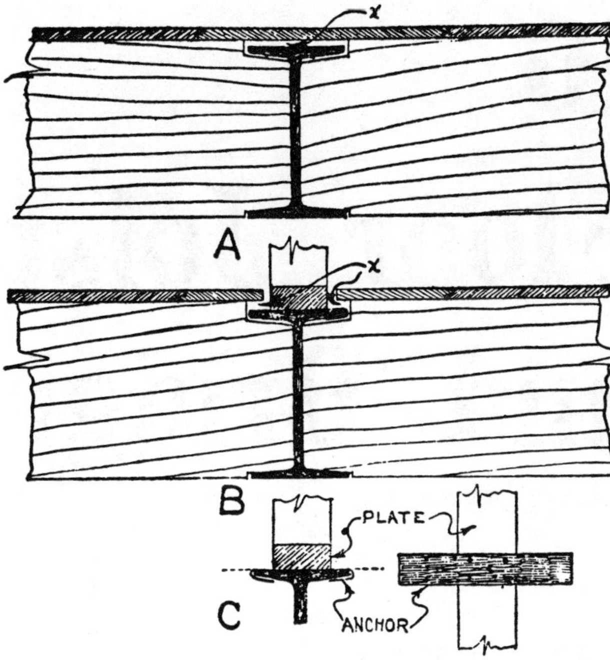

Fig. 121

In our time, when steel plays such a large part in the building world, carpenters are often called upon to frame joists onto I-beams. When this is done, care should be taken to provide ample shrinkage space between the rough floor, which should be laid diagonally, and the upper edge of the I-beam. Fig. 121 shows over *A*, at *x*, where the shrinkage space should be provided -- also over *B*, at *x*. If this space is not allowed, the floor will have a hump in it over the I-beam, as soon as the joists shrink. At *B* we show how a partition should be set, when it comes directly over the I-beam. It should never be set on the rough flooring in such cases. The practice of blocking under the rough flooring, to support the partition, is bad -- even though it holds the partition, it does not prevent the hump in the floor when the joists shrink. At *C* we are illustrating how the plate should be fastened to the flanges of the I-beam. A strip of heavy sheet-metal, nailed to the bottom of the plate at regular intervals, will, by bending the ends around the flanges, as shown, anchor the plate into position. To the right is shown the anchor nailed to the bottom of the plate, and to the left, the plate in position with the ends of the anchor bent around the flanges is shown.

A flat-grained board, or a board sawed almost in line with the annual rings of the wood, will shrink more than one that is sawed at almost a right angle to the annual rings. There is less shrinkage in the direction of the medullary rays of wood, than there is in the direction of the annual rings, and therefore the amount of shrinkage for any given piece of board depends on how the board was sawed from the log.

10 Floor Joists and Bridging

Swelling Problems

Serious swelling problems can arise where forms for concrete slabs are fitted too closely against walls. This is especially true where form lumber is very dry when it is put down. For when concrete is poured, it absorbs water, and the inevitable swelling follows, pushing out the wall or walls, as the case might be. This danger is equally present where rough floors are laid too tightly to the walls of a building. We have in mind an instance where the walls of a brick building were pushed out, leaving a crack at each corner.

In connection with what we have just said, we want to call attention to the tremendous force wrapped up in the swelling characteristics of wood. A fair idea of this force can be realized by remembering that large stones have been broken by driving dry wood into holes drilled at regular intervals, and then swelled by pouring water on it. This force has been utilized by prehistoric civilizations and down through the ages to the present time.

Ceiling Joist Supports

The first illustration of this chapter, Fig. 122, is a perspective view, showing how partitions running at a right angle to the joists should be joined to ceiling joists. It will be noticed, by referring to the drawing, that the joists rest directly over the studding. We show a single plate, which is sufficient in this case. Some builders use double plates throughout the building, not so much for strength, as to provide nailing for picture moulding. Of course, all joist-supporting partitions should have double plates.

Fig. 123 shows how to join partitions to the ceiling joists when the partitions run parallel with the joists. At *a, a, a* are shown crosspieces nailed between the joists which rest on the partition. These crosspieces should be set up 7/8 of an inch, as pointed out at *b*, so as to make room for the backing board, *c*, onto which the ends of the laths are to be nailed.

Fig. 122

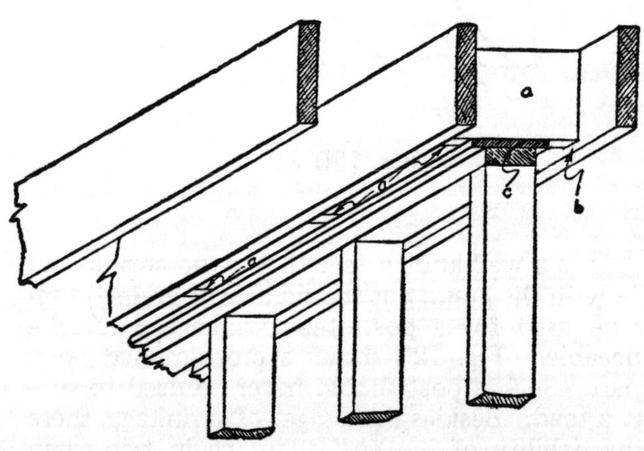

Fig. 123

Two ways of framing the ends of joists, that rest on a ribbon board, are shown by Fig. 124. The joist to the left is notched over the ribbon board, *b*, and the one to the right is sized. The purpose of sizing or notching is to make the distances, *a, a,* equal on all the joints, so that the upper edges of the joists will line up. How the joists are sized that rest on a sill, is shown by Fig. 125. In this case, also, the distances, *a, a,* are to be equal on all of the joists. At *b, b,* we point out the shoulder that takes up the difference in the various original sizes of the joists. Sizing is always necessary where the joists are used with the crowns just as they come from the yard, and for many practical purposes, this method gives satisfactory results.

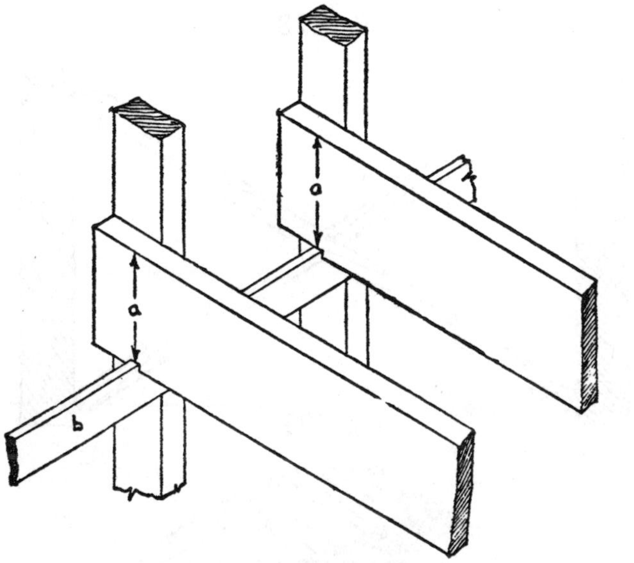

Fig. 124

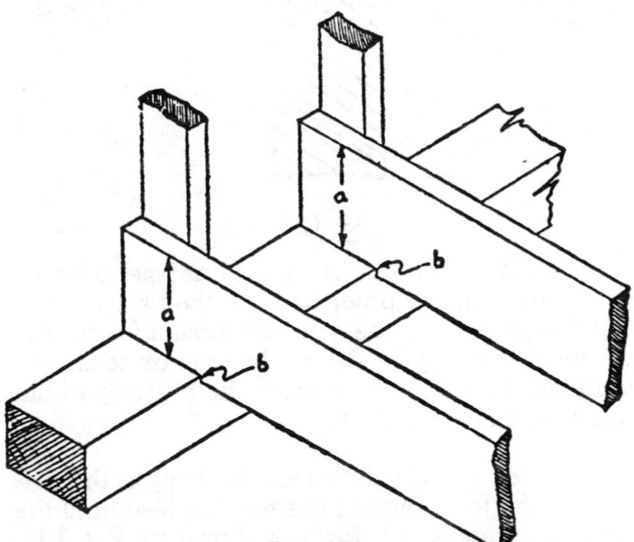

Fig. 125

In the better class of work, and where the span is rather long, the joists should be crowned, somewhat on the order shown by Fig. 126. This really means that the whole joist is sized, and therefore the ends do not need sizing again. The dotted

Fig. 126

lines indicate the original size of the timber, while the continuous lines show how the joist appears after the crowning is done. The illustration shows the crowning slightly exaggerated in order to bring out the point more clearly. The upper edge of the joist has a higher crown than the lower edge, which leaves the greater width of the joist at the center, where it should be. The lower edge of the joist should be crowned from 1/3 to 1/2 as much as the upper edge. A 3/4-inch crown for a 20-foot joist is not too much, and the amount of crown should increase or decrease, just as the length of the joist is more or less than 20 feet.

The end joists, that is, the first and the last joist, or the joist coming against the wall, should be perfectly straight. The second joist from the end should have a 1/3 crown, the third joist from the end should have a 2/3 crown, while all the other joists have full crowns. The purpose of this is to taper off the crown at the ends, so as to keep the floor around the edges of a room straight. The end joists should be anchored to the wall, so they cannot sag when the load is applied.

Bridging

Does bridging increase the strength of joists? This question is often asked, and it usually leads to a controversy. There are those who hold that bridging increases the strength of joists, while, on the other hand, there are others who claim that it does not. Misleading circumstances are responsible for this difference of opinions. For instance, an individual joist of a floor that is bridged will carry a much greater load than an individual joist of a floor that is not bridged. This fact leads many to believe that bridging does increase the strength of joists, but it does not. It distributes the load, which is to say, that if a load is placed directly over the joist, the bridging will distribute it over most, or all of the other joists. Very little, if any additional weight, uniformly distributed, will a bridged floor carry, when the bridging at the ends does not butt against something firm, like a wood or brick wall. It is true, when the bridging has a firm bearing at the ends, that a bridged floor will carry a greater load than a floor that is not bridged.

A system of cross bridging is shown by Fig. 127. The joists are shown in part, with the bridging for four spaces in place. Notice the first pair of bridging to the left. The piece joins the joist at the bottom, sloping upward, while the other joins the joist at the top, sloping downward. In the next space, the downward sloping piece is nailed at the top of the joist, directly opposite the upward slop-

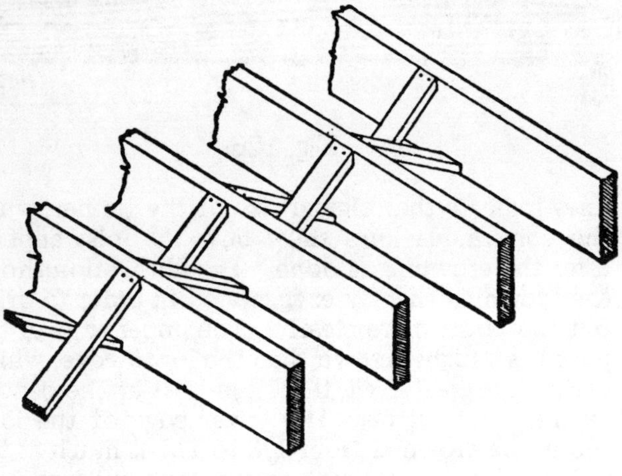

Fig. 127

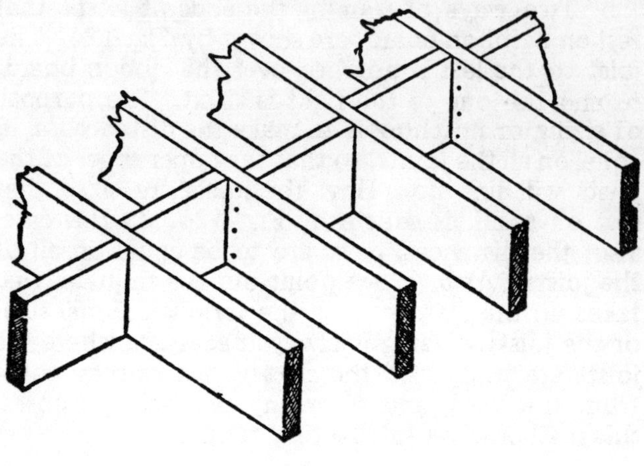

Fig. 128

ing piece just mentioned, and the upward sloping piece is nailed at the bottom of the joist, directly opposite the downward sloping piece. The pair of bridging in the third space is placed exactly the same as the first pair -- while the fourth pair is placed the same as the second pair. A little study of the illustration will show that it is necessary to reverse the position of every other pair of bridging, in order to bring the bearings on the joists directly opposite each other.

Another important thing about nailing bridging is that it should be nailed only at the top until after the rough floor is nailed down. This will bring the joist into line, and then the bottom of the bridging should be nailed. Usually the nails are stuck into both ends of the pieces beforehand, so that when they are nailed into place the carpenter can hold the bridging with one hand and nail with the other.

A form of bridging made of blocks, is shown by Fig. 128. This consists of nailing pieces of joist material between the joist in the manner shown by the illustrations. This system, sometimes called the block system, stiffens the joists and also answers for fire stops. Block bridging requires more labor to put it in place than cross bridging, and a great deal more care, for if the pieces are not cut exactly square, the joists will be thrown out of alignment. The same is true in cases where the thickness of the joists is not exactly the same from one edge to the other. But carefully done block bridging is superior to cross bridging.

A pair of cross bridging, cut square, is shown by Fig. 129. Some claim that this cut is better than the beveled cut. They say that the corners, when nailed will sink into the joists, as shown at *a a*, Fig. 130.

Our contention is that a square cut for bridging is not universally better than the beveled cut, but that in cases where the joists are of soft material and the bridging of material so hard that nails

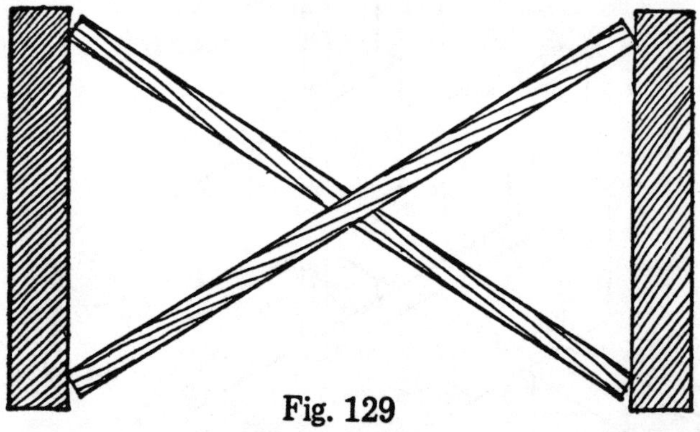

Fig. 129

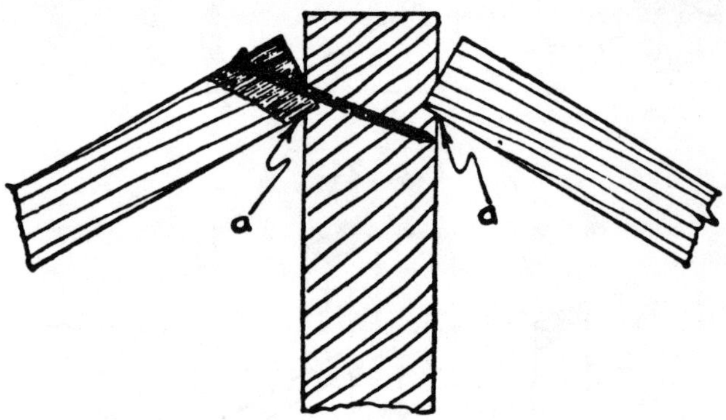

Fig. 130

will split it, the square cut has advantages over the cut to fit. On the other hand, if the bridging material is soft, and the joist material is hard, the square cut would be decidedly inferior to the established beveled cut, because the bridging would not indent the hard joist.

A method of nailing cross bridging at the bottom of the joist, leaving the top unnailed until the rough flooring is put down, is shown by Fig. 131. The bridging piece is held to the joist in an upright position, as shown, and nailed at the bottom somewhat as pointed out at *a*, both to the right and to the left. This done, the upper ends of the bridging are forced down, as indicated by the arrows in the dotted part-circles, until they come into the position shown by the straight dotted lines.

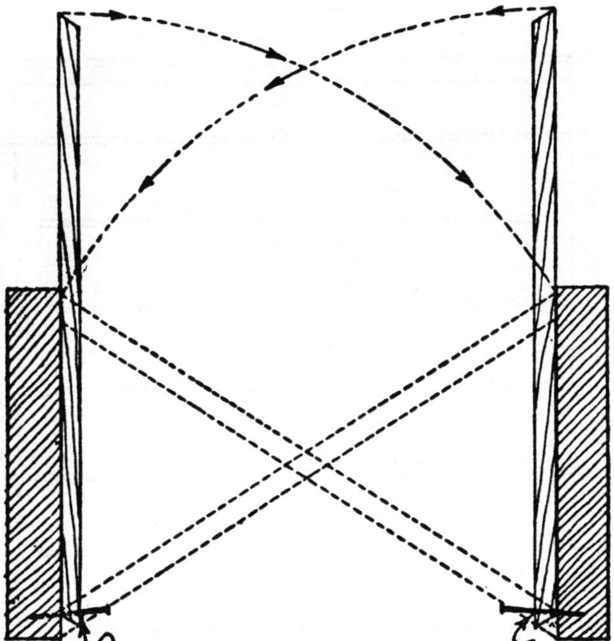

Fig. 131

Two details of the nailing, is shown by Fig. 132. At *A* the nail is shown after it has been driven, and at *B* we show the nail after the bridging has been forced down into its regular position. This method of nailing bridging can be used to advantage where the ceiling is too high to be reached by ordinary scaffolding, or where it is impossible or inconvenient to nail the bottom of the bridging after the rough flooring is down.

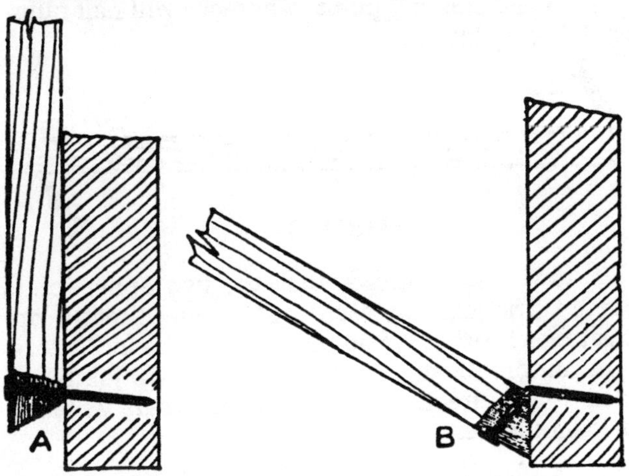

Fig. 132

A short cut-and-fit method of cutting bridging, is shown by Fig. 133. This method is quite practical in cases where the joists are not spaced equally, or for cutting bridging to replace defective or torn-out pieces. The carpenter, after cutting one end of a piece of bridging material to the proper bevel, places it in the position shown by the illustration. Then with a saw he cuts it off as indicated by the dotted line between *a* and *b*. The cut, as well as the angle of the cut, is governed entirely by the judgment of the carpenter. He has, of course, the side of the joist as a guide for the saw, that is, the saw is kept parallel with the side of the joist while the cutting is being done. A little practice will soon enable the apt carpenter to make these bridging cuts so that the bridging will fit between the joists snuggly. The position of a pair of bridging, after they have been nailed into place is shown by dotted lines.

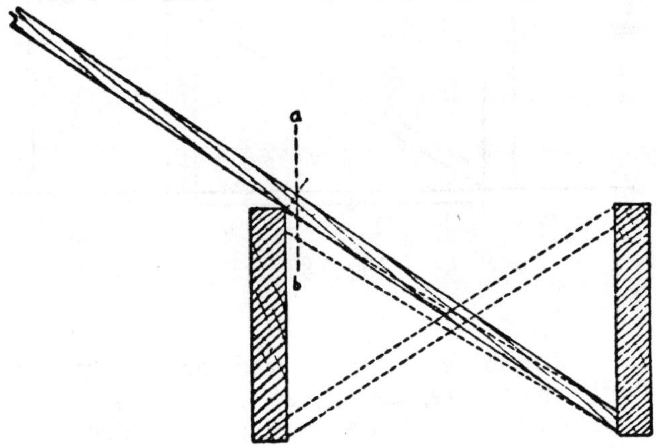

Fig. 133

A miter box for cutting bridging is shown by Fig. 134. This box is made by nailing a piece of 2x6 onto the back edge of a 2x4, as shown by the illustration. Then at point *a*, a sawkerf is cut into the 2x6, on a bevel corresponding with the bevel of the bridging cut. At a distance equal to the length of a piece of bridging from point *a*, point *b* is established and a nail stuck there to gauge the length of the bridging. Now take a piece of the bridging material, place it in the box, and with the saw in the guide at point *a*, cut the end of the piece to a bevel. This done, slip the piece forward until it strikes the nail, as shown by Fig. 135, and cut the first piece of bridging. This operation is repeated over and over until the necessary amount of bridging is cut.

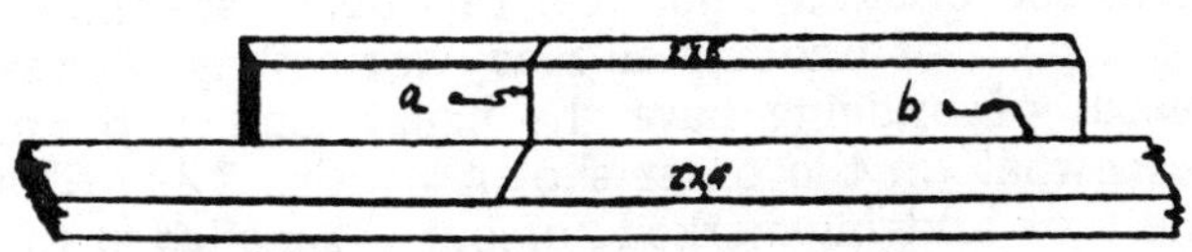

Fig. 134

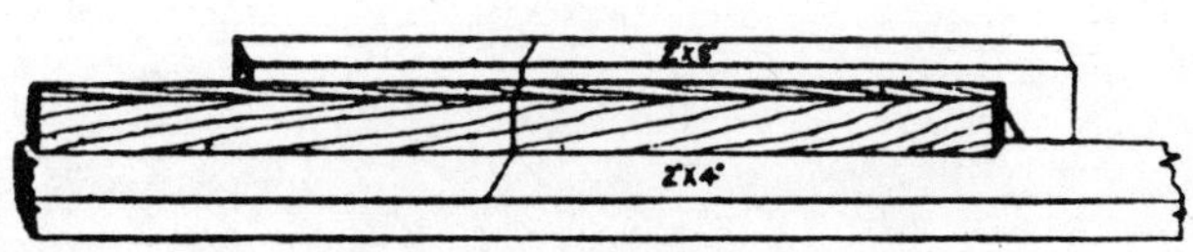

Fig. 135

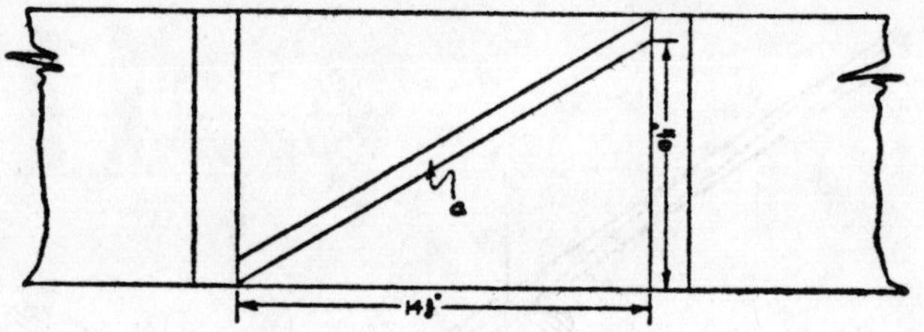

Fig. 136

To obtain the bevel and length of bridging with the square, take a piece of the joist material, and on it lay off one space, as illustrated by Fig. 136. In this case the space is 16-inches from center to center. Then draw two lines, representing a piece of bridging, as indicated at *a*. To obtain the cut, measure the distance between the bottom of the bridging where it joins the joist at the top, and the bottom of the joist, which, as can be seen, is 8½ inches. Now, by taking 14-3/8 on the body of the square, and 8½ on the tongue, and applying the square to the material as shown by Fig. 137, you will have both the bevel and the length of the bridging. The tongue of the square gives the bevel. The diagonal distance between the points, 14-3/8 and 8-1/2, will give the length of the bridging.

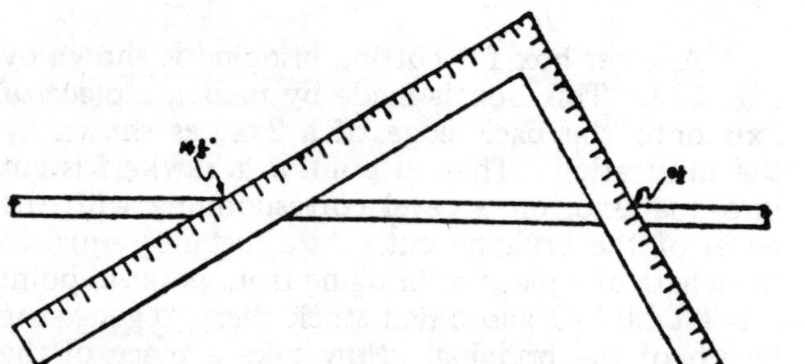

Fig. 137

A layout, showing how to nail bridging into place, is shown by Fig. 138. A line is stretched from *A* to *B*, which establishes the bridging line. The tie board, which holds the joist in position, was intentionally placed a little to the left, so as to clear the bridging line. On this board are shown eight pairs of bridging all ready for nailing. These pieces of bridging have the nails stuck in them, somewhat on the order shown by Fig. 139. The piece of bridging marked number *1*, is nailed at the top edge of the joist at point *a*, piece number *2* is nailed at the top edge of the joist, at point *b*, number *3* is nailed in the same way, at *c*, and *4*, at *d* and

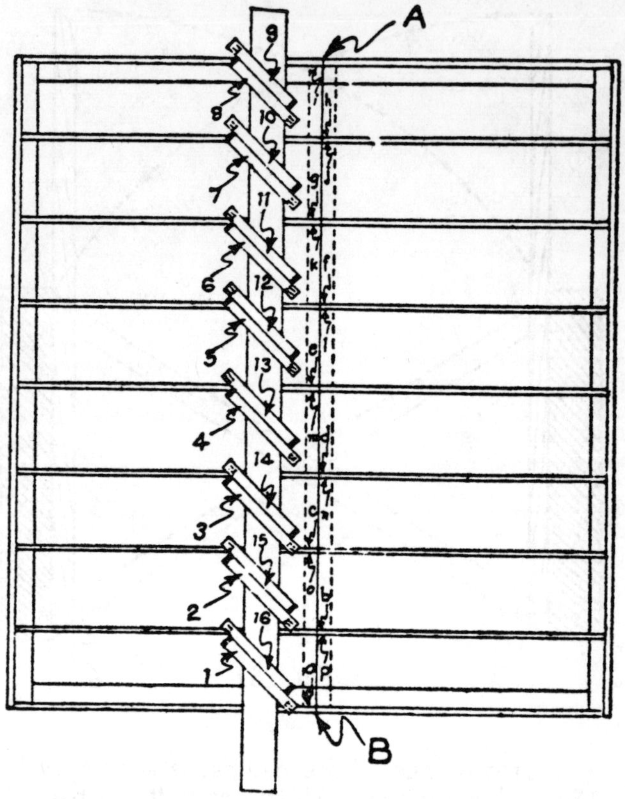

Fig. 138

so on until we reach number *8*, which is nailed at the top of the joist, at point *h*. Now we turn around and nail number *9* at the top of the joist, at point *i*, and number *10* at *j* and so on until we come back to the starting point, where we will nail number *16*, at point *p*.

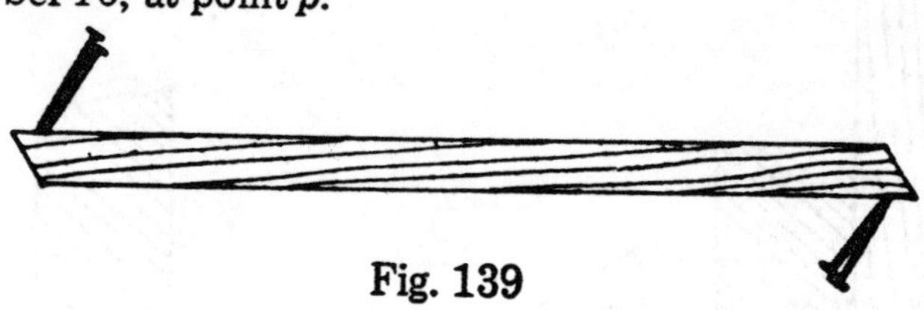

Fig. 139

It will be noticed, that the two upper joints of the bridging, on each joist, come directly opposite each other, as pointed out by the indicators. This automatically brings the bottom joints on each joist, directly opposite each other, which is as it should be.

Metal bridging is coming into use and can be used instead of wood bridging. The advantages are that there is no splitting as in wood and, if properly fastened, metal bridges will hold the joist in perfect position.

11

Rough Floors and Dropped Ceilings

Rough Floors

Rough floors are omitted only in the cheapest of buildings, and even then very little is gained by the omission. The inconvenience of working over naked joists is bound to increase the cost of labor. This extra cost, coupled with the other losses caused by omitting the rough floor, will almost, if not altogether, balance the cost of a rough floor.

It was not so long ago when rough floors consisted of shiplap, 1x6's and other kinds of 7/8-inch material, until plywood came in and gradually took the place of these materials; however, it did not completely eliminate them. Frequently we notice a house going up that is boxed with boards, or has a rough board floor, and sometimes the sheathing is also as it used to be, just plain boards.

While laying rough flooring is so simple that explanations seem almost superfluous, there are problems that arise in this work that must be solved before the workman can reach his maximum efficiency. Such problems are dealt with in this chapter.

In Fig. 140 we show an oblong floor space which is to be covered with rough flooring. The ordinary way of proceeding, especially when only two or three men are on the job, is to start at one end and put down the flooring, one board at a time, until the whole floor is laid. But in this instance we have twelve men – how are we going to proceed so as to give each man a chance to work without handicaps? Here is the way we will do it:

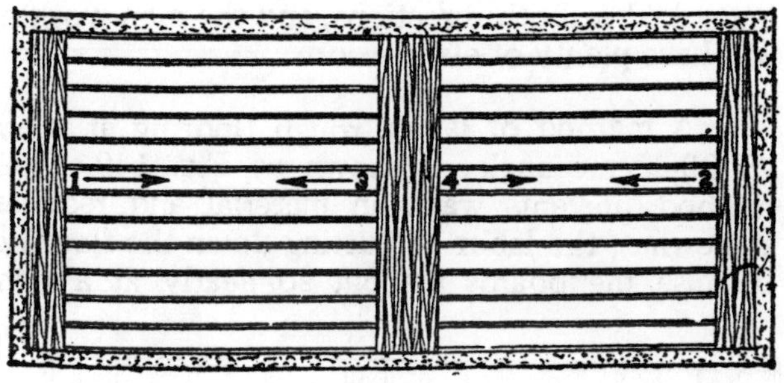

Fig. 140

Three men will work at number *1*, three more will work at number *2*, three will work at number *3*, and three at number *4*. All will be working in the directions indicated by the arrows, respectively, until the floor is down. Group 1 and group 3, when they meet, will have to rip one board in order to finish at the point of meeting. Another board will have to be ripped when group 2 and group 4 meet. In case there are less than twelve men on the job, say five or six, then we will let three work at number *1*, and two or three at number *2*, and let them work until they meet, omitting group 3 and 4.

Tying Joists

Fig. 141 shows two ways of tying joists. The dotted lines through the center show where a tie-board is nailed temporarily, in order to hold the joist in place until enough rough flooring is down to hold the joists in line. This method of tying

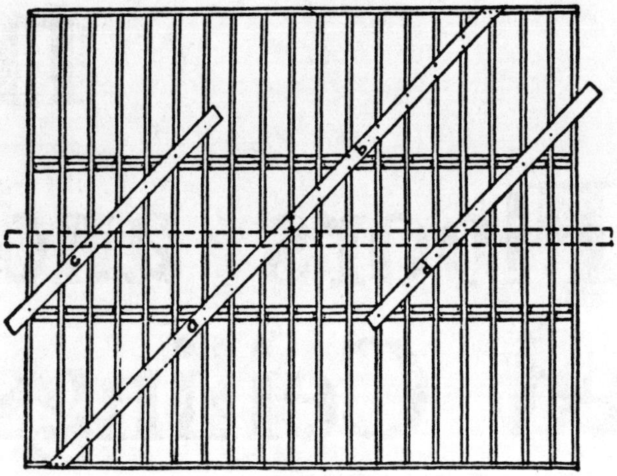

Fig. 141

joists is good when the rough flooring is laid at a right-angle to the joists. But when it is laid at a 45-degree angle, then the method shown at *a*, *b*, *c* and *d*, is better. Here it will be noticed that the boards marked *a* and *b*, are shown nailed permanently with two nails to each bearing, while the boards marked *c* and *d*, are nailed temporarily, with only one nail to a bearing. The reason for this is that the boards *a* and *b* answer both for tie boards and starting boards. Starting at these boards, the flooring is laid in both directions, and the men working will have plenty of elbow room.

A method of laying rough flooring at about an 80-degree angle, is shown by Fig. 142. This method prevents waste in material and reduces, somewhat, the labor of putting down the flooring, because the boards are cut so nearly at a right

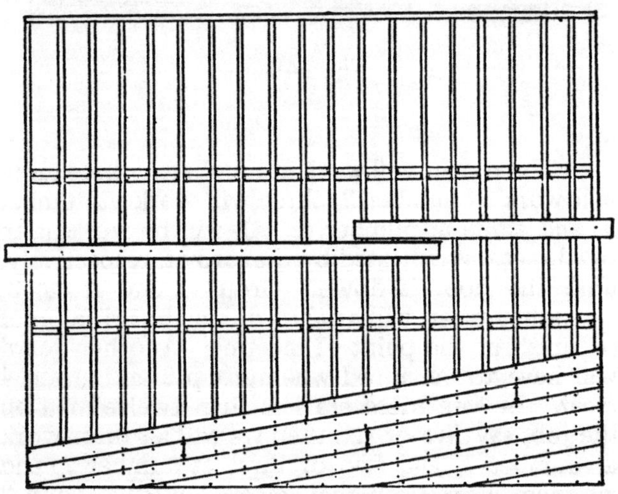

Fig. 142

angle that it will require but little extra sawing. Moreover, the boards can be cut square across, if desired, which will bring the joints slanting onto the bearings. To do this, a little more care must be taken in locating the joints, but no extra sawing will be necessary.

A simple method of laying rough flooring at a 45-degree angle, is shown by Fig. 143. The tie-boards are indicated by dotted lines at *a* and *b*. Board number *1*, marked *S* and *B*, is the starting board. This board down, one man works in the direction of the arrow shown at *A*, and puts down boards number *2*, number *3*, number *4*, and so on until he reaches number *8*, which projects one-half of its length beyond the joists. This projection is marked number *9*, and is the longest projecting part of the eight boards. Now he cuts off these projections, which are numbered, respectively, *9*, *10*, *11*, *12* and so forth until he reaches number *16*. These pieces are then placed onto the joists, where we show them numbered, *9*, *10*, *11*, *12* and so on down to number *16*. With number *16* in place the corner is done, excepting cutting off the projecting ends that remain. But while *A* has been doing this, *S* and *B* were working in the direction of the arrows shown at these points, and are probably far enough along to give *A* room to work with them, without inconveniencing any one of the three.

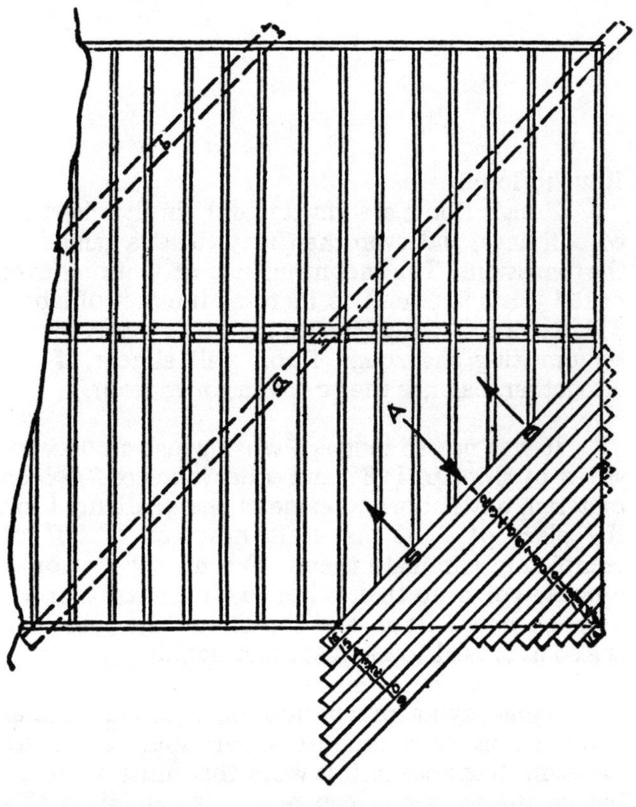

Fig. 143

Diagonal Flooring

The principal reason for laying rough flooring diagonally, is to prevent the finish flooring from running parallel with the rough flooring. And the reason why this is desirable, is:

First, the rough flooring usually is exposed to the weather, and by the time the finish flooring is put down, it is more or less warped, thus giving the finish floor an uneven or wavy surface. Second, the shrinkage of both the rough flooring and the finish flooring will bring about open joints in the finish floor at various intervals. This is illustrated by Figs. 144 and 145. The illustrations are exaggerations, or better stated, we show what would happen if perfectly green lumber were used for both the rough flooring and the finish flooring.

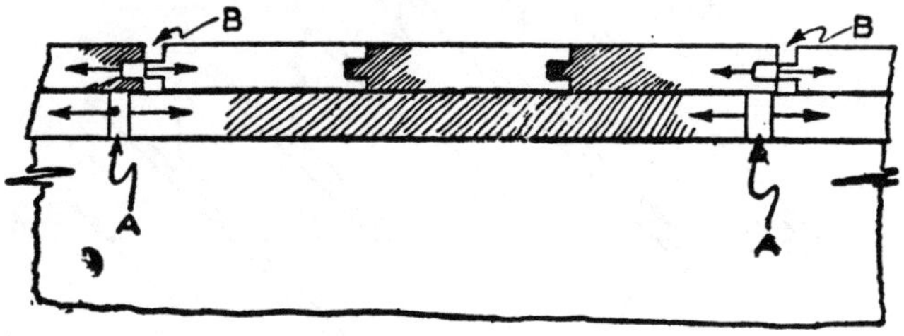

Fig. 144

At *A* and *A*, Fig. 144, we point out the open joint in the rough flooring caused by shrinkage, while at *B* and *B*, we show what would happen to the finished flooring. The shrinkage pull is indicated by arrows pointing away from the open joints. It is this pull that causes the open joints in the finish flooring. The indicators at *A*, Fig. 145, show the open joints in the rough flooring, while those at *B*, show the results in the finish flooring, exaggerated, of course.

When the rough flooring is laid diagonally, the results of shrinkage do not show up every third or fourth joint, but are distributed evenly over all the joints of the finish floor, and thus become scarecely noticeable.

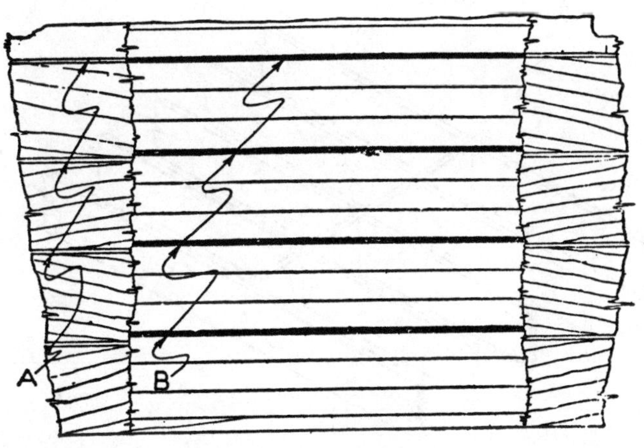

Fig. 145

Skill in Sharpening Tools

Knowing how to sharpen tools properly, is half the carpenter's trade, the other half is being able to skillfully handle them.

There is a distinction between sharpening tools and grinding tools. Even filing tools is not always sharpening. It depends on what kind of edge you are after, as to when to use the word "sharpening," and when to use "grinding," If grinding is all that is necessary to give you the edge you want, then grinding is sharpening. But if you grind a tool and then put it on the oil stone to give it a keen edge, grinding is simply grinding, and the oil-stone work is sharpening. If filing a tool gives you the edge you want, then filing is sharpening, but if you have to finish it with the oil stone, then filing is filing, and the oil-stone work is sharpening. This reasoning can be applied to all tools that from time to time need sharpening.

There is no better recommendation a carpenter can expect to have, than that of making smooth and well-fitted joints. This is true, not only in finishing, but it is also true in all rough work, and very important. A well-fitting joint does not only look better, but it is a better joint, and stronger, than the ill-fitting and ugly one, such as one frequently sees. That is the reason we mentioned tool sharpening. It is utterly impossible to make good joints with poorly sharpened tools. But we do not want to be misunderstood in this. Good judgment and common sense are assets that no carpenter can easily afford to dispense with. There are times and conditions under which it would be stupid to work over a joint in order to make it well-fitting, just as there are times and conditions that enter into the tool sharpening. To know when to do these things and when to forego doing them, is a fine shade of distinction that every mechanic must make many times during his whole career.

Floor Openings

Fig. 146 shows an opening for a stairway, or as it is called, a wellhole or well. Here we have *a*, *a*, trimmers, *b*, *b*, carrying joists, *c*, *c*, *c*, tail joists, and *d*, header. It will be noticed that the carrying joist to the left answers also for the left trimmer of the well. This construction is not an unusual one; however, there are many stairways on which it cannot be used, and a trimmer, such as is shown to

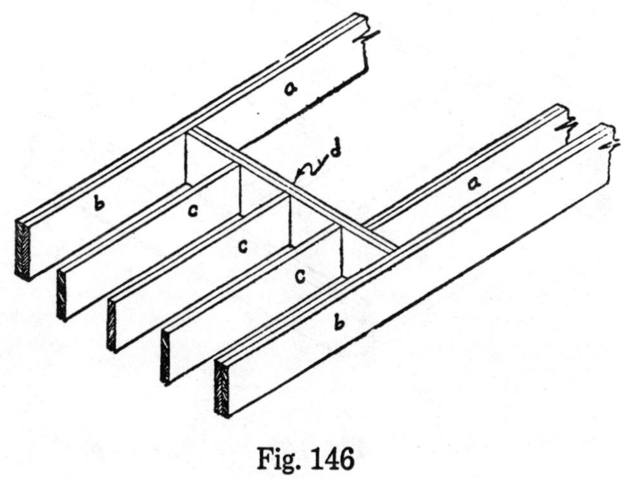

Fig. 146

the lower right, is required on either side of the well. The joints, as we show them, are held together with nails, and for most residential work that is sufficient. In cases where the header, or even the trimmers and the tail joists, are subjected to a great deal of strain, joist hangers or stirrups should be used for reinforcing. Joist hangers can be obtained on the market, or they can be ordered made of wrought iron.

A wellhole in which the carrying joists, *a*, *a*, rest on a brick wall, is shown by Fig. 147. The dotted line represents the inside face of the wall. The header, *b*, in this case is on the side and runs parallel with the run of the stair, whereas in the previous figure the header was on the end of the well. Three of the tail joists in this figure are marked *c*, while the absence of the mark on the others, means ditto. What was said about stirrups in the explanation of the previous figure, will apply throughout this chapter.

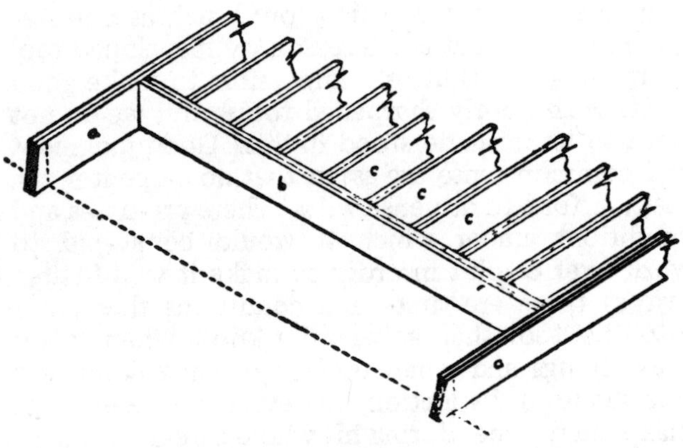

Fig. 147

A plain perspective view of the construction of a fireplace opening in a floor, is shown by Fig. 148. The dotted lines represent the face of the well.

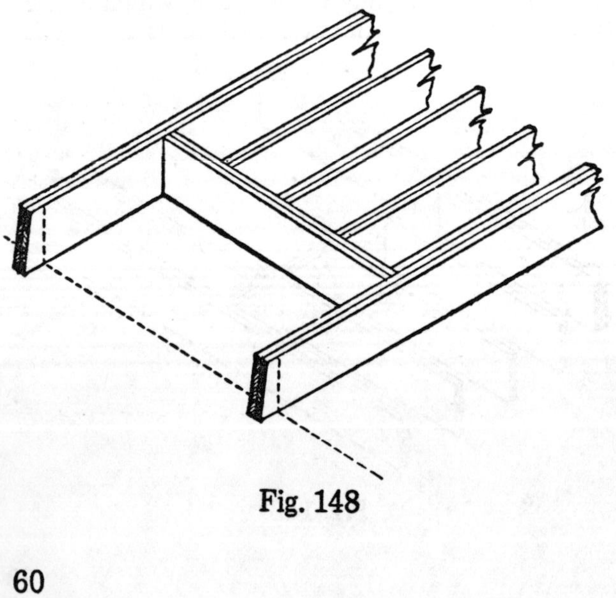

Fig. 148

How to carry joists where there are flues built into the joist-supporting wall, somewhat on the order shown by dotted lines, is illustrated by Fig. 149. Sometimes a chimney breast is built where such flues occur, in which case the construction is the same, excepting that the carrying joists are doubled.

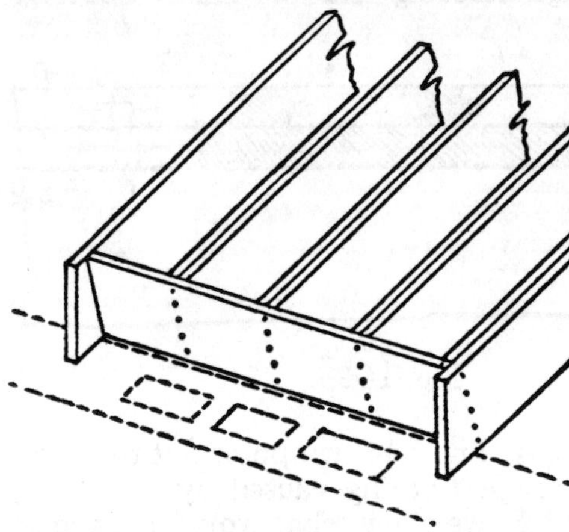

Fig. 149

A method of construction that should be used cautiously, is shown by Fig. 150. It shows how, in case of a shortage of full-length floor joists, pieces can be utilized by framing a header in between two carrying joists, as shown. Onto this header the tail joists, for that is what they are, are nailed. We repeat, this construction should not be used where the floor is subjected to much strain, unless there are joist-supporting partitions underneath, located in such a way as to make the floor above perfectly safe.

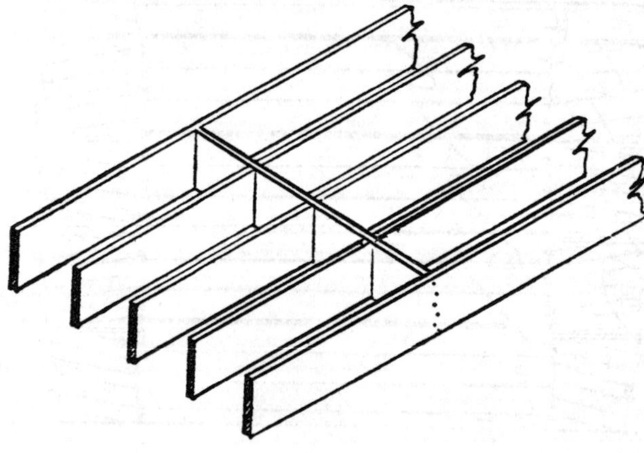

Fig. 150

The construction of an opening for a chimney, is shown by Fig. 151. The trimmers are shown at *a*, *a*, the header at *b*, *b*, the tail joists at *c*, *c*, and the carrying joists at *d*, *d*. The upper left trimmer is also the carrying joist. This construction is all

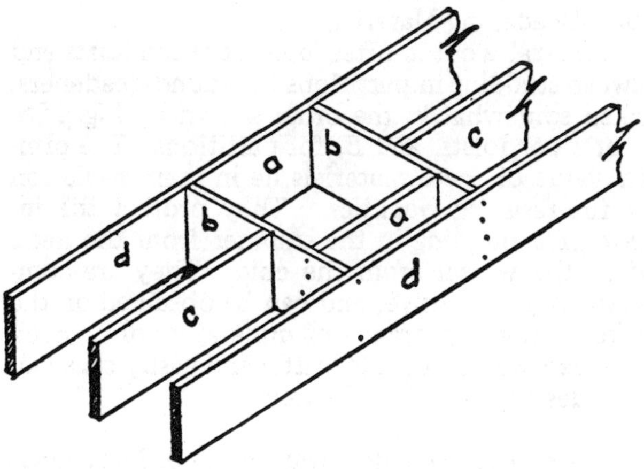

Fig. 151

right when the chimney comes right for it, but in many cases this is not the case. Then two trimmers, one on either side, like the one to the right, must be put in.

Sound-Proofing

Much as folks are anxious to hear, there are times when they want to be where it is quiet. And to accomplish this, sound-proofing is necessary. Bedrooms, offices, churches, hospitals, schools, jails, theaters and other buildings or parts of buildings, often must be made sound-proof to be satisfactory.

There are a number of sound-deadening materials on the market, felt paper, sound-deadening felt, asbestos sheathing, Cabot's Quilt, etc. In fact almost any building paper has some sort of sound-deadening qualities. The most satisfactory, however, are those which are at the same time fireproof and vermin-proof.

A method of sound-proofing a floor by means of some kind of sound-deadening felt or quilt, is shown by Fig. 152. Onto the rough flooring is laid (not nailed) a layer of deadening material, and onto it 2x2 strips (also not nailed) and then the flooring is nailed onto the 2x2's in such a way that no nail will enter or pass through the deadening material. This is a good method, and if fireproof deadening material is used, it can hardly be improved upon. The 2x2 strips, and the extra space necessary under ordinary circumstances would hardly be noticeable.

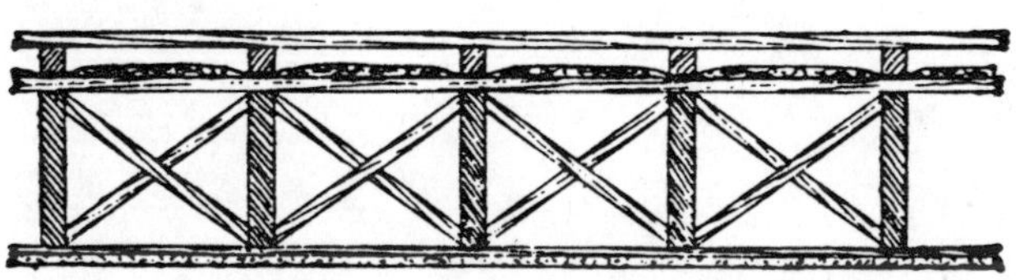

Fig. 152

An air-space method of sound-proofing a floor, is shown by Fig. 153. Here two sets of joists are employed. One set to support the floor, and another to carry the ceiling. In addition to sound-proofing the floor, this method prevents the vibra-

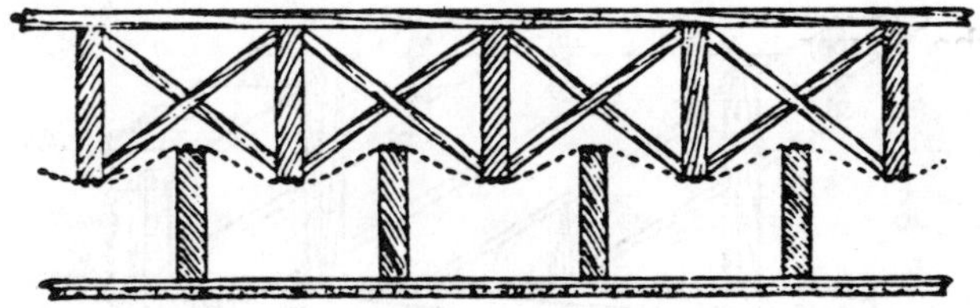

Fig. 153

tion of the floor from being transmitted to the ceiling. This is desirable where the floor is to be used for dancing and similar purposes, and the ceiling is plastered. The dotted line shows how a sound-deadening material can be added to this construction, not only making it more soundproof, but more nearly fireproof, provided a fireproof material is used and carefully placed.

Without the fireproof sound-deadening material between the two sets of joists, the arrangement is a dangerous fire-trap. The fire could spread in every direction with fuel enough to keep it going, while access for extinguishing it would be difficult. These things should be kept in mind when sound proofing is considered.

A similar method of sound-deadening for partitions is shown by Fig. 154. The irregular line between the two sets of studding represents sound-deadening material of some kind. What we said about fire, in the explanation of the previous figure, can be applied to this also.

Fig. 154

Dropping Ceilings

Fig. 155 shows how to strip joists, that for some reason are not spaced right to receive, either the lath or other ceiling material, whatever that might be. The stripping, of course, must be governed by the requirements of the material used. How to drop the ceiling some distance below the joists, is shown by Fig. 156. Fig. 157 shows how to drop the ceiling when the joists are spaced right for the ceiling material. In this case it will be noticed, that the nailing strips run parallel with the joists, whereas, in the previous figure they ran crosswise.

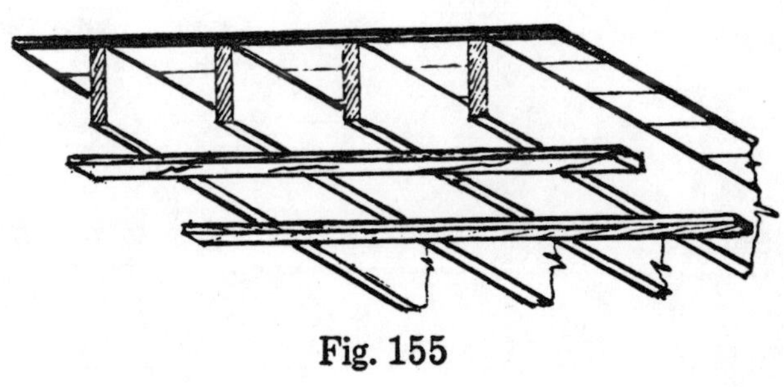

Fig. 155

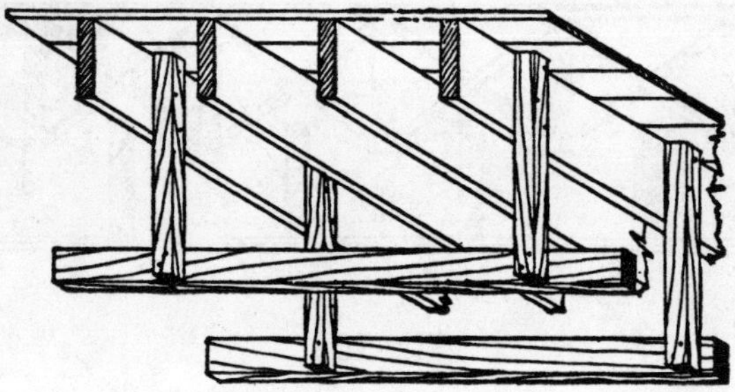

Fig. 156

A stripped or a dropped ceiling helps somewhat to deaden sound, the fire danger, though, is greatly increased. The access for extinguishing fires in dropped ceilings is difficult, and fire can spread in a few moments throughout the space between the ceiling and the floor, eating up hangers and joists for fuel.

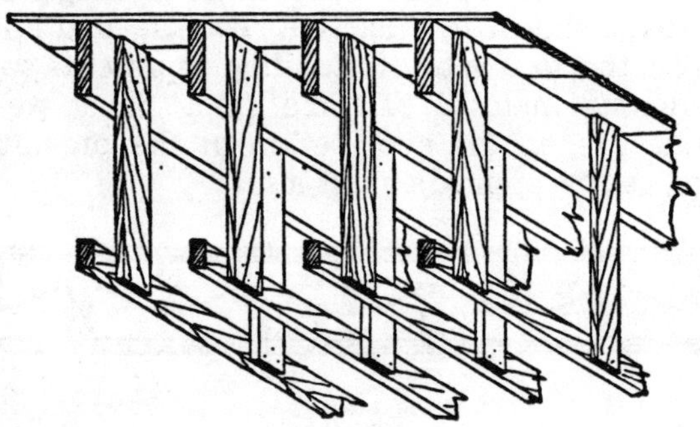

Fig. 157

Sound-Deadening Materials

Mineral wool is often used between joists and between studding in partitions for sound-deadeners, packed somewhat in the order shown by Fig. 158, *A*, for floor joists, and *B*, for partitions. The principal value of these materials lie in their insulation and fire-resisting qualities. They protect the interior of a building in the summer from the heat, and in the winter from the cold. They are comparatively inexpensive, and can be obtained on the market. The appearance of mineral wool is much like sheep wool, but it is brittle and easily crushed, which destroys much of its value.

In order to make doorways more or less soundproof, two doors should be used, leaving an air-space between the doors when they are closed. When the doorway is used for heavy traffic, one door can be left open, but when traffic is light, and sound-proofing is desired, then both doors should be kept closed. In cases of outside doorways, a storm door will answer at the same time for a sound-deadener. The same can be said of windows. Interior windows with two sets of sash, leaving an air-space between, will prevent, to a great extent, the transmission of sound. For outside windows, a storm sash added to the regular window will answer both for sound-deadening and against cold.

Fig. 158

12

Skeleton Walls

Marking for Studding

Marking for studding, joists or for openings, is not a difficult job, however, it must be done with care, if the proper degree of accuracy is to be attained.

One of the first requirements, is a good pencil, well sharpened, with lead not too hard and not too soft. A pencil that is too hard will not leave enough mark to be readily seen, and one that is too soft will wear away too fast. The most essential tool is the steel square. The figures and graduation marks should be clearly visible. A good way to bring out the figures and marks on an otherwise good square, is to clean out the grooves thoroughly, and then apply a coat of white enamel. Allow this to dry until it will cut without smearing and then with a sharp chisel shave off the enamel. This will leave the figures and graduation marks filled with enamel and easily read.

Studding Pattern

Fig. 159 shows at *a*, an edge view of a studding pattern lying on a studding to be marked, *b*, shows a top view of the same layout, and *c*, shows an enlarged detail of the gauge block fastened to one end of the pattern studding.

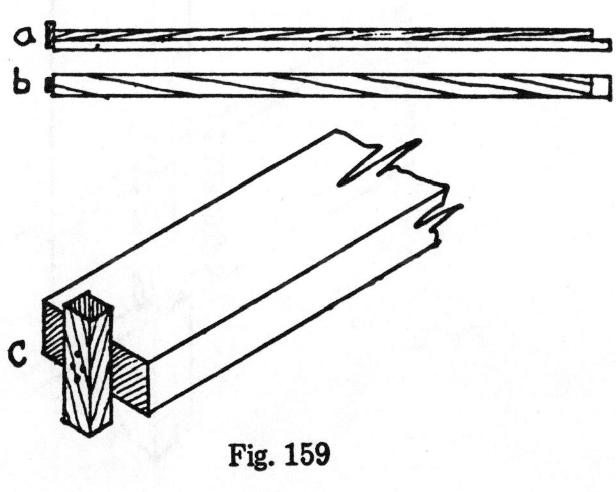

Fig. 159

Skeleton Walls

Fig. 160 shows at *B*, a very common method of laying off plates, which are shown lying side by side, with the studding marks on them. At *A* is shown the bottom plate nailed in place and the top plate nailed onto the studding, ready to be raised. When the wall is up, the studding are toe-nailed to the bottom plate. Fig. 161 shows another method of doing the same thing. At *B* the plates are side by side, and laid off. At *A* the wall is shown lying on the floor, with both the top and bottom plates nailed onto the studding, ready to be raised. When the wall is up, instead of toe-nailing the studding to the plate, as in the other instance, the plate is nailed to the floor. Fig. 162 shows the same wall

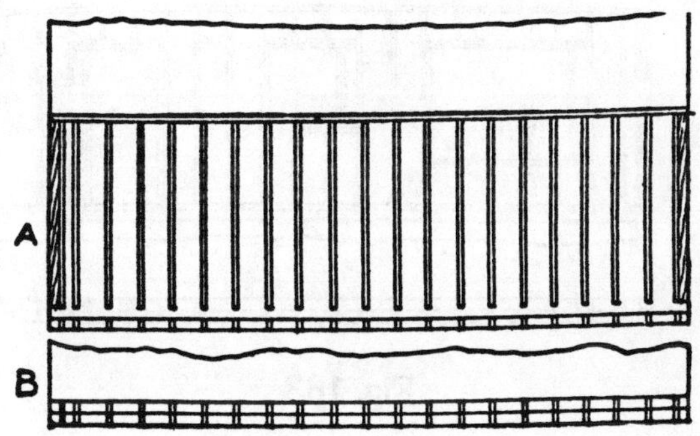

Fig. 160

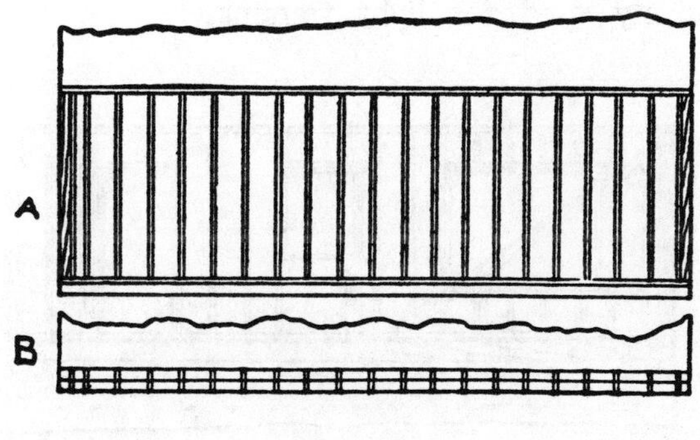

Fig. 161

after it has been raised and nailed in place. To the left can be seen the sway brace.which holds the wall in a plumb position, one way. The dotted lines indicate rough openings.

Fig. 162

Still another method of putting together a skeleton wall, is shown by Fig. 163. At *B* are shown the top and bottom plates, marked for the studding and for the trimmers of the openings. The latter are indicated by the use of an X-mark. At *A* is shown the wall lying on the floor with both plates nailed onto the studding, and the openings framed. Fig. 164 shows this wall raised and braced. Fig. 165 shows another necessary brace in order to keep the wall perfectly plumb, both ways.

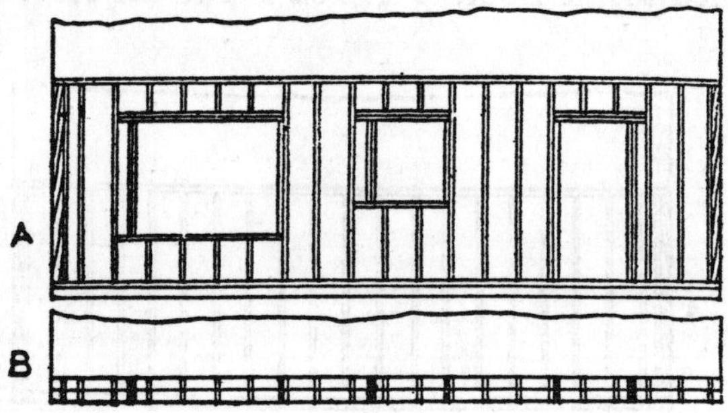

Fig. 163

It will be noticed, that in the method of framing openings shown here, one side of the opening is framed in such a manner that by doubling the studding it answers for a trimmer. This method is extensively used for light framing.

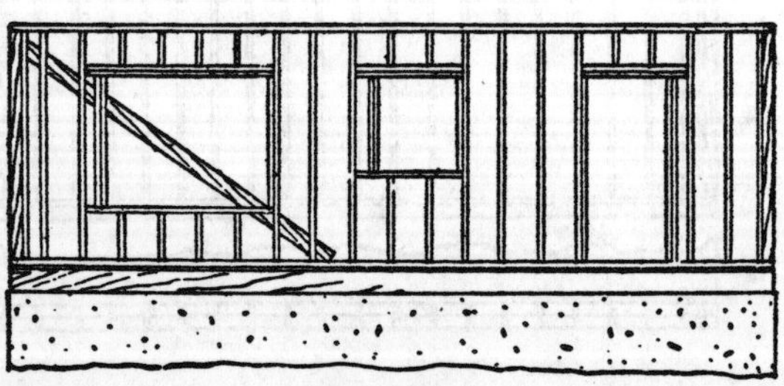

Fig. 164

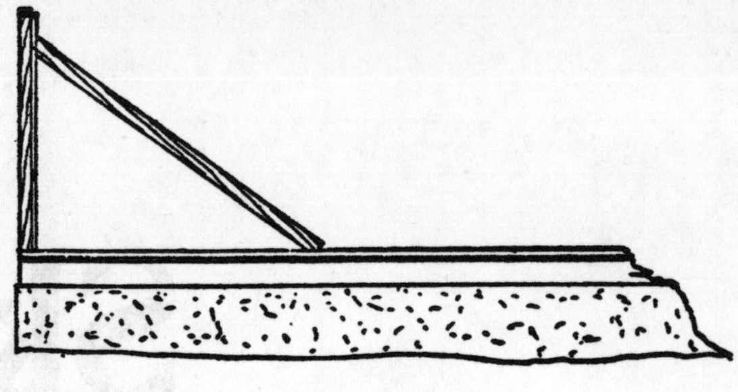

Fig. 165

Two-Story Skeleton Walls

The corner posts of two-story skeleton walls are nailed in place first and braced, as shown in Fig. 166, *c, c.* At *d* we show a brace holding a corner post. The center studding, number *10,* is then erected. After the ribbon boards *a* and *b* are in place, the studding numbered *1, 2, 3, 4* and so on will be set. A good way to do this is for one man to handle the studding, another nail the ribbon board and a third toe-nail the studding to the bottom plate. Fig. 167 shows how the studding are notched to receive the ribbon board, and Fig. 168 shows the skeleton wall completed, including the double top plate.

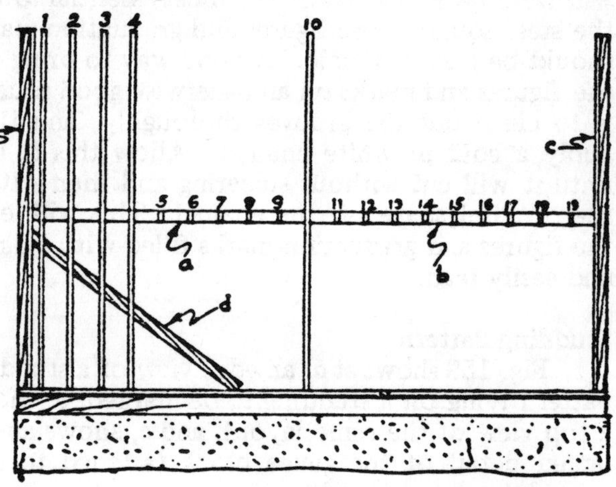

Fig. 166

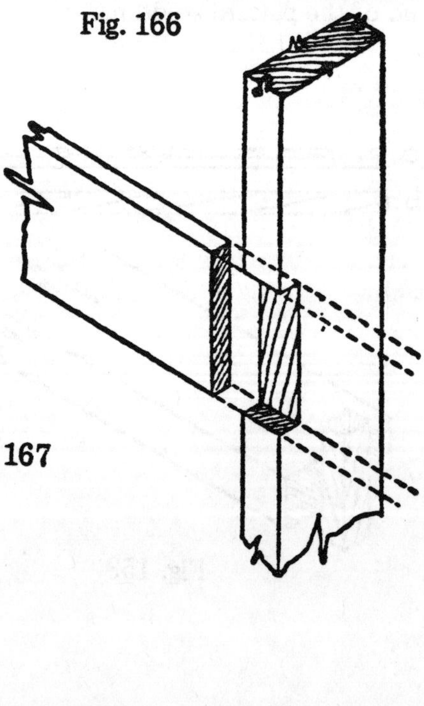

Fig. 167

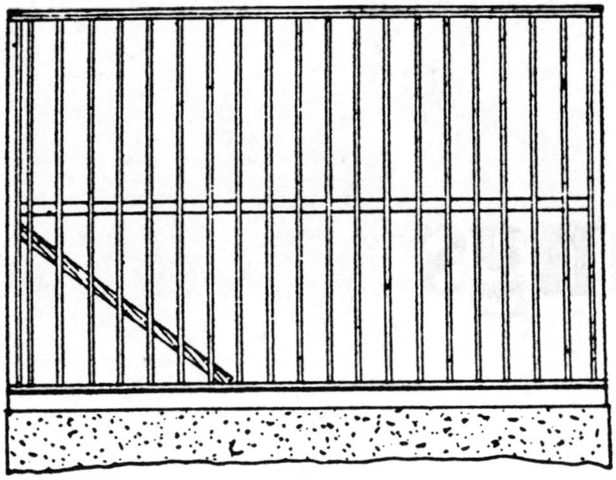

Fig. 168

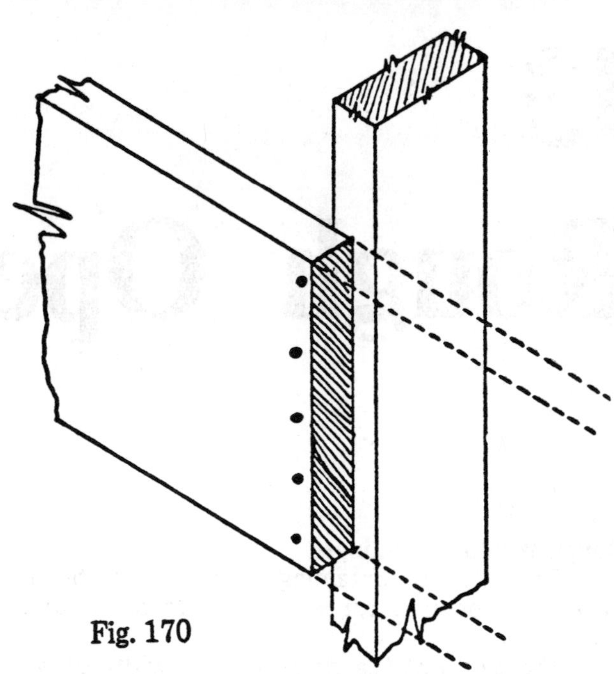

Fig. 170

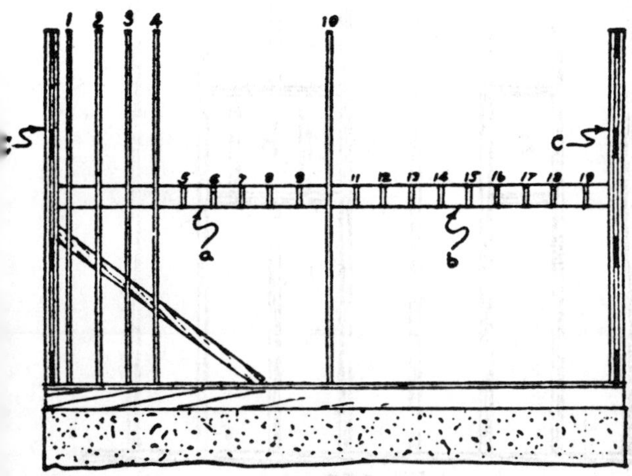

Fig. 169

Fig. 169 shows how to proceed on the sides where the studding are nailed to the end joists. As in the other case, the corners marked *c*, *c* are erected first. Then the joists marked *a* and *b* are nailed in place, as shown. With a man nailing the joist to the studding and another man toe-nailing the studding to the bottom plate, proceed to erect the studding in the order shown by numbers, *1*, *2*, *3* and so on. It will be noticed by referring to Figs. 168 and 169, that the braces shown are nailed on the inside, in order to keep them out of the way when the studding are erected. Fig. 170 shows how the joists are nailed to the center studding, which in Fig. 169 is marked number *10*. Fig. 171 shows the skeleton wall completed.

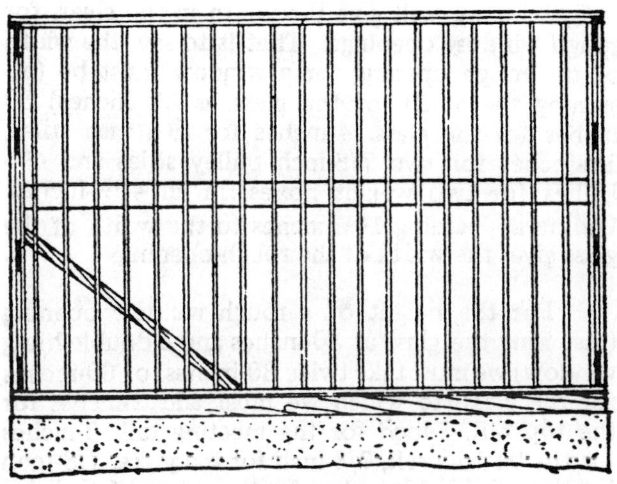

Fig. 171

Bracing

In order that the reader will not be misled, it should be said that the braces shown in the illustrations are given merely to show how braces should be nailed. The corners should be braced two ways on every story, and the walls should be braced enough to hold them in proper alingment. As a rule, proper bracing is governed by circumstances. No hard-and-fast rule can be laid down, other than the rule of substantial construction.

13 Rough Openings

Rough Window Openings

Formulas for obtaining the width and height of rough openings for windows, are as follows:

The width of the glass, plus the width of two sash stiles, plus the thickness of two pulley stiles, sometimes called jambs, plus the width of two weight boxes, will give the width in the clear, for rough window openings. That is to say, the width of the rough opening for a window must be (assuming the width of the glass as 30 inches) 30 inches for the glass, 4-inches for two sash stiles, 1¾-inches for two 7/8-inch pulley stiles and 4¾-inches for two weight boxes, in all 40½-inches. Ordinarily, adding 10½-inches to the width of the glass gives the width of the rough opening.

For the height of a rough window opening (assuming the glass as 30-inches and a double-hung window) we must take twice 30-inches, or 60-inches for the glass, to which we must add 2-inches for the top rail, 1-inch for the meeting rail, 3-inches for the bottom rail, 7/8-inch for the yoke, 7/8-inch for the subsill, 1¾-inches for the sill and 1-inch for clearance, in all 70½-inches. For ordinary purposes, adding 10½-inches to the height of the glass will give the height of a rough opening for a double-hung window.

Taking up the illustrations, Fig. 172 shows at *a*, the simplest kind of rough opening, consisting of a top and bottom header nailed between two studding. At *b*, we show the same size opening which is framed so as to be located at a specific place, hence a studding had to be cut out, and in addition to top and bottom headers, two trimmers are necessary. Fig. 173 shows at *A*, a commonly used method of framing rough openings for windows in very cheap work. This method is all right where it is not necessary to locate the window at a given point. At *B* is shown a somewhat better construction, which is merely a modification of what is shown at *A*.

The methods of framing rough openings for windows shown in Fig. 174 are good, and are commonly used on residential work. Attention should

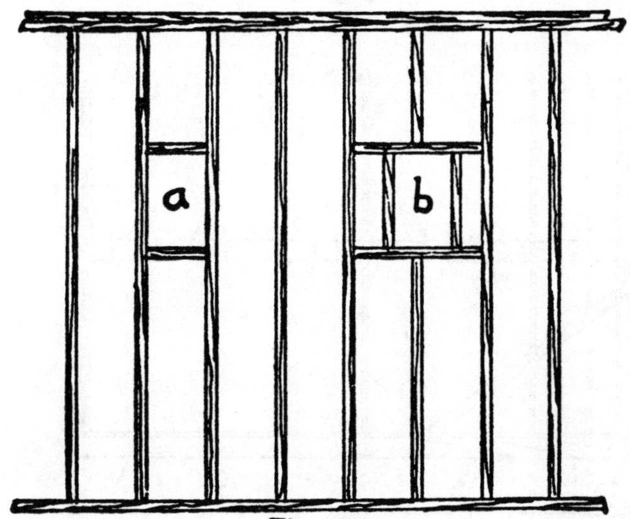

Fig. 172

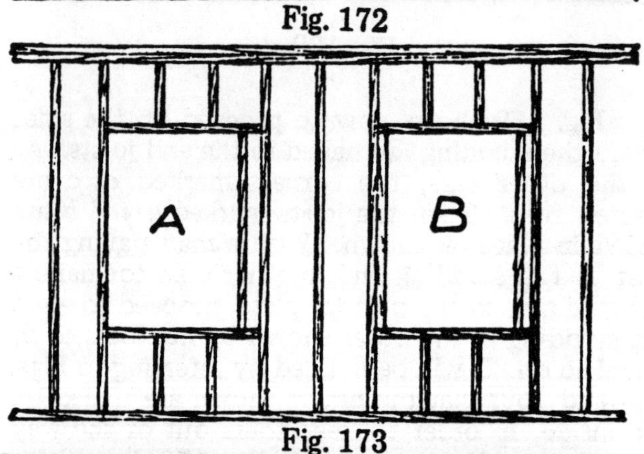

Fig. 173

Fig. 174

be called to the construction of the top headers. At *a*, we show a plain doubled header, which is all right, but will not support as much weight as the header shown at *b*, where the 2x4's are set on edge, with a lath between to bring them to the width of the studding. The former of these construction costs a little less in labor and material than the latter, but for ordinary purposes, especially where boxing is on the studding, one construction is as good as the other. One reason, but not the only reason, for doubling headers and trimmers in rough openings, is to provide nailing for the trim, or finish, as it is sometimes called.

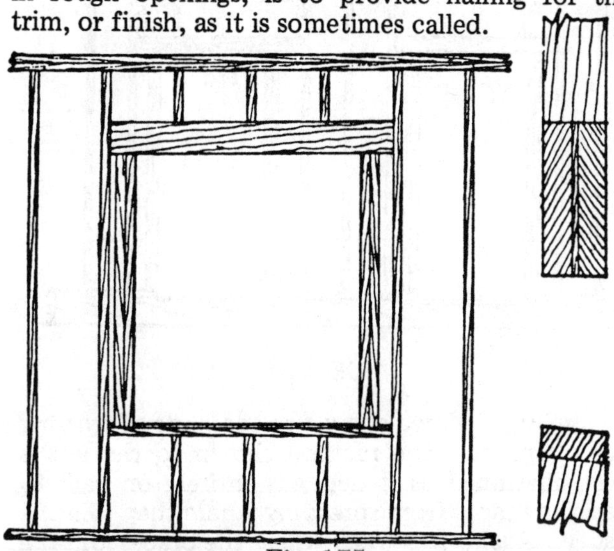

Fig. 175

Where a wall must support a great deal of weight, as is often the case, the construction shown in Fig. 175 is very good. The top header, shown in detail to the right, is built up of 2x6's or 2x8's set on edge. The trimmers are doubled 2x4's, while the bottom header is sloped outward, as is shown by the detail to the right. Fig. 176 shows a rough opening for a twin window. The top header is built up of 2x6's or 2x8's, with laths on the inside to bring it out in alignment with the edges of the studdings, as shown by the detail to the right. The bottom header is doubled, as shown. This makes a substantial construction, and provides amply for nailing.

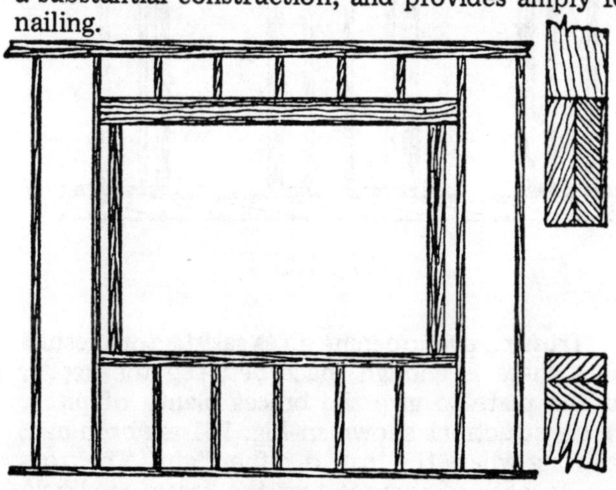

Fig. 176

As we explained in a previous chapter, some builders frame the rough openings before raising the skeleton walls, others raise the walls and cut and frame the openings afterward. We have used both methods, and like them. The former is a labor saver, while the latter, perhaps, makes possible a larger degree of accuracy.

We show by Fig. 177 a templet for marking studding to be cut out for window and door openings. The templet is made of a regular studding, as shown to the left. To the right, is shown the templet set against a studding ready for marking the top and bottom cuts. The upper crosspiece is so

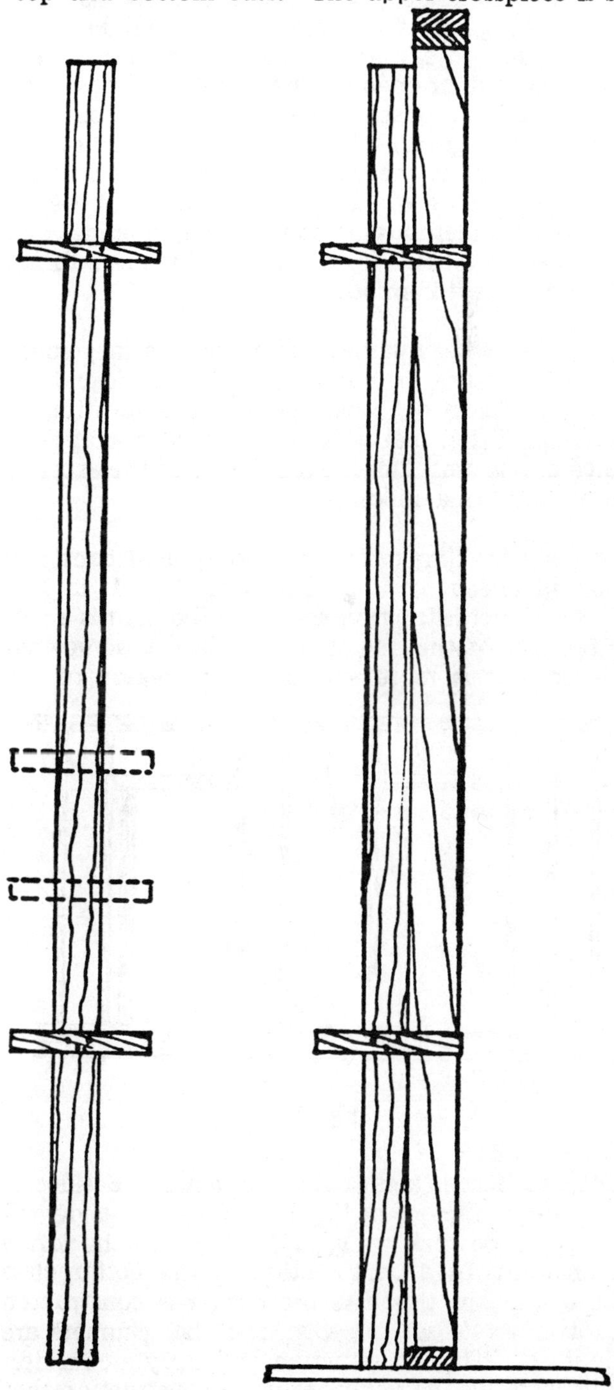

Fig. 177

located on the templet that when the cutting is done and the header is in, the height of the opening will be right. The same can be said of the bottom crosspiece, it must be placed in such a way that allowance will be given for the header. For doors, only the top crosspiece is necessary. The dotted lines, shown on the templet to the left, indicate how additional crosspieces can be placed for small-openings. The reason the crosspieces extend both ways, is to make it possible to mark either side of the studding, for it is not always convenient to do the cutting from the same side.

We are dealing with doors in particular, and while the journeyman carpenter often does not have much to say about the construction or the location of door openings, he sometimes does. The architect is supposed to locate the door openings, and usually governs himself by the needs of the room, which is the way it should be. However, there are times when the carpenter must be informed on methods of constructing openings and of locating them. In locating doors, these things should be kept in mind:

The owner's wishes, the intended arrangement of the furniture, the convenience for traffic and the relative effects on other rooms. The location of openings often makes or mars the general appearance of a room. This is equally true of the exterior appearance of a building.

A very simple, but commonly used construction in cheap work, is shown to the left in Fig. 178. The construction shown to the right is a modification of the one at the left, and is somewhat better -- it is more substantial by reason of the

Fig. 178

doubled header and doubled trimmers. Besides, it provides sufficient nailing both for the outside and inside woodwork. Fig. 179 shows to the left a commonly used construction in the better class of work. In this case the header is constructed of two 2x4's set on edge, and the trimmers are doubled. This construction will carry a considerable load. To the right is shown a construction that will carry a heavy load. The header and trimmers are doubled, but the header is reinforced by a blind header placed directly under the plate. This blind header can be either single or double, whichever the case may require. More than that, it can be made of 2x4's, 2x6's or 2x8's, depending on how much weight must be supported, and on the span of the opening. That fundamental requisite of good judgment is as important in framing rough openings as it is in building any other part of a structure.

Fig. 179

In Fig. 180 we show two methods of trussing over openings. The method shown to the left is weak, inasmuch as it depends entirely on nails to keep the braces from spreading where they join the header. Again, it is weak where the braces join the studding, for here there is nothing to keep them from slipping but the nails. In supporting a heavy load, nails should never be depended on entirely. For light loads, this construction will give fairly good results, but it is a faulty form. The method shown to the right is good. The studding are framed afterward and nailed to the braces as shown.

Fig. 180

Trussing over opening gives satisfactory results where there is enough space between the header and the plate to give the braces plenty of pitch. The constructions shown in Fig. 181 are both bad, excepting where the load is rather light. The trouble with both of them is that the braces are likely to spread when the load is applied. A blind header, as shown to the right in Fig. 179, would give much better results.

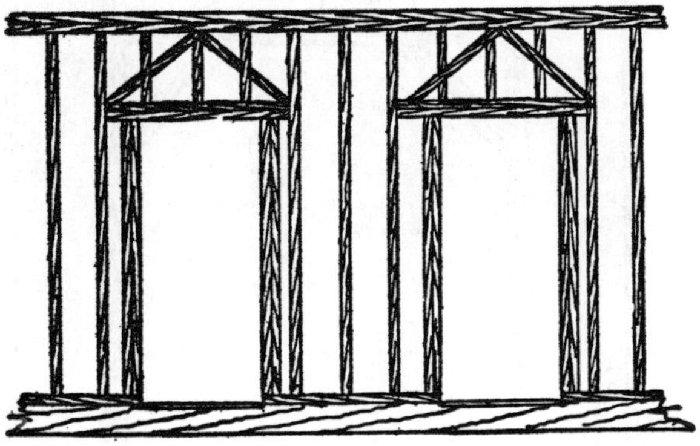

Fig. 181

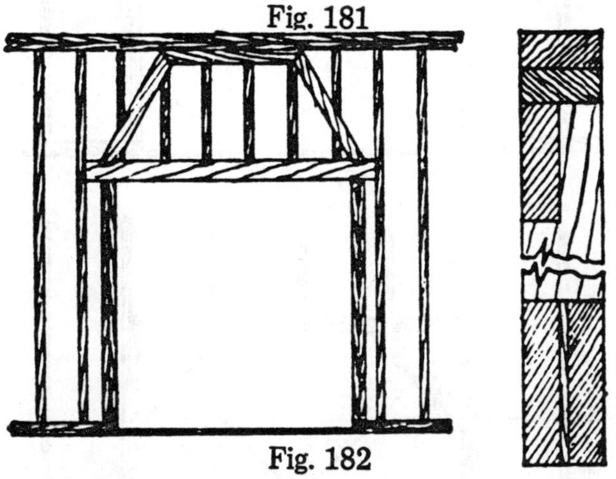

Fig. 182

A good method of trussing over a wide opening is shown by Fig. 182. In this case we have a header set on edge, with the braces of the truss notched into it. The detail to the right shows the construction of both the header and the top member of the truss. We show the truss constructed of a single piece, but if necessary, it can be doubled throughout. A modification of this truss is shown in Fig. 183, which is a better form than the one just considered. The braces are notched into the

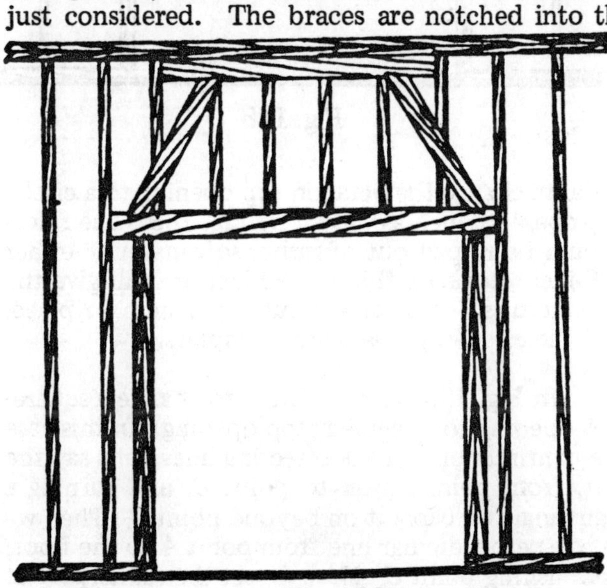

Fig. 183

header and also into the top member of the trusses. This will support a heavy load, provided the trimmers of the opening are substantially supported. The bottom support of door openings is even more important than the top, for if the bottom gives way, the top will have to come down, no matter how well the frame is made.

In Fig. 184 is shown, with figures, the size of a rough opening for a door measuring 3 feet by 7 feet. It will be seen that the rough opening must be 3-inches wider than the door, and 3-inches higher, assuming that the door jamb is made of 1-1/8-inch material. If 7/8-inch material is used for the jamb, 2½-inches added to the width of the door will be enough. The height, however, should be 3-inches higher than the height of the door, measuring from the rough floor to the bottom of the header.

Rough openings should be framed with straight material without defects. The measuring, marking and cutting of the lumber should be done painstakingly, which of course, will mean that the joints will be good, the headers will be level, and the trimmers plumb.

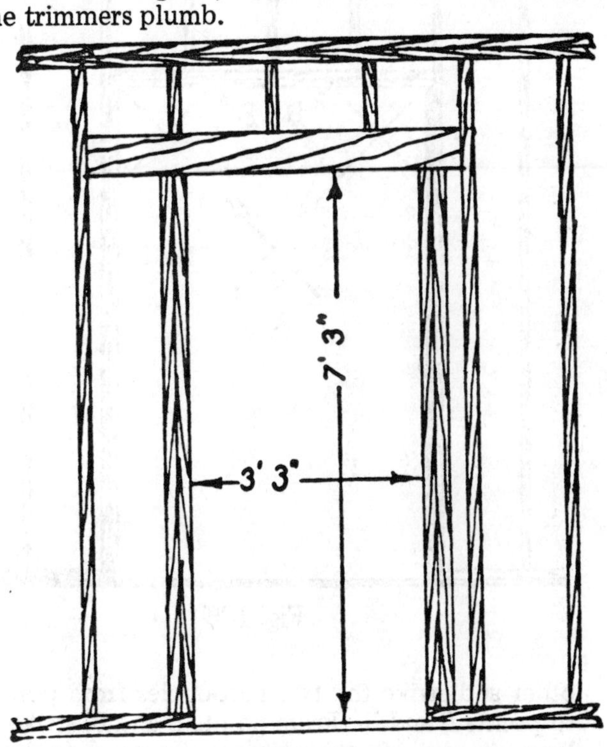

Fig. 184

A square-top rough opening for a door, is shown by Fig. 185. The same opening is shown in Fig. 186 where we show how to proceed to turn a square-top opening into an octagon-top opening. First bisect the two corners by striking two lines on a 45-degree angle, as shown by dotted lines, a little farther than half-way across the opening, or beyond the point marked *A*. Having this point, set the radius pole, first at one corner and then at the

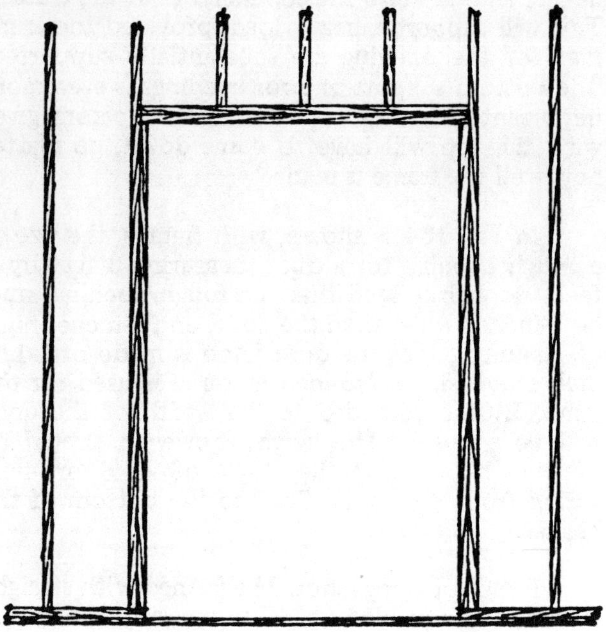

Fig. 185

Fig. 187

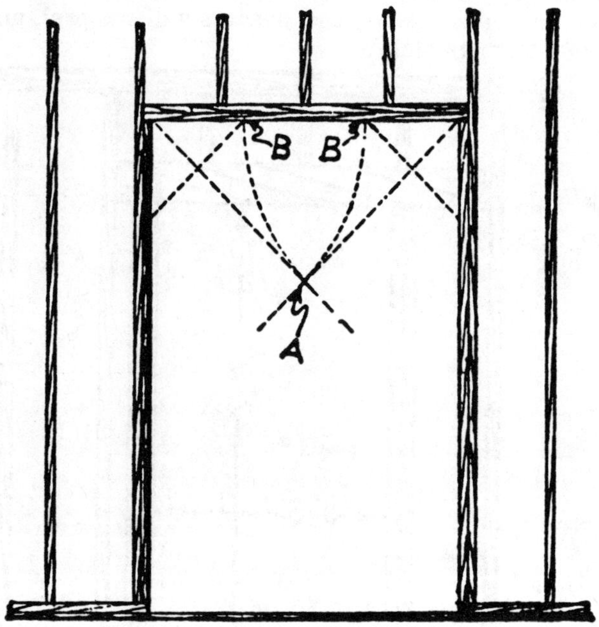

Fig. 186

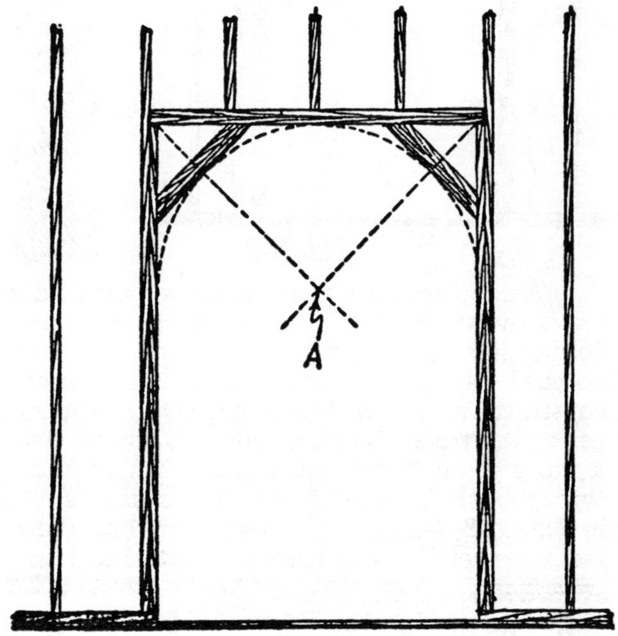

Fig. 188

other and strike the two part-circles from point *A* to points *B* and *B*. From points *B* and *B*, strike the lines cutting off the two corners on a 45-degree angle, as indicated by dotted lines. These points obtained, we can proceed to complete the octagon-top opening by framing the angle pieces and nailing them into place. The opening is shown completed in Fig. 187.

But suppose we want to frame a circle-top opening -- in that case we will have to place the radius pole at the point where the two bisecting lines cross, or at point *A*, Fig. 188, and strike the half-circle shown by dotted line. The half-circle shows how much must be filled in at each angle, in order to change the octagon-top opening to a circle-top opening, such as shown by Fig. 189. The fillers should be ripped out of rather soft material, either 7/8-inch stock or thicker, whichever will give the best results. The fillers cut, nail them in place, and the circle-top opening is complete.

In Fig. 190 we show how to change a square-top opening to a segment-top opening. In this case we continue one of the bisecting lines, let's say the left, from point *A*, on to point *B*, and turning a right angle we carry it on beyond point *C*. Then we drop a perpendicular line from point *A* to the floor, establishing point *C*. Now we set the radius pole at point *C* and strike the segment in such a way that

Fig. 189

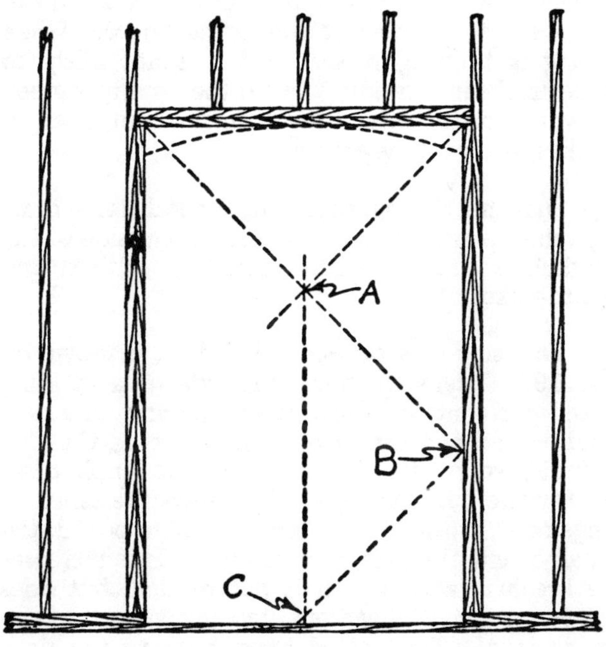

Fig. 190

Fig. 191

the highest point will contact the bottom of the header, as indicated by the dotted part-circle. The space between the part-circle and the header must be filled in, which will make the completed opening appear similar to what is shown in Fig. 191. The fillers should be cut out of soft material, and nailed into place, keeping them flush with the edges of the studding and header.

It has been said that the radius for a true segment must be one and one-half times the width of the opening. On this principle our diagram is based.

14
Boxing

Boxing with Boards

There are many reasons why a frame building should be boxed -- with boards -- or with other materials. The first reason is that it encloses the building.

Boxing brings the studding into alignment, and holds them there. It keeps the walls in a rigid position by preventing vibration of the studding. Well nailed boxing, whether it is placed horizontally or diagonally, will, in most cases provide all the bracing necessary in ordinary residential work. It should be remembered, however, that a wall cut in various ways by large openings, or too many openings, loses much of its bracing value. In such cases special bracing should be provided. Moreover, local conditions must be taken into consideration in determining the amount of bracing necessary for any building.

In localities visited by earthquakes, the bracing of buildings should meet the demands of those localities. Where the buildings are subject to a great deal of wind pressure, the bracing should be done in keeping with the needs of such conditions. And where the earth is treacherous, so that it is almost impossible to put in foundations that will stay put, there the bracing should be made to meet the local requirements. But where good foundations are possible, and there is no danger of earthquakes or high wind pressure, much of the otherwise necessary bracing can be dispensed with.

Factory buildings, mills, or buildings that for various reasons are subjected to vibrations and strains, should be provided with special bracing which will meet the requirements of such conditions.

The purpose for placing boxing diagonally is often misunderstood, both by carpenters and by laymen. While we are aware that placing boxing diagonally gives a building additional bracing value, that is not the principal reason for doing it. A building that is finished on the outside with lap siding should be boxed diagonally throughout, not merely on the corners, as we see it done so frequently. The boxing, as it seasons, will shrink and thus cause the siding to crack where the edge joints of the boxing occur. Sometimes these cracks run from one corner of a building to the other.

In cases where the siding is somewhat cross-grained, the cracks will appear at the nails and run in toward the center of the siding board. When boxing is thoroughly seasoned it is not likely to cause cracks in the siding, but if the boxing is green or water-soaked when the siding is put over it, cracks are almost inevitable.

Diagonal boxing, excepting for additional bracing value, is not necessary where the outside of the building is finished with shingles or with stucco and the like.

One side of a one-story building, is shown by Fig. 192. Here we have one side of a building boxed horizontally, showing an opening for a twin window and one for a door. This boxing, if nailed as indicated by the dots, will provide ample bracing for the building. Fig. 193 shows the same arrangement, with the two lower corners boxed diagonally, and the rest horizontally. This increases the bracing value, which is not needed, but does not prevent the siding from cracking when the boxing shrinks. The dotted lines, both to the right and to the left, show how some builders box the sides diagonally up to the corners of the openings, and then fill in between with horizontal boxing.

Fig. 192

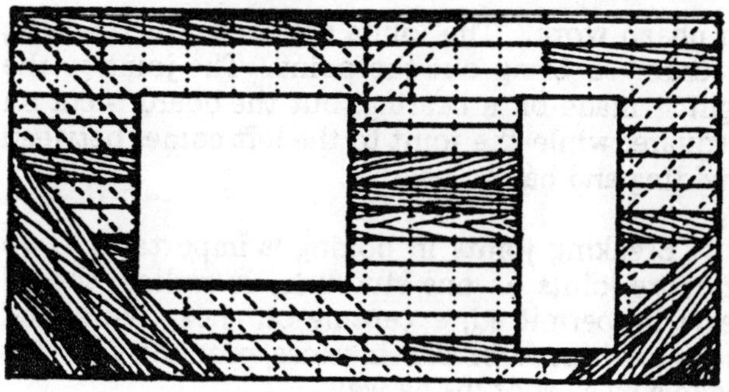

Fig. 193

Other builders continue the diagonal boxing from both ends until they meet somewhat in the order shown at the top center by dotted lines. This prevents the siding from cracking, but the joints coming on the studding next to the window opening make a bad construction, and will probably cause

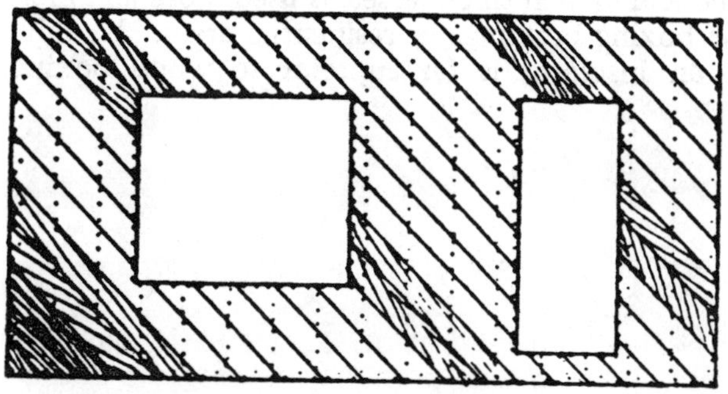

Fig. 194

the plastering to crack. A much better way to apply the boxing diagonally is shown by Fig. 194. This method not only gives full bracing value, but it prevents shrinkage cracks from occuring in the siding. In case of a gable roof, the diagonal boxing should be extended so as to cover the gable at the same time.

How end joints are sometimes made in horizontal boxing, is shown by Fig. 195. This method, where the studding are spaced 16-inches on center,

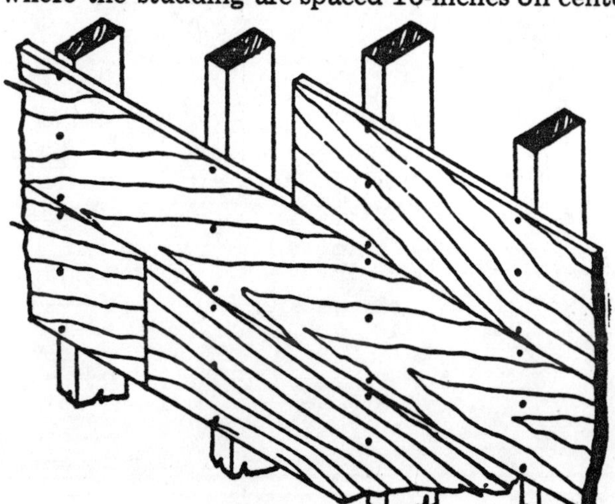

Fig. 195

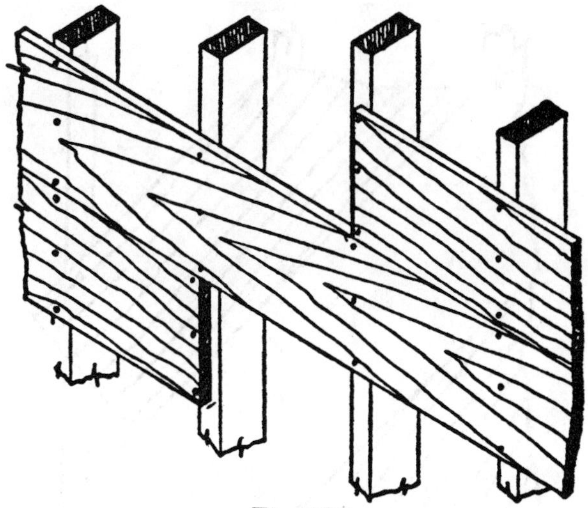

Fig. 196

does not make a bad joint -- however, it is seldom used in the better classes of buildings. Fig. 196 shows the approved method of making end joints. Every joint is made on a bearing, and both ends are nailed. This requires a little more labor and material than the former, but it cannot be improved upon. In the former method, some builders who use it place nailing blocks on the inside of the joints in order to hold them. When that is done, we feel that it would be cheaper to use the approved method, for what is saved in labor and material on applying the boxing, is lost again when the nailing blocks are put in place.

The approved joint for diagonal boxing, is shown by Fig. 197. This joint is made on a bearing and well nailed. Fig. 198 shows joints often used

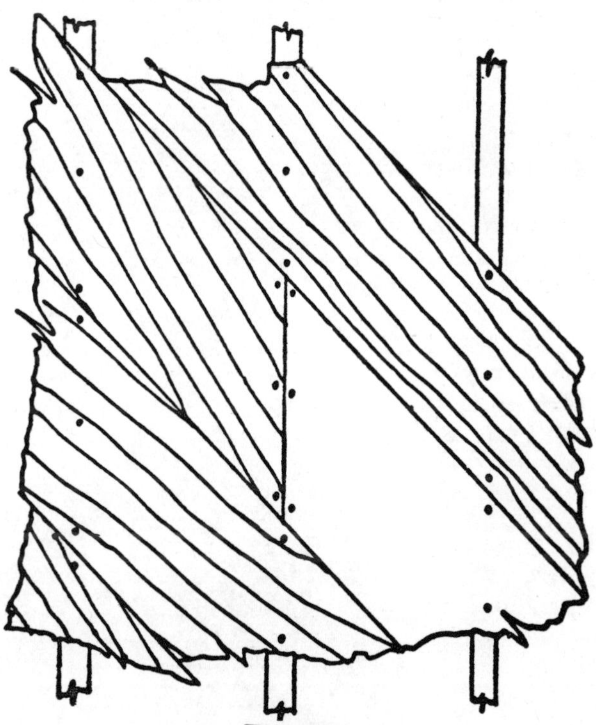

Fig. 197

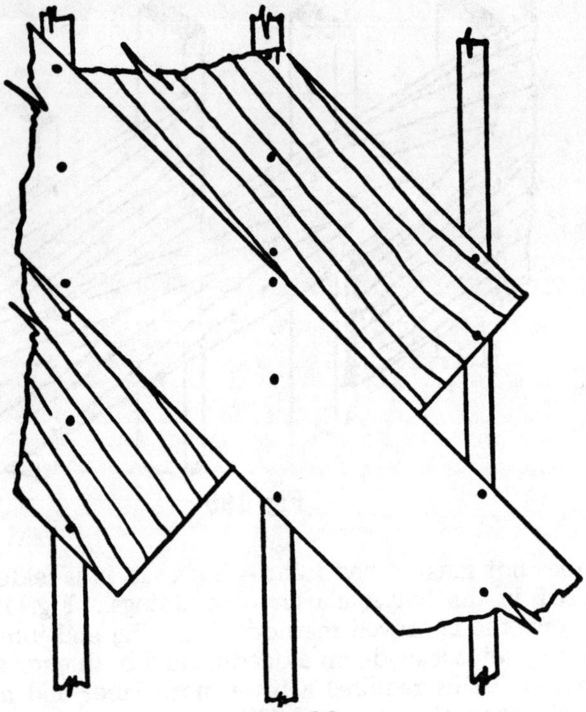

Fig. 198

on cheap work. The joints are shown incomplete, in order to bring out the point. The joint to the right is made on a bearing, but the board is cut on a square, while the joint to the left comes between studding and has no bearing.

Breaking joints in boxing is important. Two or more joints on one studding, unbroken, should never be permitted, excepting on cheap work, and many one-space breaks in close proximity always make a weak spot in the wall.

We were speaking on the basis of boxing with boards, which answers for much of the bracing necessities under ordinary conditions. Boxing with plywood or other manufactured material can be put on an equal basis, relative to its bracing value of a building.

In recent years celotex, which comes in ½-inch and ¾-inch thicknesses, is used more and more for boxing buildings in residential areas. Its value as an insulator is far superior to wood boxing.

15

Rough Porch Construction

The fundamental parts of porches are more or less the same, in most porches, but the designs vary greatly. In this chapter we shall deal with the skeleton part of porch platforms, ceilings, chords and lookouts. The lookout cornice, though, is passing out of use, and in its place we have the various forms of open or bungalow cornices, which to some extent are giving way to a style that we are going to call a dehorned cornice, which is more nearly borrowed from the past than modern.

Method of Procedure

Fig. 199 shows a simple method or procedure. The four beams, supported with temporary struts at *a, a, a, a* are joined to the main building by means of nails or anchors, somewhat on the order shown by the details in Figs. 203 and 204, which we will explain more fully in another place. *A* detail of the struts is shown to the right of the drawing. A 2x4 nailed to the main building between beams, flush with the top edge of the beams, serves in carrying the ends of the porch flooring abutting the building. This is pointed out by indicators on the drawing. The joists, which are supported by the beams, are cut equal in length for each section, and nailed as shown. It will be noticed that the joists are staggered from one section to the next.

The purpose of this is to make it convenient to nail through the beams into the ends of the joists. The dotted line shows how a line should be stretched, locating the outside of the face joists. The beams must be cut off the thickness of the joists shorter than the dotted line shows, so as to allow for the outside joists. The platform shown is for an 8-foot porch, but it will be seen that the figures show the framing of the rough work only 7 feet, 9 inches, which allows 3 inches for fascia, moulding and projection. Three inches is ample for this, in fact, 2½-inches will do. Fig. 200 shows the same layout with these exceptions:

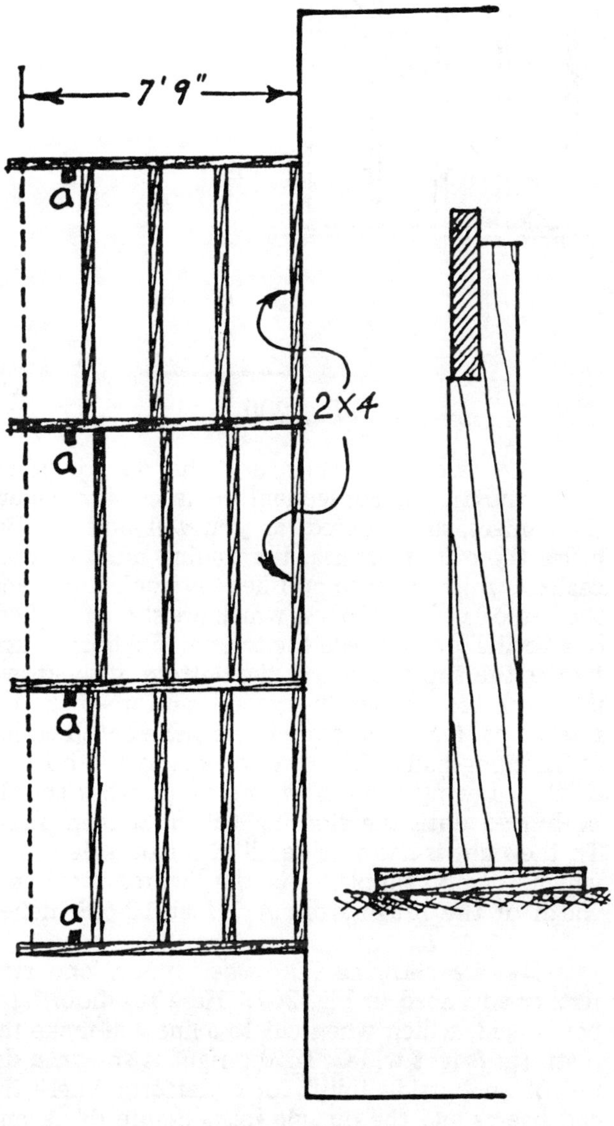

Fig. 199

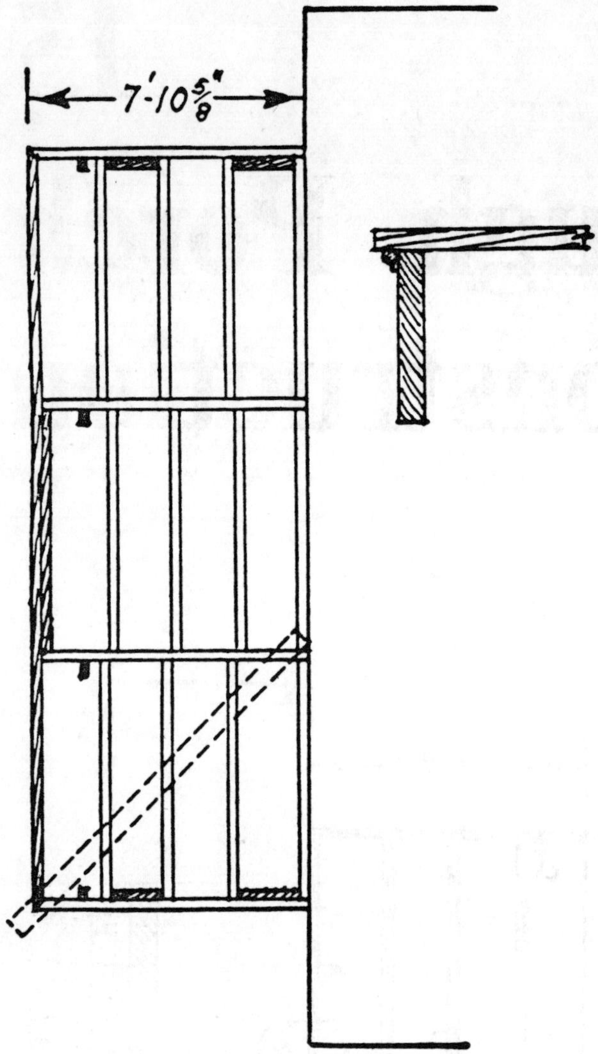

Fig. 200

The two end beams and the outside joists make the finish, consequently the corners where these meet, are mitered, as you will notice. Because there is no fascia, the nailing must be concealed, which is accomplished by toe-nailing and the use of nailing blocks, which are shown shaded, one to the left between the two middle beams, and two at the top and two at the bottom of the drawing. The blocks merely provide reinforcement for the toe-nailing. The corners are nailed with casing or finishing nails. The dotted lines at the bottom of the drawing show how the rough work should be braced until the flooring will hold it in place. To the right is shown a detail of the outside finish, which is very simple. As the figures show, the width of the rough work is 7 feet 10-5/8 inches.

We are showing the same layout, one step farther advanced in Fig. 201. Here the flooring is partly laid, which when cut to a line will make the platform 8 feet wide. To the right is shown a detail of an outside finish for a platform where the end beams and the outside joists are doubled, and a fascia is used for finishing. This construction is commonly used, and conforms with what is shown in the next figure. In the layout shown in Fig. 202 all of the beams are doubled as well as the outside joists. This is a good construction, especially for very large porches, and of course, will require a fascia for finishing. The only objection we have to this layout, is that the bulky beams will not dry out readily, and thus sooner or later will rot.

Fastening Beams

Fig. 203 shows a good method of fastening porch beams to the main building. It will be noticed that the boxing is omitted where the porch joins the building, thus making it possible to nail directly into the sill. Fig. 204 shows, perhaps, the best method of fastening porches to the main structure. In this case the boxing runs to the foundation wall, but the nailing is reinforced with anchors extending in to the sill, as shown. These anchors can also be used with the method shown in Fig. 203.

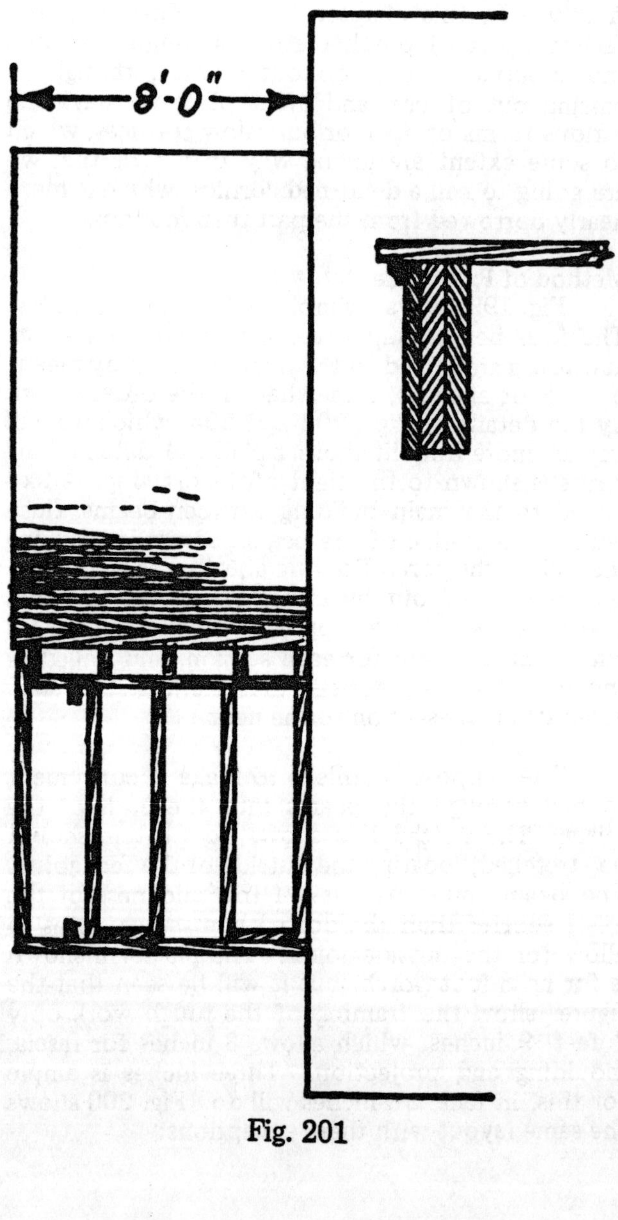

Fig. 201

Fig. 202

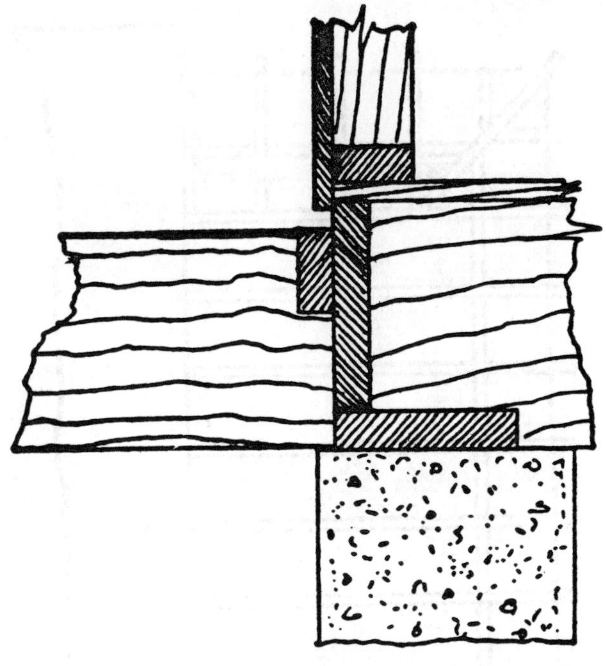

Fig. 203

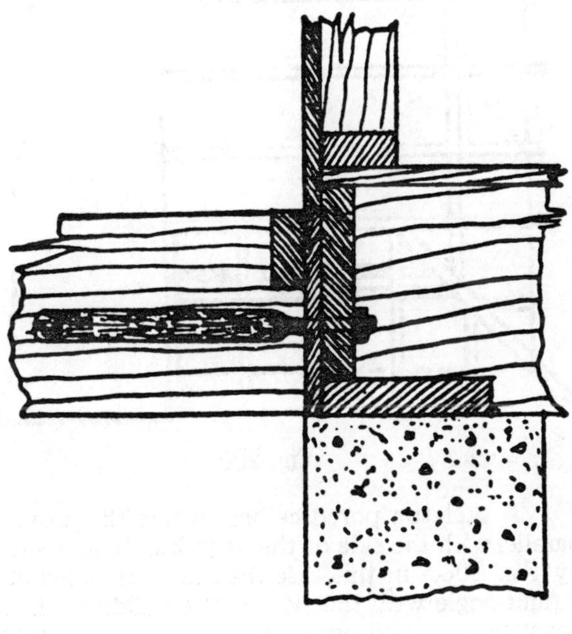

Fig. 204

Porch Ceilings

The constructions of porch ceilings shown by the illustrations are simple, and therefore practical. Fig. 205 shows a plan of a skeleton ceiling, where the ceiling runs parallel with the side of the main building. It will be noticed that we show, shaded, blocks between the lookouts at either end of the porch. These blocks are necessary -- for onto them the ends of the ceiling boards are nailed. In case of a cornice other than a lookout cornice, these blocks are not necessary, but in their stead a ceiling joist is placed. When the ceiling joists and lookouts are nailed in place as shown by Fig. 205, a line should be stretched where the lookouts are to be cut off, this is shown by dotted lines on the drawing. The width of the cornice should be established in such a way that a definite number of full-width ceiling boards will fill the space without ripping the last one. This not only makes a better job, but it reduces the cost of labor and waste in material. After the lookouts are cut off, the fascia is nailed on and the cripple lookouts shown by dotted lines at each corner are nailed into place. This done, the rafters are framed and put in place.

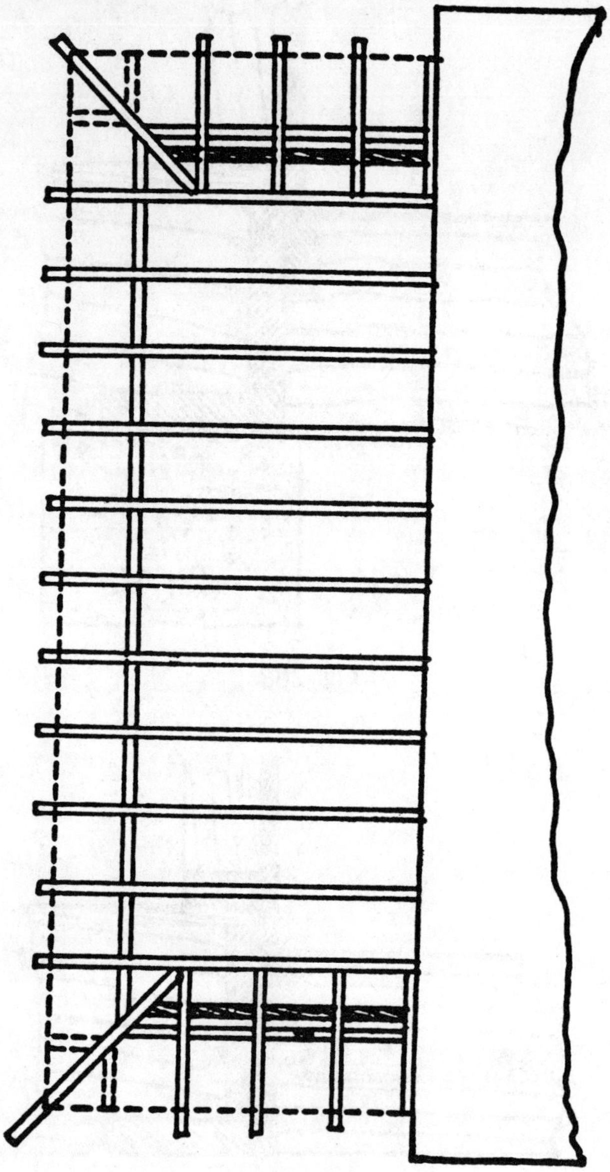

Fig. 205

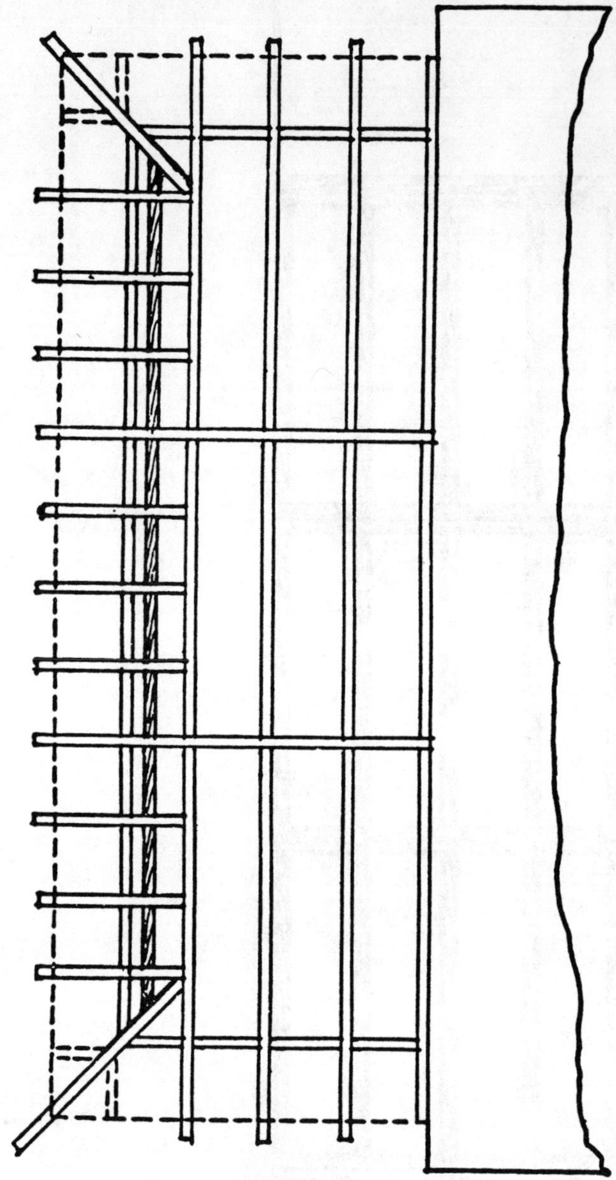

Fig. 206

A skeleton porch ceiling, where the joists run parallel with the side of the main building, is shown by Fig. 206. In this case the ceiling is nailed on at a right-angle with the side of the building. In this construction, two beams are necessary to carry the ceiling joists through the center of the porch. Here, the nailing blocks between the lookouts are placed at the front of the porch, rather than at each end, as shown by Fig. 205. The lookouts are marked and cut off the same as explained in the previous figure. The roof, also is a duplicate of the other.

Cornice Chord

Fig. 207 shows a detail of a cornice and a chord which are suitable for the two preceding ceiling constructions. We will take up chords and cornices in the next chapter, where we intend to show different designs. In this case we are using the lookout cornice to support the method we have been dealing with. The finished chord is

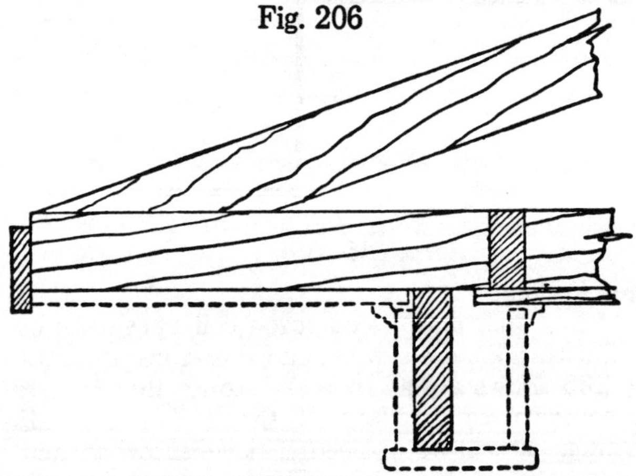

Fig. 207

shown by dotted lines -- also the cornice ceiling. The porch ceiling is shown where it joins the nailing blocks, and the fascia is shown to the left, shaded. We also show how the rafter joins the lookouts, but if the method of joining the rafters to the lookouts is used, the lookouts must be so spaced that the rafters will come directly over them.

How to fasten the rough chord to the main building, is shown by Fig. 208. At *A* is shown a side view, while at *B* is shown a view looking from the inside of the main building. This makes a substantial fastening, which will not come apart.

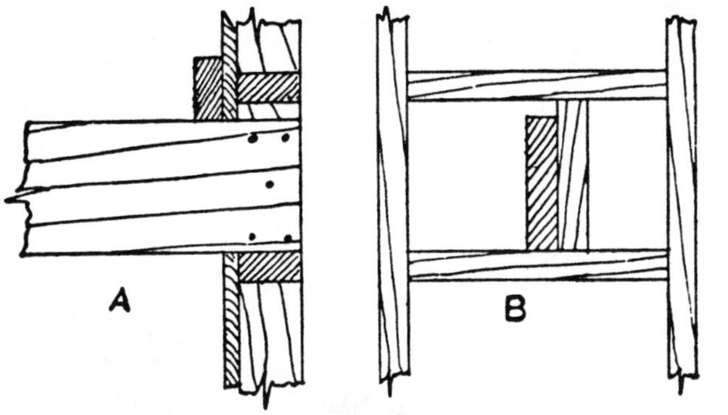

Fig. 208

Fig. 209 shows at *A* a method of fastening the rough chord to the main building, which is often used, but it is a weak construction. It would not be so bad, if a nailing block were placed inside of the boxing, as shown at *B*, but for an A-Number-One fastening an anchor should be used in addition to the block, as shown in the drawing.

While we are using 5/8-inch ceiling in explaining constructions of porch ceilings, we are aware that other materials are often used, such as plastering of various kinds, plaster-boards, pressed wood, and sheetrock. Whatever kind of material that might be used instead of 5/8-inch ceiling, plywood is probably the most commonly used of them all. When other materials are used, the construction shown for ceiling boards, will answer the purpose, with few, if any, changes.

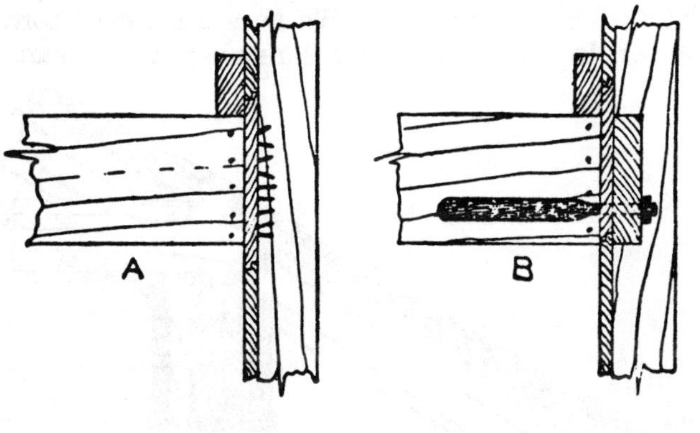

Fig. 209

Where the ceiling boards are run parallel with the side of the building, joints are often necessary, especially in large porches. The approved method is to make the joints on bearings, and not too many on the same joist. Another method is to allow the joints to come wherever they will between bearing, thereby holding the cost of labor and waste of material to a minimum, which of course, means that it is usually used on cheap work. The joints at angles can either be miter joints or lace joints. The former is shown at *A* and the latter at *B* in Fig. 210.

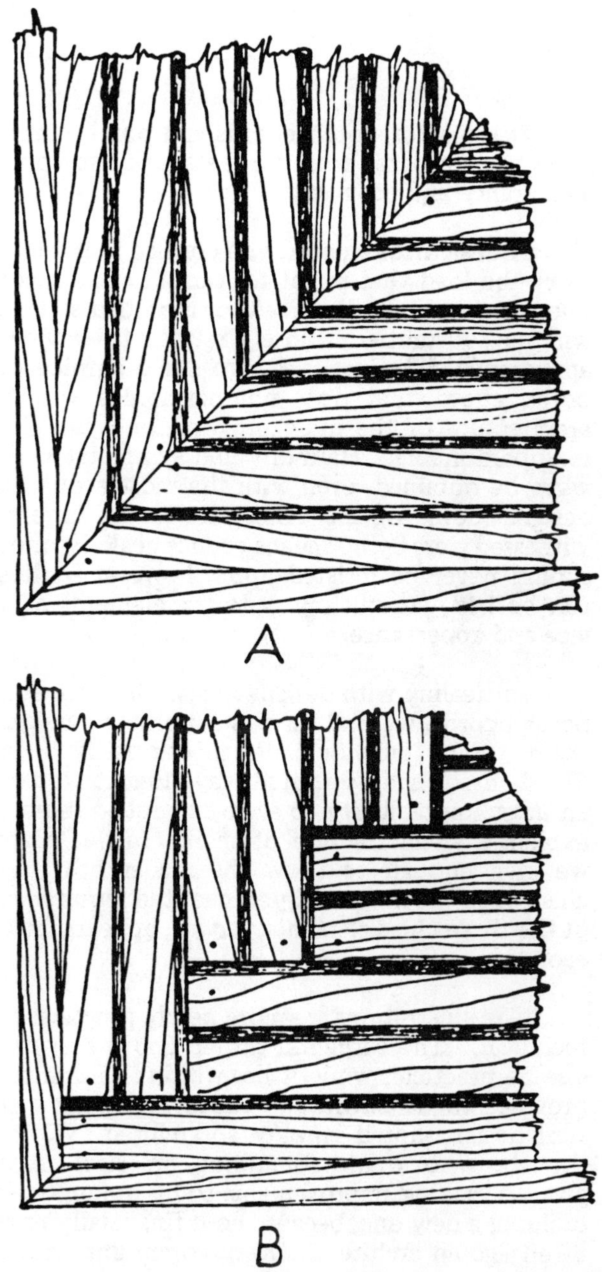

Fig. 210

16 Cornices

Designs for porch chords and porch cornices must be considered from three standpoints, service, appearance and economy.

A design for a chord or a cornice that will not carry the load that might for various reasons come upon it should be discarded, or modified so that it will be substantial. Closely related to service, is appearance. To use a design for a cornice or a porch chord that does not harmonize with the architecture of the main building, or is unsightly, is unpardonable. Harmony and symmetry can always be obtained, even with the most economical design. Some of the ugliest designs we have ever witnessed were by no means economical. Economy should never be insisted upon, however, unless it can be had without injury to the elements of service and appearance.

In dealing with designs for porch cornices and porch chords, we find there is a limitless number of them. Each individual design can easily be modified in so many different ways that it would be an utter impossibility to even attempt to show and explain a small fraction of them. In this chapter we give enough to show the reader how easily changes can be made and yet meet the requirements of the three chief elements, service, appearance and economy.

The illustrations are more nearly practical than technical. The carpenter should above everything else be practical, while it lies within the architect's province to be technical. Those who want a classical design should employ an architect. No one, though, who wants something practical, should give up making improvements to his home, or even building a new one, because he is financially unable to engage an architect. Every community has carpenters who can do a well-appearing practical job, without the services of an architect.

In Fig. 211 is shown a lookout cornice especially suitable for a hip roof. The design is similar to the one shown in the previous chapter, Fig. 207.

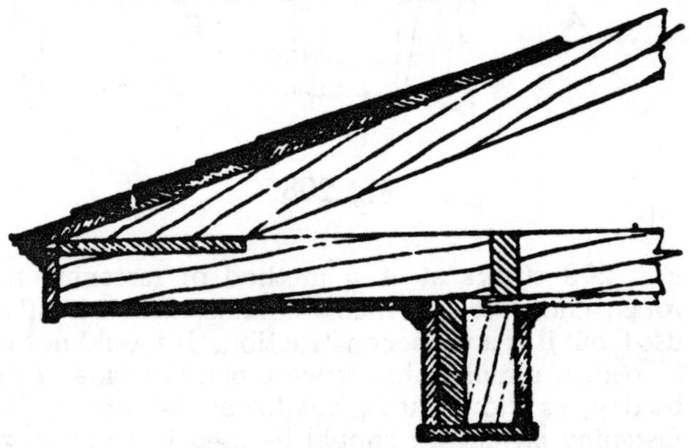

Fig. 211

In this we show more details. The rafters rest on a toe-board, which gives the cornice more rigidity during construction, and added strength to the job when it is finished. The toe-board makes possible an independent spacing for the rafters. That is to say, the rafters do not have to come directly over the lookout, as was the case in the design that had no toe-board.

The design for the chord is an old familiar one, with a nosed soffit. A single beam, which is blocked out to give the finished chord the proper width, carries the load. We have a similar chord shown in Fig. 212 with a 2-member rough beam.

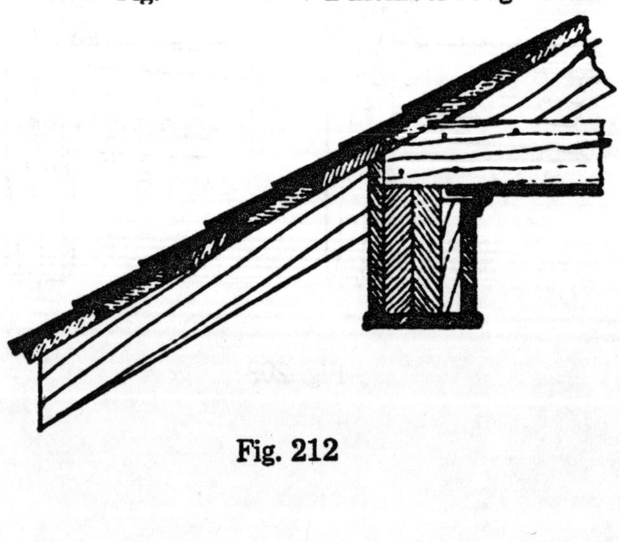

Fig. 212

This beam is also blocked out to give the chord its width. The cornice is an open cornice, very simple in construction. 6-inch matched lumber is used for sheathing, which at the same time answers for the bottom finish of the cornice. Neither bed moulding nor crown moulding is necessary with this design.

A good design for an open cornice, is shown by Fig. 213. Here we have a bed moulding and a small fascia, which answers for the crown moulding. A 2-member rough chord, blocked out, makes the framework onto which the finished chord is nailed. The two sides extend a little below the soffit, which makes a pleasing finish. This design lends itself readily to a panelled soffit.

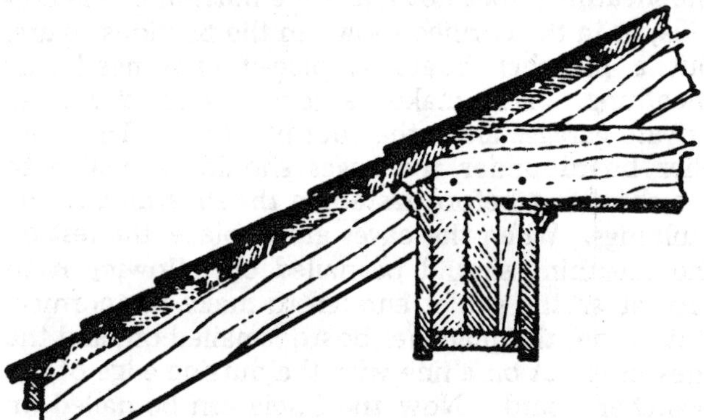

Fig. 213

A boxed cornice with a compound chord design is shown in Fig. 214. This has both a bed moulding and a crown moulding, while the fascia has a beveled bottom edge, and is in a vertical position. The plancher board must also be beveled where it joins the fascia. A little different design is shown by Fig. 215. This is also a boxed cornice, but the fascia is nailed at a right angle to the pitch of the roof. It will be noticed that the rough beam makes the outside finish of the chord. The soffit shows a reveal at the outside corner and a nosing at the inside corner. A plancher board finishes the bottom of the cornice.

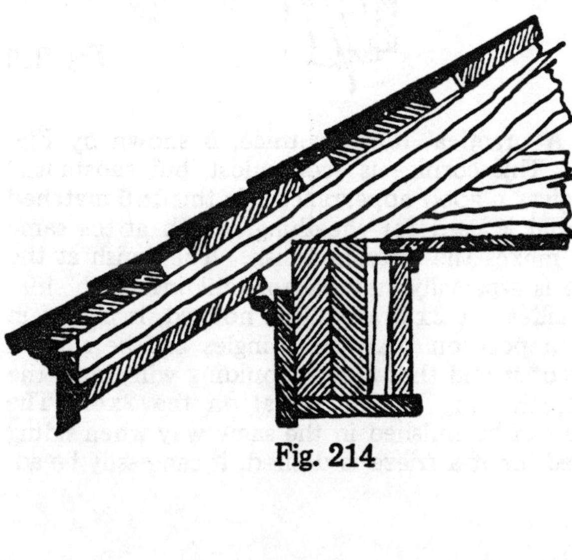

Fig. 214

Fig. 215

Fig. 216

A self-supporting chord is shown in Fig. 216, in which the rough chord is omitted. The two sides are held in place by blocks, onto which the soffit is also nailed. Fig. 217 gives a diagram, showing the construction of the chord and how the rafters are cut. The dotted lines show where the rafters come when in place. The regular sheathing, placed solid, makes the bottom finish of the cornice. This design gives good results where economy is desired, and where the supports for the chord are not too far apart. The writer has used it with good results.

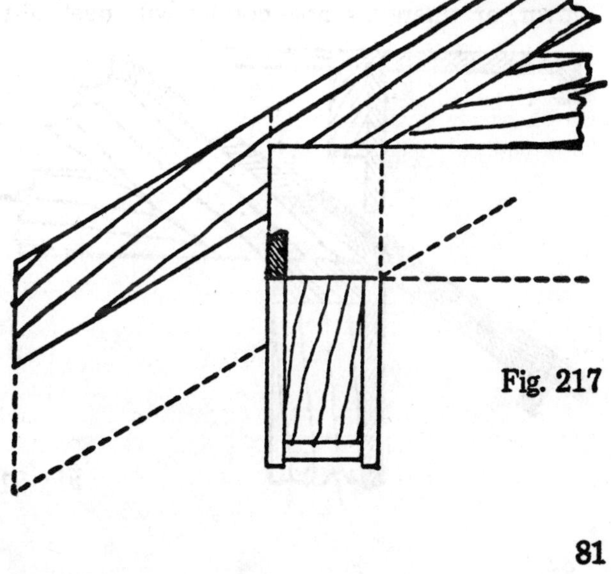

Fig. 217

All the cornices shown so far can be adapted to main roofs, with little, if any changes. From here on we will show cornices for main roofs, which at the same time can be used on porches. We are doing this in order to show as many designs as possible. When we take up roof-gutters, we will show them with still other designs for cornices.

The chords we have shown are all of ordinary size, yet all of them could either be increased in size or decreased. The width, especially, of a chord should be governed by the size of the columns, or on the other hand, where the size of the chord is established, the size of the columns should be governed by the width of the chord. We recall one job where an extremely wide chord, with a panelled soffit, was supported by rather small columns. In another instance, and this is more often the case, an ordinary sized chord was supported by extremely large columns. We will deal with this further when we take up porch columns.

The styles of cornices, especially house cornices, are as various and go from extreme to extreme. Each of these extremes has its advantages and its disadvantages. The wide cornice is more expensive, but protects the sides of the house against the effects of the elements, such as rainfall, hail, snow, sunshine and so forth. While this protection is not one-hundred percent perfect, it nevertheless protects. On the other hand, the dehorned cornices does not protect against any of these things. It has the advantages of being less expensive, and whereas it does not protect the sides of the building against rain-fall, it permits the rain to wash the sides of the building as it falls against it. And whereas it does not protect against the effects of the sunshine, it makes the effects of the sunshine uniform from top to bottom, which is not the case with the wide cornice.

We have a simple cornice shown in Fig. 218. This cornice is suitable for sheds, cheap garages and other buildings where the expense must be held down, or where a simple cornice will meet all the

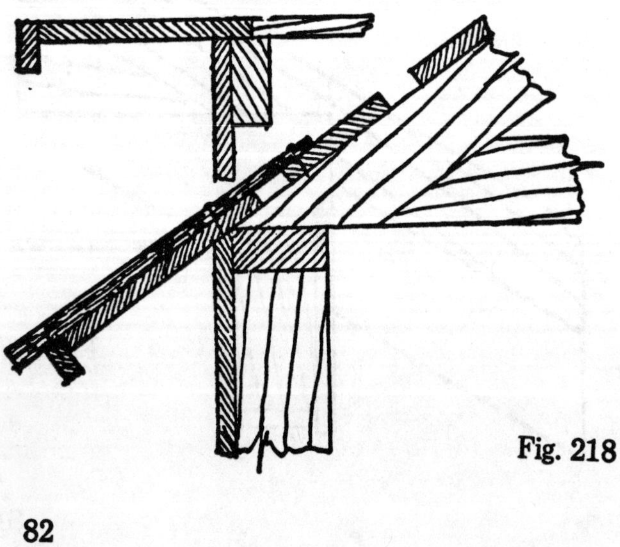

Fig. 218

requirements. The upper sketch shows the layout for the gables, while the main drawing gives the eaves. We show the double course of shingles nailed on, with the nails so located that the shingles will reinforce the cornice. The fascia is important, for it not only helps to give the cornice a more pleasing appearance, but it stiffens the edge of the cornice and keeps it in alignment. Depending on the width of the cornice, the fascia can be from 1½-inches to 3 and even 4-inches wide. However, an 8-inch cornice with a 2-inch wide fascia is probably the most practical for this style of cornice.

A similar, but much more substantial, cornice is shown by Fig. 219. Here it will be noticed that the sheathing does not make the bottom of the cornice, as in the cornice shown in the previous figure, but a plancher board is placed underneath the sheathing. This makes a good cornice for small private garages and other out-buildings. The plancher board under the eaves should be nailed to the sheathing before fastening the sheathing to the buildings. When the eaves are in place, the rest of the sheathing should be nailed on, allowing it to project at the gables enough to make the cornice. This done, the plancher board is nailed on, and the sheathing cut on a line with the outside edge of the plancher board. Now the fascia can be nailed on and the job is ready for shingling. The doubled course of shingles, as we show, should be nailed in such a way that the shingles will reinforce the cornice.

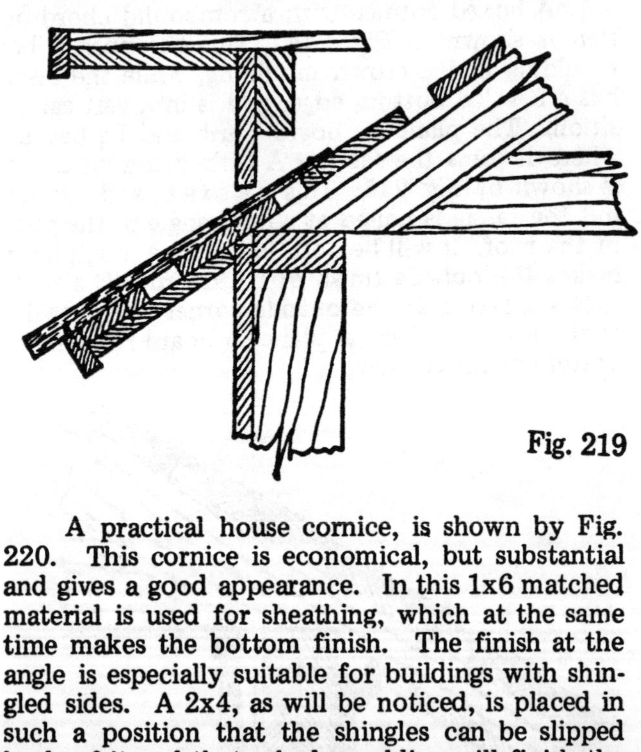

Fig. 219

A practical house cornice, is shown by Fig. 220. This cornice is economical, but substantial and gives a good appearance. In this 1x6 matched material is used for sheathing, which at the same time makes the bottom finish. The finish at the angle is especially suitable for buildings with shingled sides. A 2x4, as will be noticed, is placed in such a position that the shingles can be slipped back of it and that a bed moulding will finish the angle, showing a little reveal on the 2x4. The angle can be finished in the same way when siding is used, or if a frieze is desired, it can easily be ad-

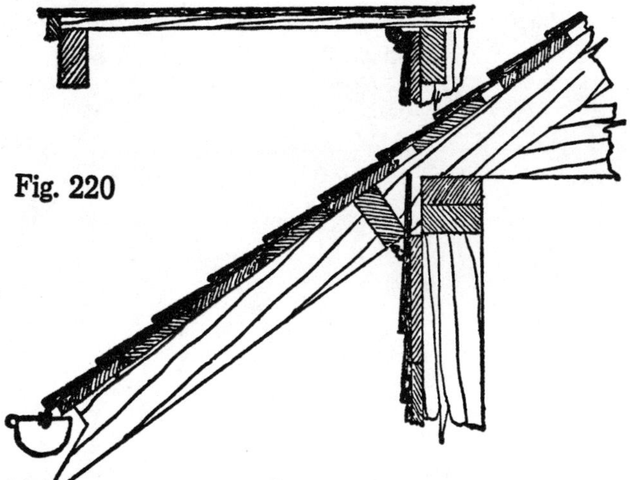

Fig. 220

justed so as to meet the requirements. In the upper drawing, we show a cross section of the cornice at the gable ends. A half-moon gutter can be utilized with this cornice, without injury to the appearance.

A cornice that is a step toward what is called a swayed cornice, is shown by Fig. 221. The rafters do not project beyond the main building, but lookouts are fastened to the rafters, set at a lower incline, which carry the cornice. With this design it is possible to use better material for the lookouts, such as cypress or white pine, which of course would add value to the appearance.

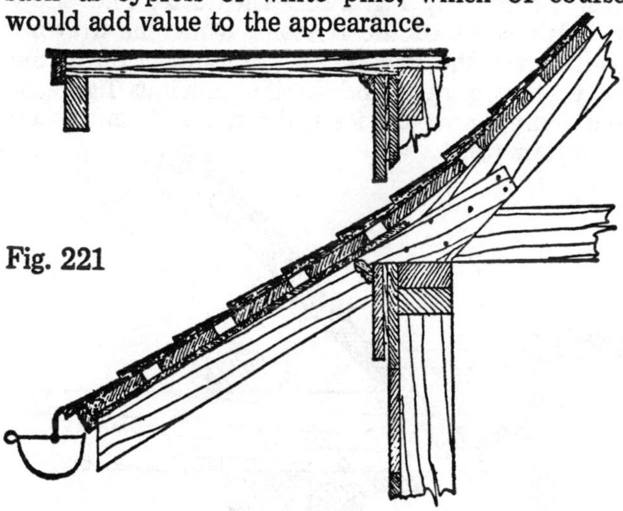

Fig. 221

A swaying cornice is shown by Fig. 222. Such cornices were extensively used in many localities, in the early part of this century. A crown-mould gutter, similar to that shown on the drawing, works well with this design.

By Fig. 223 is illustrated a cornice construction, that works well with a hip roof, and on which a crown-mould gutter can be used. A similar cornice, but different in design is shown by Fig. 224. The dotted lines represent the ceiling joists, but in many instances where this method is used, the ceiling joists are placed below the plate to increase the attic room, however, the ceiling joists can be placed higher or lower than the position indicated by dotted lines.

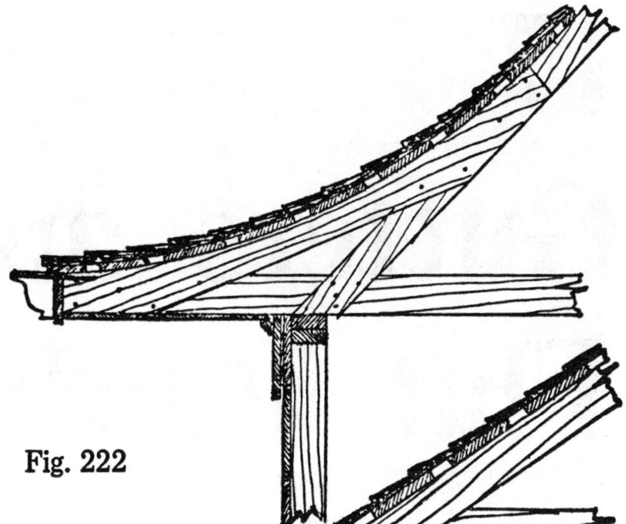

Fig. 222

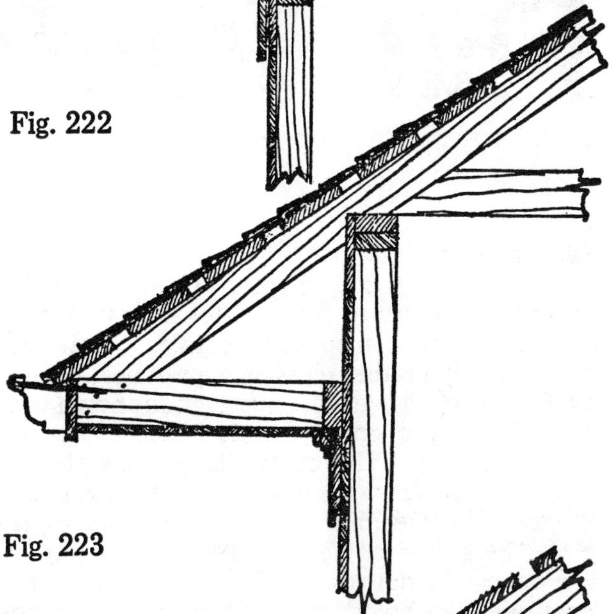

Fig. 223

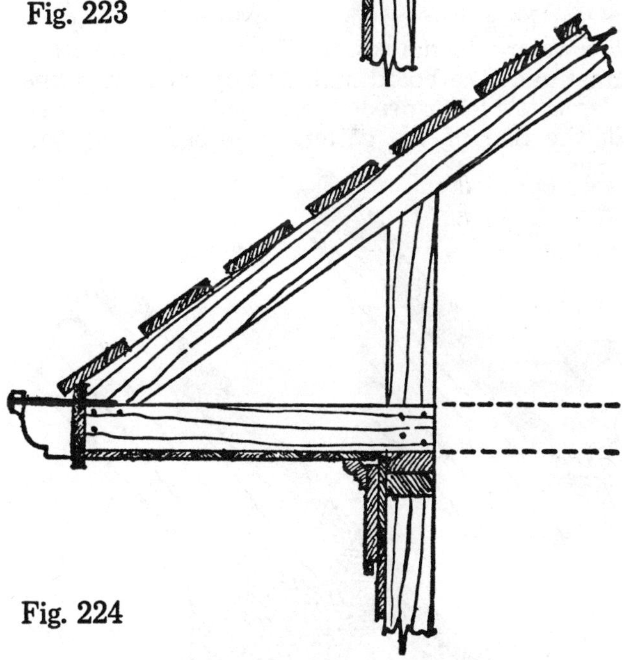

Fig. 224

It will be noticed that matched ceiling is placed between the sheathing and the lookouts, which makes the bottom finish of the cornice. The angle finish is also different from what was shown in the previous figure. The edge of the cornice is finished with a fascia, and the lookouts are finished at the end with a plumb cut. A half-moon gutter will work well with this cornice design. A suggestion for the cornice at the gable is shown above the main drawing.

17 Gutters and Flat Roofs

Dehorned Cornices

We show in this chapter designs of semi-dehorned and dehorned cornices, together with gutter designs. A shortened open cornice construction which makes a rather pleasing appearance, is shown by Fig. 225. The gutter construction shown in the main drawing is simple, and is sometimes called a Yankee gutter. The dotted lines show the high point, in case an extra board is used for a bottom to give the gutter fall. The best way, though, is to place the gutter on the eaves in such a way that the water will run toward the downspout, without the use of the board indicated by the dotted lines. Very little fall is needed, provided there are no sags in the line of the gutter. The best lining for a built-up gutter, such as shown in this chapter, is copper, but galvanized sheet metal will do, if it is kept well painted. A little different design is shown in the upper drawing. The dotted lines to the left show how a crown moulding can be reshaped and used with good results.

A built-up gutter with a good carrying capacity, is shown by Fig. 226. The dotted lines show the high point of the bottom, while the drawing itself shows the low point. The objection to the gutter shown here and in the previous figure, is that sliding snow and ice will catch on it, and often cause damage.

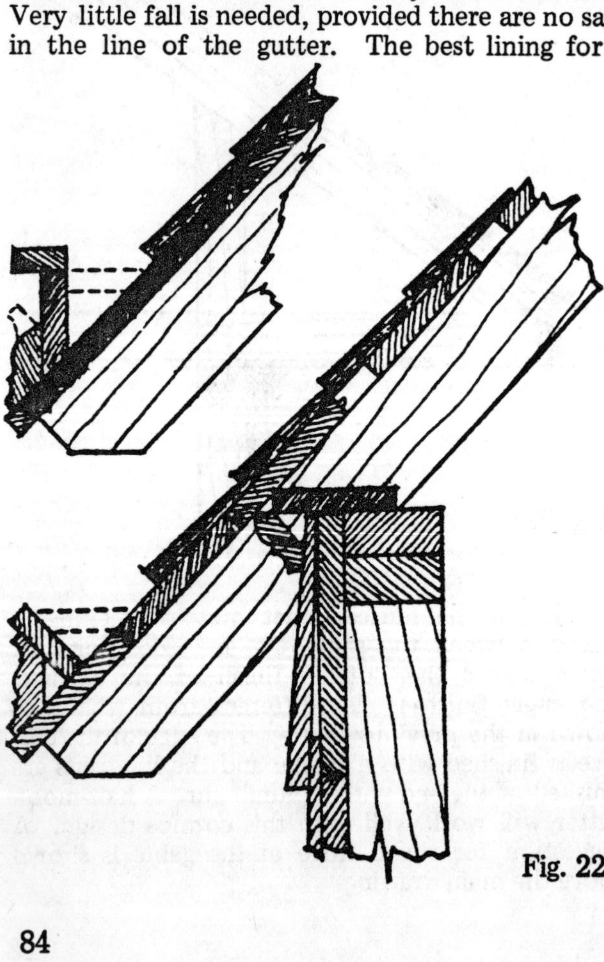

Fig. 225

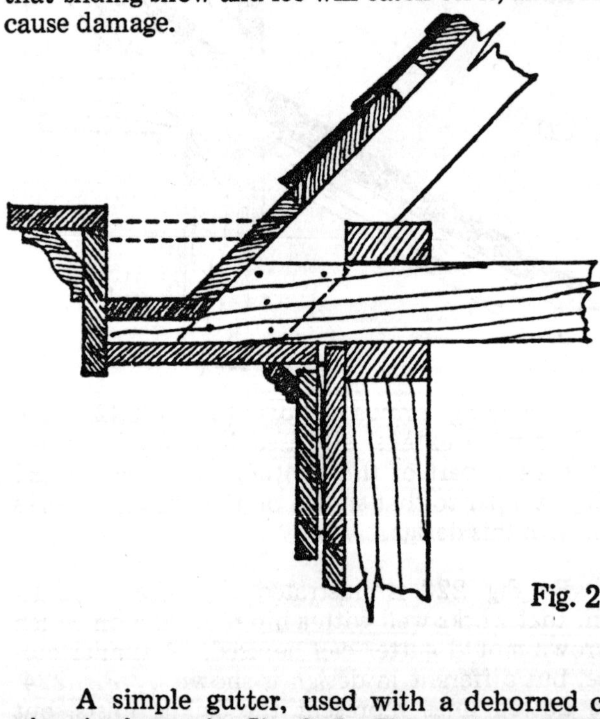

Fig. 226

A simple gutter, used with a dehorned cornice, is shown in Fig. 227. This gutter will allow snow to slide over the outside edge, and yet has a good capacity for carrying water. The fall must be obtained when the bottom is put into place. If the bottom is kept perfectly straight, very little fall is

Fig. 227

needed, especially if copper lining is used. Copper will not rust, even if the water does not drain off readily.

Two designs of wooden gutters are shown by Fig. 228. The one in the main drawing is used with a dehorned cornice, while the one to the left is shown with a regular cornice. The joints on wooden gutters are made by inlaying a strip of sheet lead over the joint, embedded in white lead and oil. A wooden gutter, if kept well painted, will give service for a long time, and it has the advantage of holding its shape well, if properly put on and cared for.

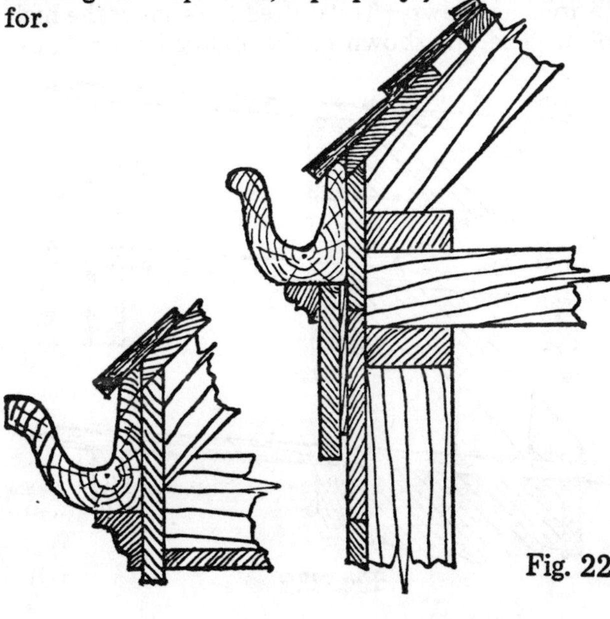

Fig. 228

A gutter for a dehorned cornice, with a concealed downspout, is shown by Fig. 229. This gutter makes a good appearance, and will allow snow to slide over it without damaging the edge of the gutter. Because of its location, it should be lined with copper, and the downspout should be either copper or cast iron soil pipe. The dotted lines show the concealed location of the downpout.

Fig. 229

A gutter very much like the one just explained is shown by Fig. 230. The form of this gutter is simpler than what is shown in Fig. 229, and has a larger capacity. The horizontal dotted lines show how a bottom can be placed, and sloped in such a way that the water will drain toward the downspout, which in this case is to be a concealed downspout.

All the designs we have shown, either of the cornice or of the gutter, can easily be modified to meet the needs of the various situations under which they might be used. The capacity of a gutter must needs be governed by the amount of roof surface there is to take care of. The same thing is true of the fall. A short gutter, with not much roof surface, will not need as much fall or as much capacity, as a long gutter, with a great deal of roof surface draining into it.

While copper lining for gutters is rather expensive in the first place, in the long run it is the

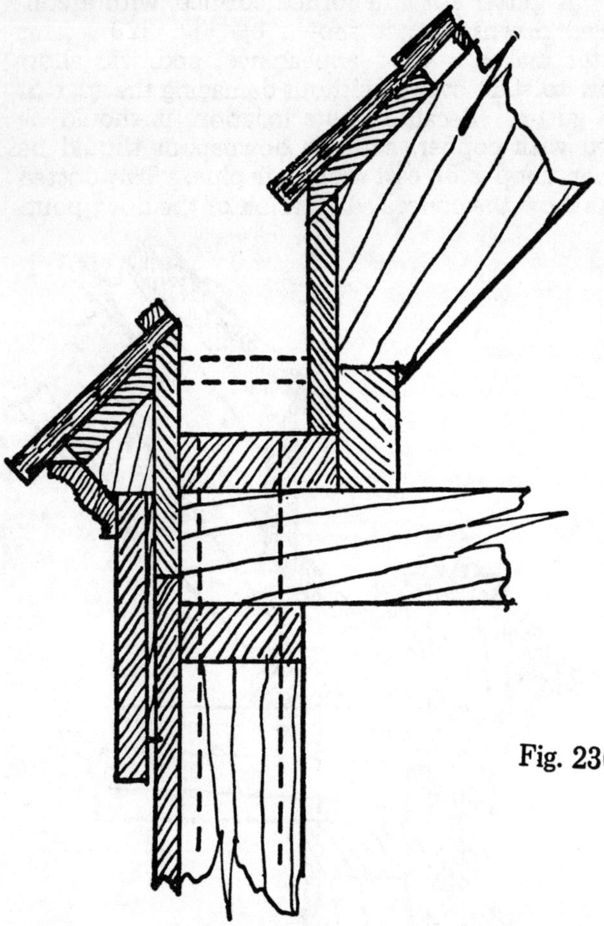

Fig. 230

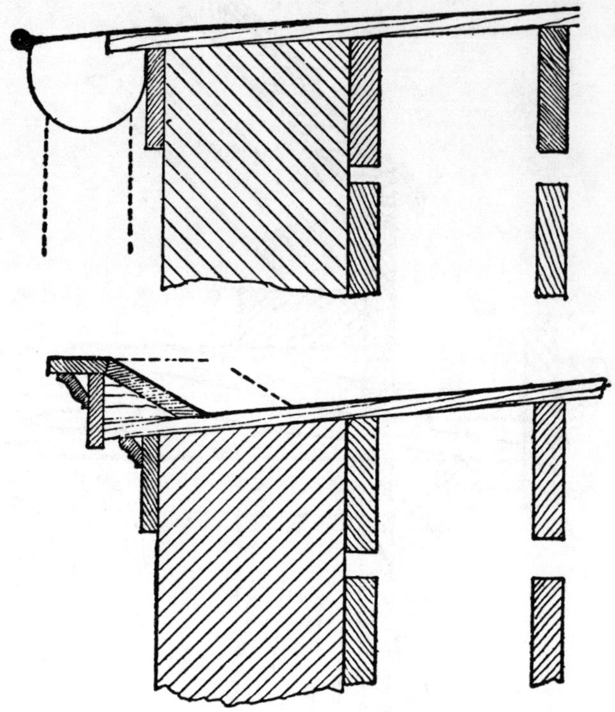

Fig. 231

cheapest. For if it is placed properly and not damaged by puncturing or tearing, it will last as long as the building on which it is used. Galvanized sheet metal is cheaper in the first place, but unless kept well painted will cost a great deal more than copper in the life of a building, speaking of permanent buildings. Tin, if it is of a good quality, and is kept well painted, will give good service. It should be painted on the bottom before it is put into place, which is usually done. The trouble is, that after it is on, in many instances, so long as it does not leak, it is forgotten, and when it begins to leak, it is too late to save it with paint.

Flat Roofs and Gutters

The flat roof for business houses, shops and store buildings, especially in smaller towns and cities, is with us permanently. Not much change in the construction of the flat roof for such buildings has taken place, excepting where they are made strictly fire-proof -- but even then, the means of carrying away rain and snow water are much the same as those found on the ordinary flat roof.

Two flat-roof gutters, and their relation to a flat roof, are shown by Fig. 231. The upper one is a half-moon metal gutter, which is perhaps the simplest and cheapest gutter for flat roofs to be found. The gutter itself is the principal part of the cornice, if it can be called a cornice. The perpendicular dotted lines indicate the position of the downspout. The bottom design has a little more cornice, and makes a better appearance than the other. The dotted lines show the form of the gutter at the high point. The improvement in appearance, though, adds to the cost of construction. One of the things that should always be kept in mind in any gutter, and especially one for a flat roof, is the support. If a gutter is well supported and straight, not much fall is required to carry the water.

Three simple and serviceable gutters are shown by Fig. 232. The one shown at the top is made by ripping a 4x4 or 4x6 diagonally, and fastening it to the roof as shown. The dotted lines show the high point. The one shown at the middle is similar, ex-

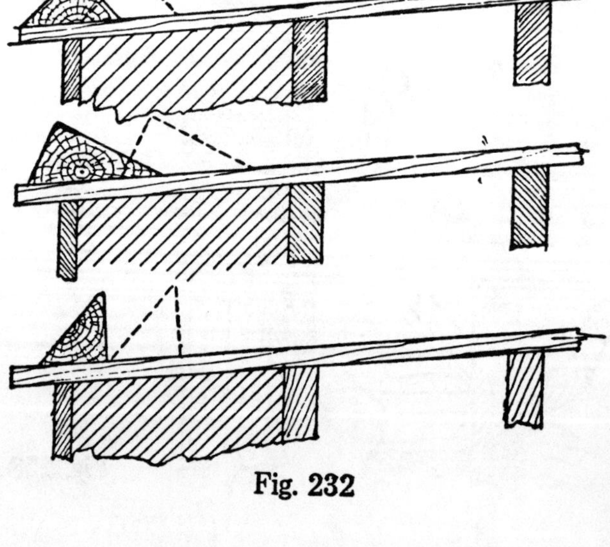

Fig. 232

cepting that a larger timber is ripped diagonally, say a 4x6, while the bottom one is formed by placing the ripped 4x4 in such a way that the gutter will be at a right angle with the roof. The roofing material makes the lining for these gutters.

The gutter forms shown by Fig. 233 is suitable for a roof with a little more pitch than the ordinary flat roof. The upper one is formed by nailing a lookout to the foot of the rafter as shown. The downspout is indicated by two perpendicular dotted lines. The low point of the gutter is also indicated by dotted lines. This cornice construction is both pleasing and economical. At the bottom is shown how the same form can be made by framing the rafter so it will make the lookout. The dotted lines at the bottom show the part of the rafter that is cut off and transferred to the upper edge

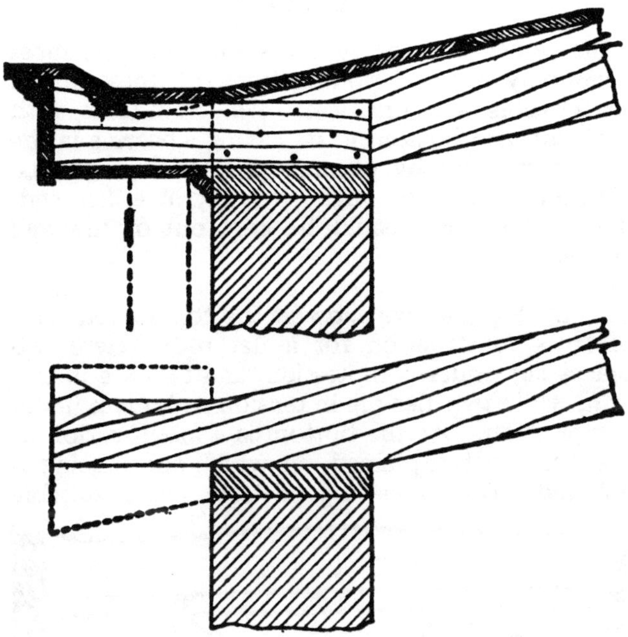

Fig. 233

of the rafter, where it is nailed and forms the bearing for the gutter sheathing. The dotted lines show how much of this piece is cut off in framing. The low point is also indicated by a dotted line. Otherwise the construction of this gutter and cornice is the same as the one shown at the top.

In Fig. 234 is shown a plan of a flat roof on which the gutters shown by Figs. 231 and 232 would be suitable. The arrows show the direction of the fall. The position of the half-moon gutter is shown with continuous lines, while the gutters shown by Fig. 232 are indicated by the dotted lines, and the dotted arrows show the direction of the fall. A little study of this drawing with the gutter designs referred to will make this clear.

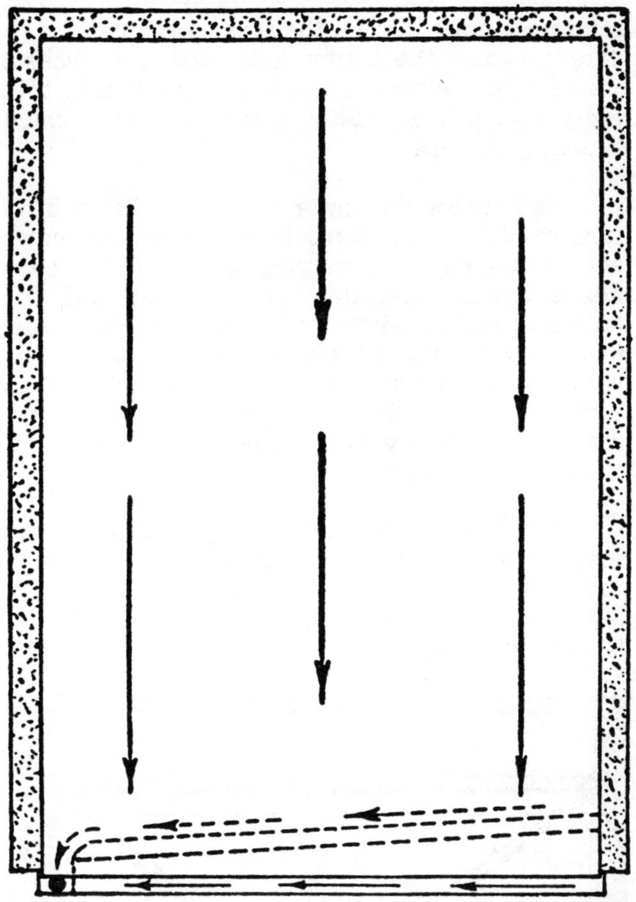

Fig. 234

Perhaps the most used flat-roof construction, is shown by Fig. 235. At the top we show the highest roof joist, trussed and tied to the ceiling joist directly below it. At the bottom is shown the lowest roof joist, trussed and tied to the ceiling joist below. The truss shown at the top will carry a load much greater than will ever come on it under ordinary conditions, but the bottom truss is not strong enough to support its load without

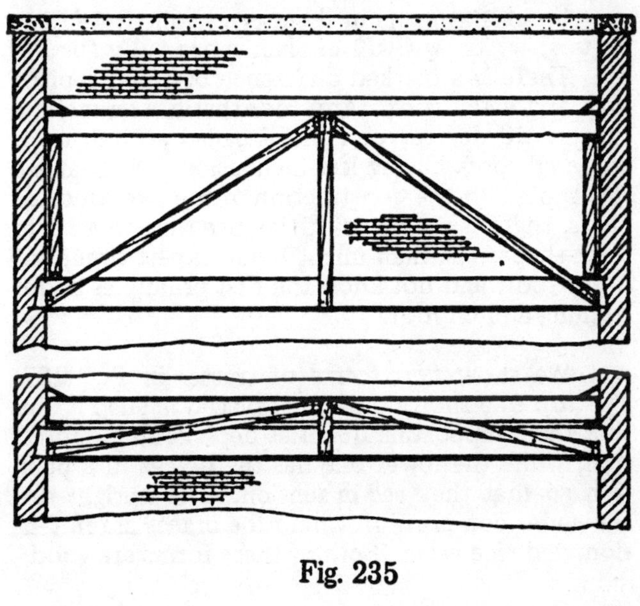

Fig. 235

reinforcement. In fact, the strongest truss of such a roof carries the lighter load, and the weakest truss carries the heavier load -- we are assuming that every roof joist is trussed in the same way, which is usually the case.

Sometimes the lower part of a flat roof is reinforced by support-partitions, which eliminates entirely the danger of sagging. But when there are no partitions under this part of a flat roof, the trusses should be reinforced, either with extra bracing, or by putting the trusses at closer intervals. It is not an uncommon thing to find a flat roof so badly sagged at the lower part that it will not drain, consequently water pools there and soon develops leaks in the roof.

We show by Fig. 236 an inverted truss for flat roofs. As a rule, there is little difference between the strength of this truss and the one shown in Fig. 235. They both depend entirely on nails for their holding qualities.

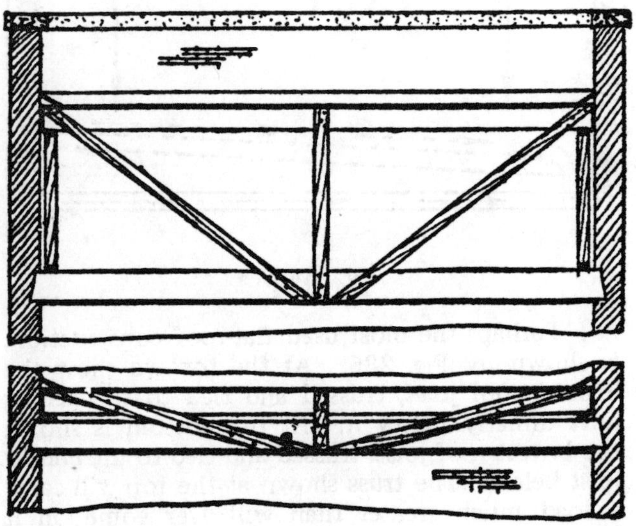

Fig. 236

Some flat roofs in themselves constitute huge gutters, which we will explain more fully further on. There is a marked difference between a pitch roof and a flat roof. Any roof that has an inclination of 20 degrees or more is called a pitch roof, while all roofs having less inclination than that are flat roofs. In the construction of the two kinds of roofs, there is even more difference than there is in their shapes. A man might be an expert framer of a flat roof, and not know the first principles about framing a pitch roof.

We show two forms of trusses in Fig. 237. The top one shows the braces placed in such a way that their supporting qualities depend on compression, while the lower one has the braces in a position so that they are in tension. The upright ties act under compression where the braces are in tension, and vice versa. Both of these forms are good -

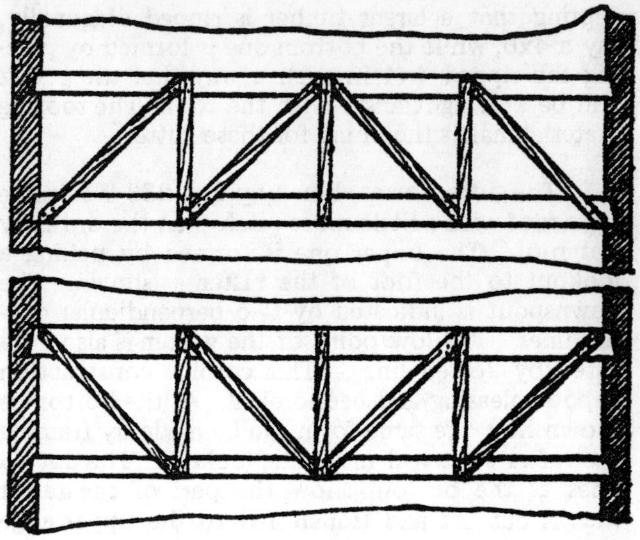

Fig. 237

- the upper one, however, is probably the most commonly used. Modifications of these forms are possible and practical. We show them with both top and bottom joists having a bearing in the side walls. But by placing extra ties or supports, whichever the case might require, at either end, one of the joists could be kept out of the wall entirely.

By Fig. 238 we show at the top a good form of truss construction for a flat roof. Here two braces act under compression, and two are in tension, while two ties are in tension and one is under compression. At the bottom is a modification of the truss just explained, in which the bracing is doubled. This makes a good truss, and is suitable

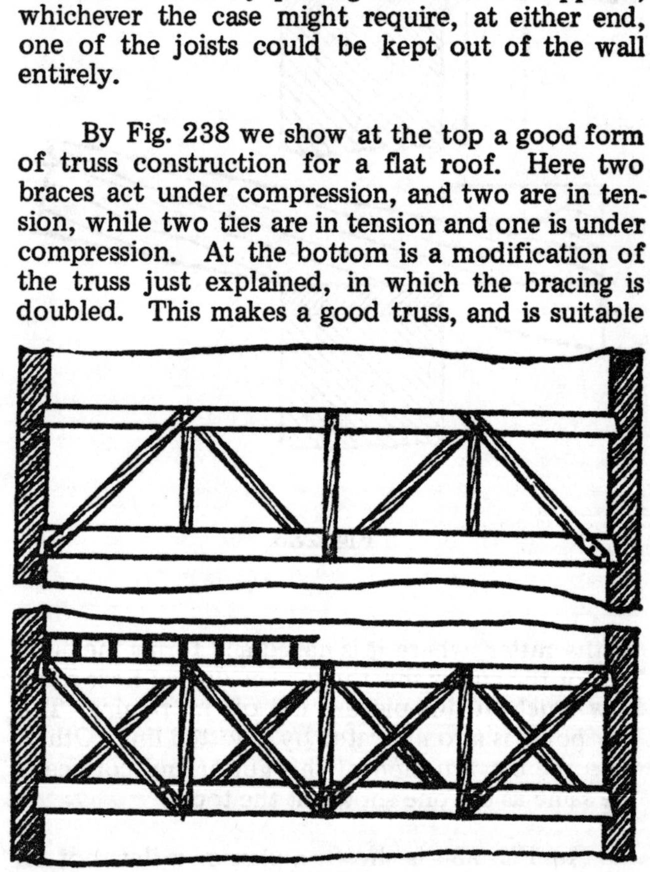

Fig. 238

for carrying sections of joists, as shown, to the left and top.

In Fig. 239 we show a plan of a roof which slopes somewhat in the order indicated by the arrows. This roof is more nearly on the order of a

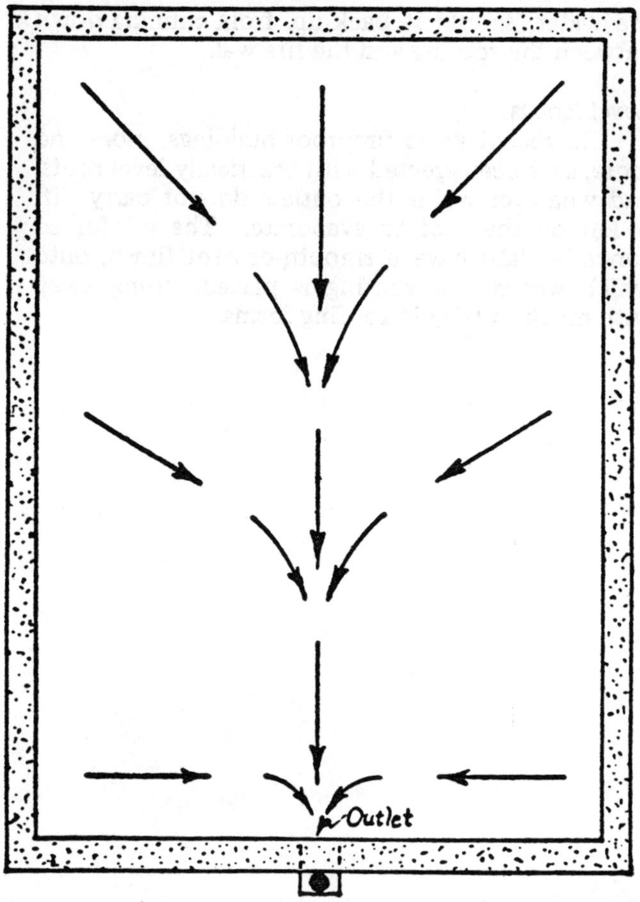

Fig. 239

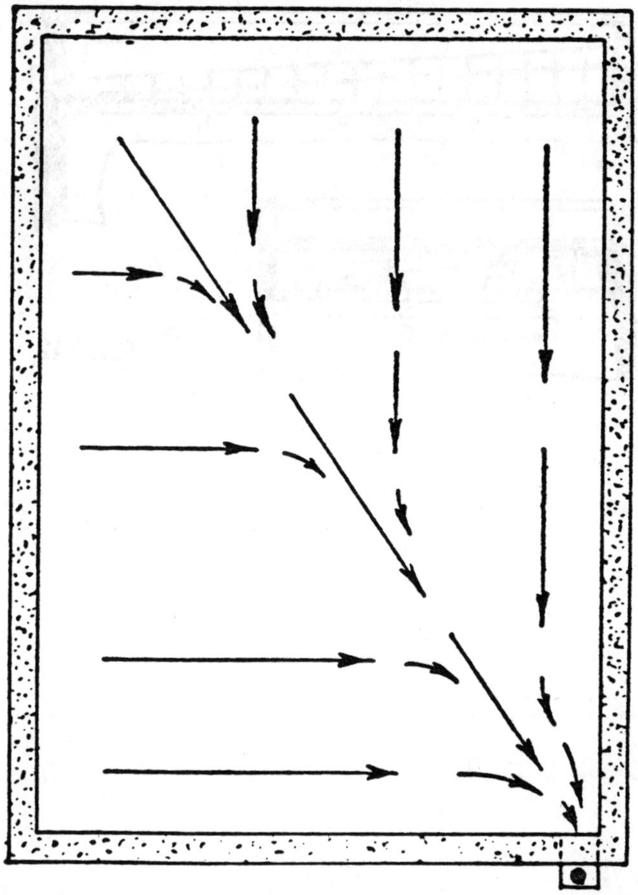

Fig. 241

huge gutter, as we mentioned before. It slopes to the center, and the center inclines to the low point, carrying the water to an outlet through the fire wall, as shown on the drawing. Fig. 240 gives two sections of this roof. At the top the roof is shown cut in two at the center. To the right is shown the sheathing on, covering six joists. The rest of the joists to the left are uncovered, showing how the inclination of them is gradually reduced until the high point is reached at the extreme left, where the joists are on a level. The bottom drawing shows a cross section with the side and center supports in place and the two highest joists set. The inclining dotted lines show the position of the joists at the lowest point of the roof. The reinforced concrete base should be noticed in these illustrations.

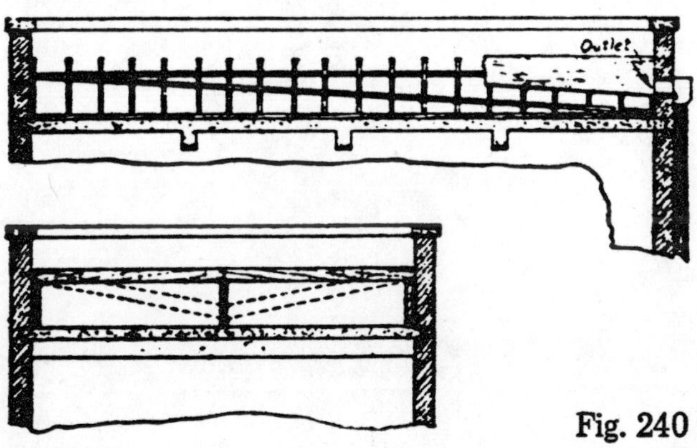

Fig. 240

It will be noticed by studying the two sections that the side supports are the same in height the full length of the roof, while the center support inclines gradually toward the low point, or the outlet for the water through the fire wall. This construction for a flat roof is suitable for buildings that have a good base to work on, especially through the center toward which the water drains. We show this roof built over a reinforced concrete slab, with beams, which makes a good base. The outlet should be large, and the downspout should have a good capacity, so it will easily take care of the water as fast as it might come to it.

A plan for a somewhat different flat roof construction is shown by Fig. 241. This gutter runs in a diagonal direction, as indicated by the arrows, to the low point at the corner where the outlet through the fire wall takes the water to the downpout. The precaution given in the explanations of the previous figure, hold good there. Fig 242 shows two sections of this roof construction. The upper one shows how one half of the roof inclines to the outlet, while the lower one shows the joists in place at the high point. By dotted lines the inclination of the joists at the low point is indicated. The inclined angle support is pointed out by arrows. The roofing near the outlet should extend up the side of the fire wall far enough to prevent

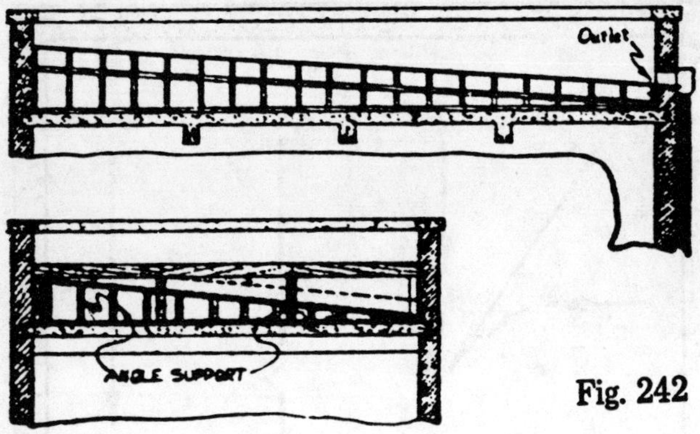

Fig. 242

the water, should it back up, from running down between the roofing and the fire wall.

Level Roofs

In recent years fireproof buildings, more and more, have been erected with practically level roofs, and whatever water the outlets do not carry off, is left on the roof to evaporate. The reinforced concrete slabs have a smooth cement finish, onto which waterproof roofing is placed, doing away with much of the old roofing forms.

18

Columns

Columns

Speaking of temporary supports, we have props, shores and struts. While the term "column" could be applied to temporary supports, the application would be rather far-fetched. When we speak of permanent supports, the term "post" is often used instead of "column", especially when referring to porches with small solid supports. While we like to use the word "column" when speaking of porch supports, there seems to be a logical basis for applying the term "post" to solid supports with less than a 6-inch diameter or thickness. This is our own conclusion, based on experience and on the following quotations:

"A stick of timber, a bar of iron, etc., when used to sustain end loads which act lengthwise of the pieces, are called columns, posts or struts, that is, if they are so long that they would bend before breaking. When they are so short that they would not bend before breaking, they are called short blocks."

A few definition of the word "column" follows:

(1) "A solid, vertical body of greater height than thickness, generally serving as a support." (2) "A cylindrical or slightly tapering body set vertically upon a stylobate, and surmounted by a spreading mass which forms its capital." (3) "A kind of pillar, a cylindrical or polygonal support for a roof, ceiling, statue, etc., somewhat ornamental, and usually composed of base, shaft, and capital." (4) "A vertical shaft or pillar; a prop or support." (5) "Anything resembling in form or position, a column in architecture; an upright body or mass, a shaft or obelisk."

The last quoted authority leaves the field of architecture, where we stopped quoting, and enters other fields in which the word "column" is used to convey meanings that do not in any way pertain to carpentry. The quotations we have given, however, will give the reader enough light so he can know what constitutes a column, in the realm of architecture.

In Fig. 243 we show three of the simplest porch posts in use, or porch columns, as we sometimes call them. To the left is shown a plain 4x4 post, which makes a substantial support, and if properly trimmed, does not make a bad appearance. In the center is a 4x4 chamfered post, and to the right is a turned column, made out of a 4x4. While these columns are out of date, they are still used on cheaper classes of work. One frequently sees them on back porches, when the front porches have more nearly up-to-date columns.

The designs shown are as simple as they can be. There are variations and modifications that can be made, so that an unlimited number of designs is possible. The corners of the square post can be rounded or chamfered, and in this way change the design. Changes in the designs of chamfers are possible by changing the depth or the length of them -- or by changing the beginnings and endings to ogees, concaves and so forth. The variations of small turned columns are unlimited, but they are out of use almost altogether. The square and chamfered 4x4 posts can easily be made to fit into modern construction.

Details giving the relationship of these columns to the porch, are shown in Fig. 244. The simplest one is shown to the left. The center shows how sometimes a metal shoe is used to prevent the column from rotting at the base. The top shows a very simple way of joining the column to the chord. To the right is shown a more modern way of joining 4x4 posts to porch chords. Both the cap and the base are nailed to the 4x4 column, as is indicated by the dotted lines and dots. They do not add to the strength of the column; in fact, the base will shorten its life, because the joint will not dry out as readily as it would without it, and therefore will rot much sooner. If it is well painted on the back before the base is put on, and kept well painted afterwards, the results will be quite satisfactory.

The purpose of the cap and of the base is to obtain a better appearance. It will be noticed that a bed moulding is shown at the bottom and a cove

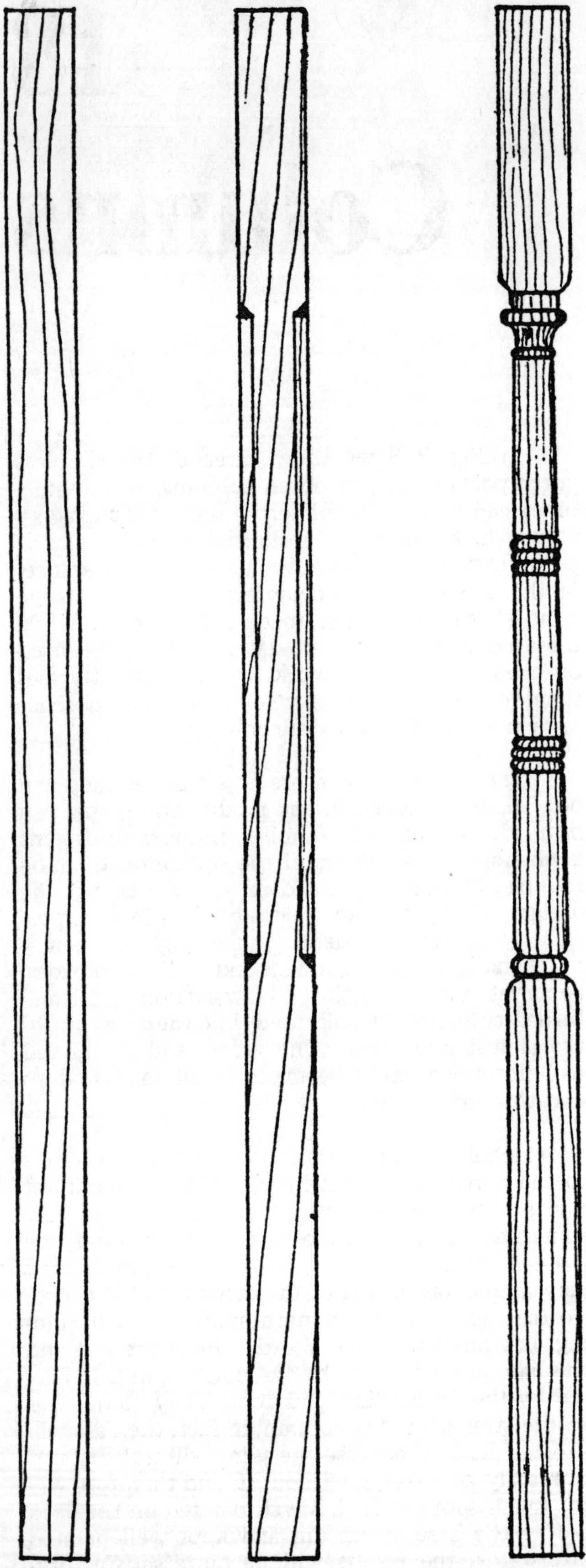

Fig. 243

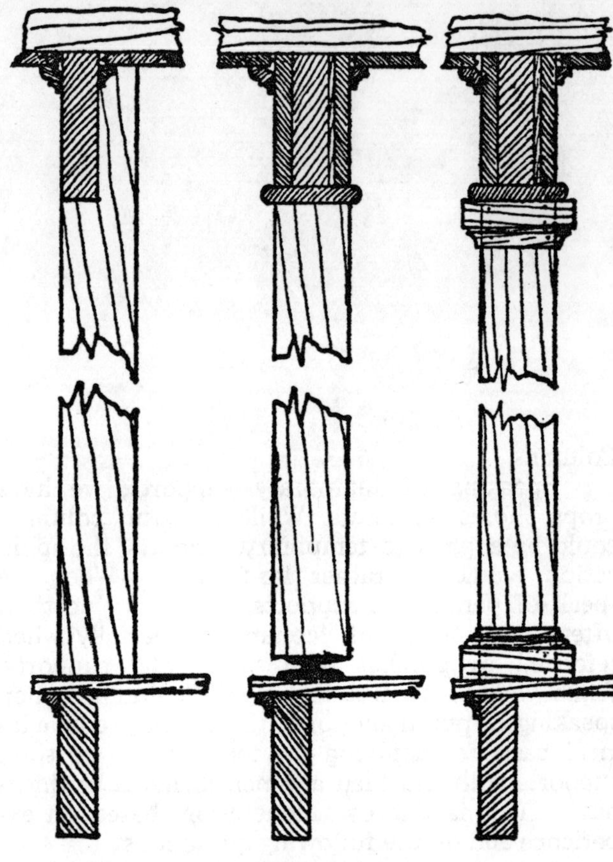

Fig. 244

at the top to taper off the base and the cap, respectively. In the same way many variations in the design can be obtained by changing the style of moulding used.

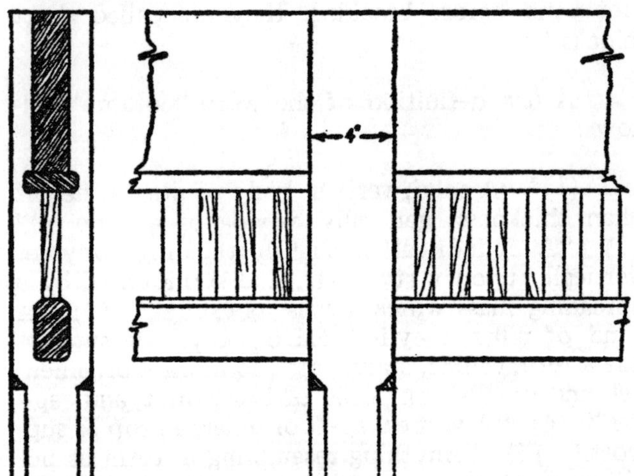

Fig. 245

A method of joining a 4x4 column to the chord, is shown by Fig. 245, which was much in use more than a score of years ago. The detail shown is somewhat modernized. Where square balusters are shown, the old timers used turned spindles. Fig. 246 represents one end of a shed-roof for a cheap porch, on which is shown the construction

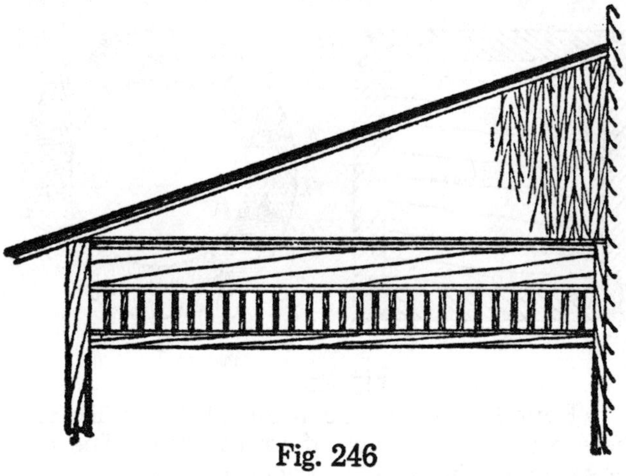

Fig. 246

given in detail in the previous figure. The ends of such roofs were frequently finished with beaded ceiling, as we are indicating to the right with shading.

Half columns were fastened to the building where the chord joined the house, to receive the siding and give additional support to the chord. Square or chamfered half columns were placed directly against the boxing, and the siding was fitted to them, but the turned half columns had to have a board placed onto the boxing to receive the siding, and onto this board the half column was nailed.

One of the most commonly used constructions, is shown by Fig. 247. Also shown is a plain

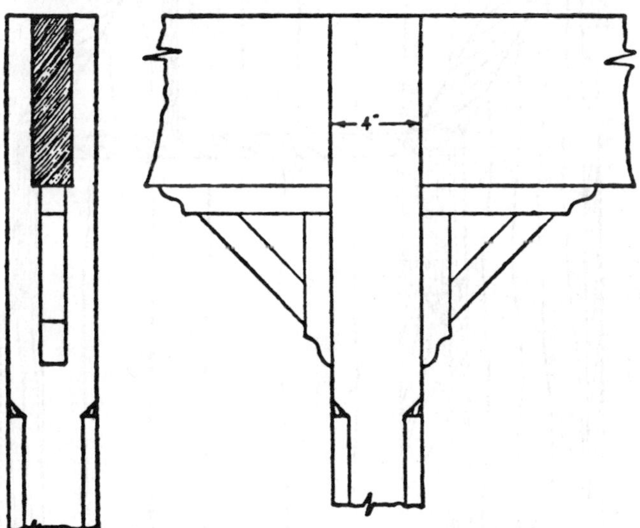

Fig. 247

straight-lined bracket, for simplicity, but in the days of our fathers, such brackets represented curlycues of innumerable descriptions. Fig. 248 shows the simple effect that the bracket construction gives on a cheap porch. Such porches are still to be seen, minus the brackets, on some new buildings, especially where economy is a necessary factor in the choice of construction.

Brackets as well as turned porch posts are in disfavor nowadays, since many home owners want

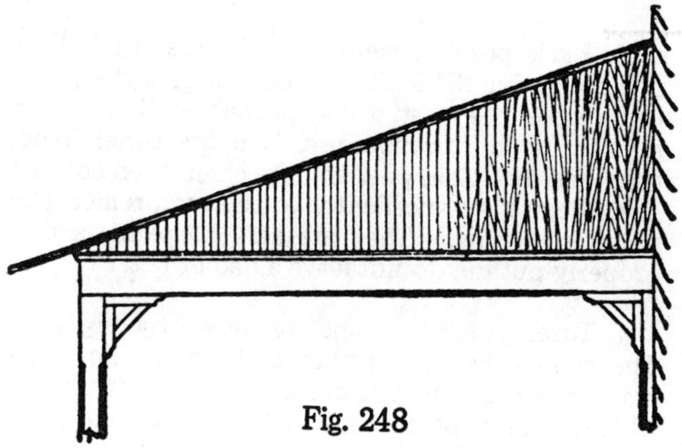

Fig. 248

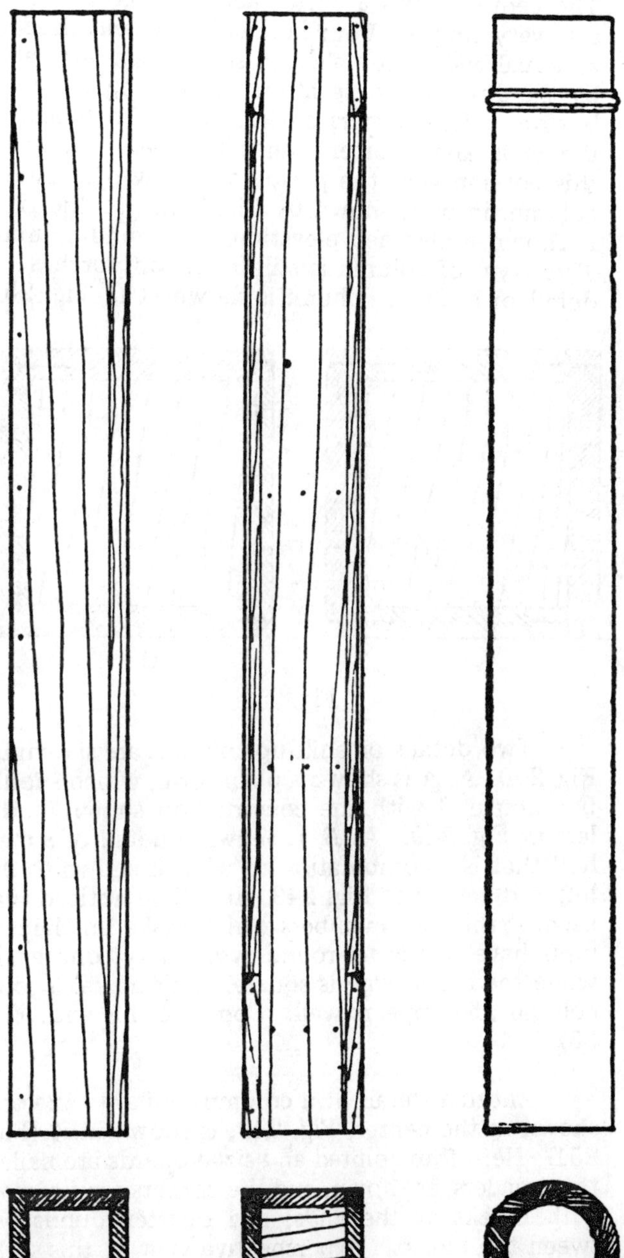

Fig. 249

their back porches screened. Besides that, they make painting difficult, and do not give the pleasing appearance that was expected of them when they were in common use. On the other hand, both the 4x4 plain post and the chamfered column lend themselves excellently to the requirements of porch screening. They are easy to paint, and if properly put up, do not leave a bad effect.

Three porch columns are shown by Fig. 249. The one to the left is a plain built-up column. The plan shown at the bottom gives the construction, which is perhaps the simplest that can be used. Four boards jointed and sized to an equal width are nailed together as shown by the illustration. The center shows a better appearing column, but also very simple. Four boards joined and sized to an equal width are nailed to spreaders as indicated by the dots on the elevation and by the plan at the bottom. The corners are finished with blocks at the ends and quarter-rounds between. Compare this column with the perspective view of a similar column, in part, shown by Fig. 252. To the right is shown a plan and elevation of a round column. This style of column is still in use on porches. A detail of a round column is shown at *B*, Fig. 251.

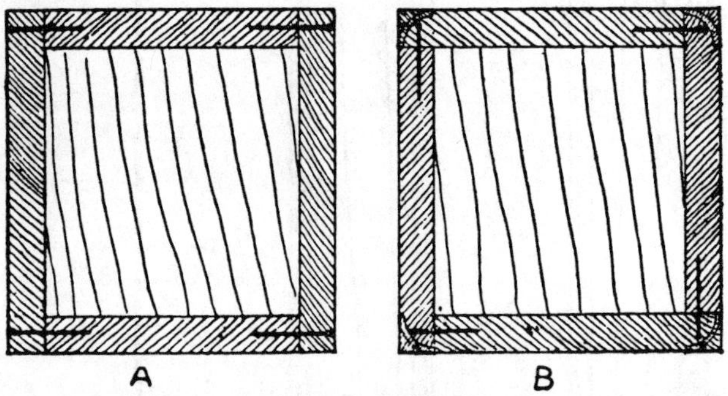

Fig. 250

Two details of built-up columns are shown by Fig. 250. At *A* is shown a plain form, which should be compared with the construction shown to the left in Fig. 249. At *B*, is shown a detail of a method that is a combination of what is shown to the left and center of Fig. 249. It will be noticed that each of the four members which make up the column have a quarter-round worked on one edge, while the other edge is square. This makes a good column and appears well. Compare this with Fig. 252.

Another detail of a column similar to the one shown at the center, Fig. 249, is shown at *A*, Fig. 251. Here four jointed and sized boards are nailed to spreaders, as shown, and the corners are finished with blocks at the ends, and quarter-rounds between the blocks. A perspective view of this style of column is shown in part by Fig. 252. Fig. 253 shows two perspective views of the corner blocks. The heavily shaded parts indicate two methods of tapering the end that joins the quarter-round.

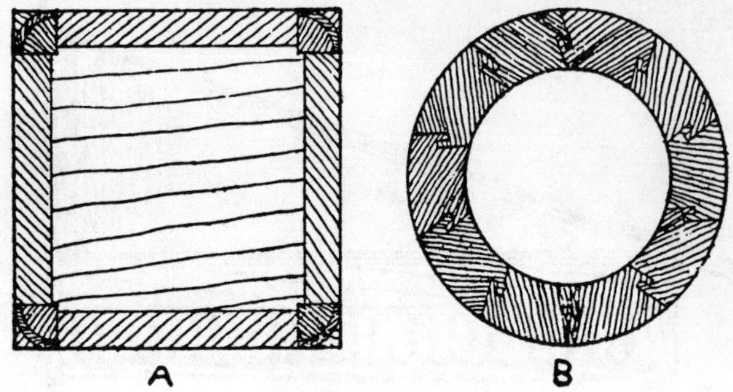

Fig. 251

Round columns are made at the mills, and there are various kinds of forms used. The one shown at *B*, Fig. 251, is a good one, but it is not the only good construction for such columns.

A cap of a round column, is shown by Fig. 254 at the top -- at the center is shown a base for a round column, and at the bottom we show a base for a square column. The cap for a square column is usually the same as the base, speaking of columns that are made on the job. The dotted lines indicate that the column extends to the bearing, and that the base is merely a band nailed around the column. The same is true of the cap.

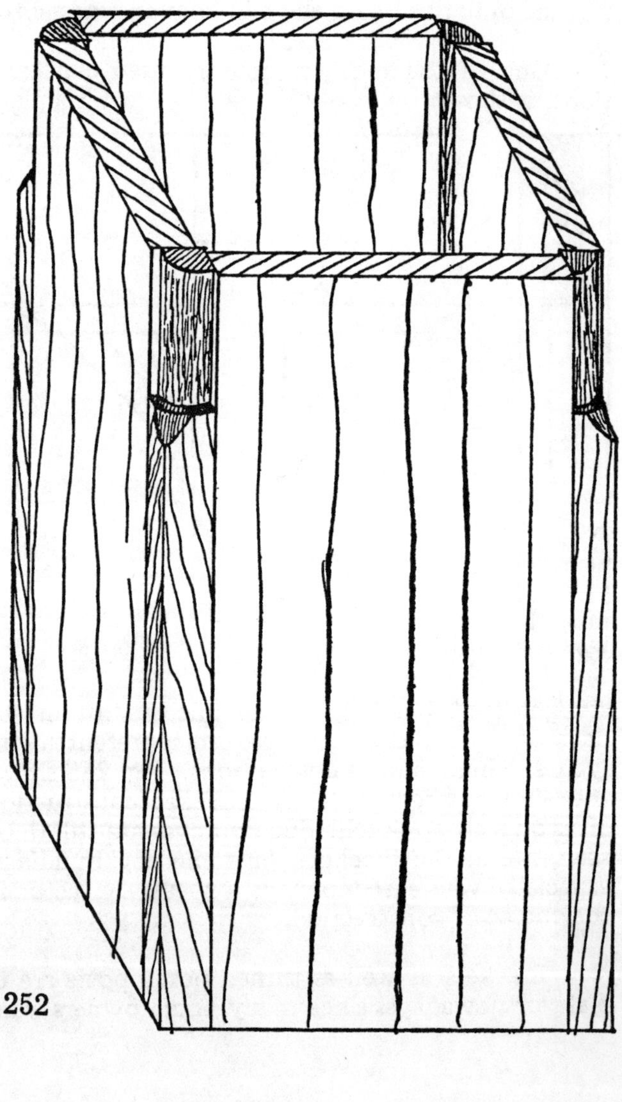

Fig. 252

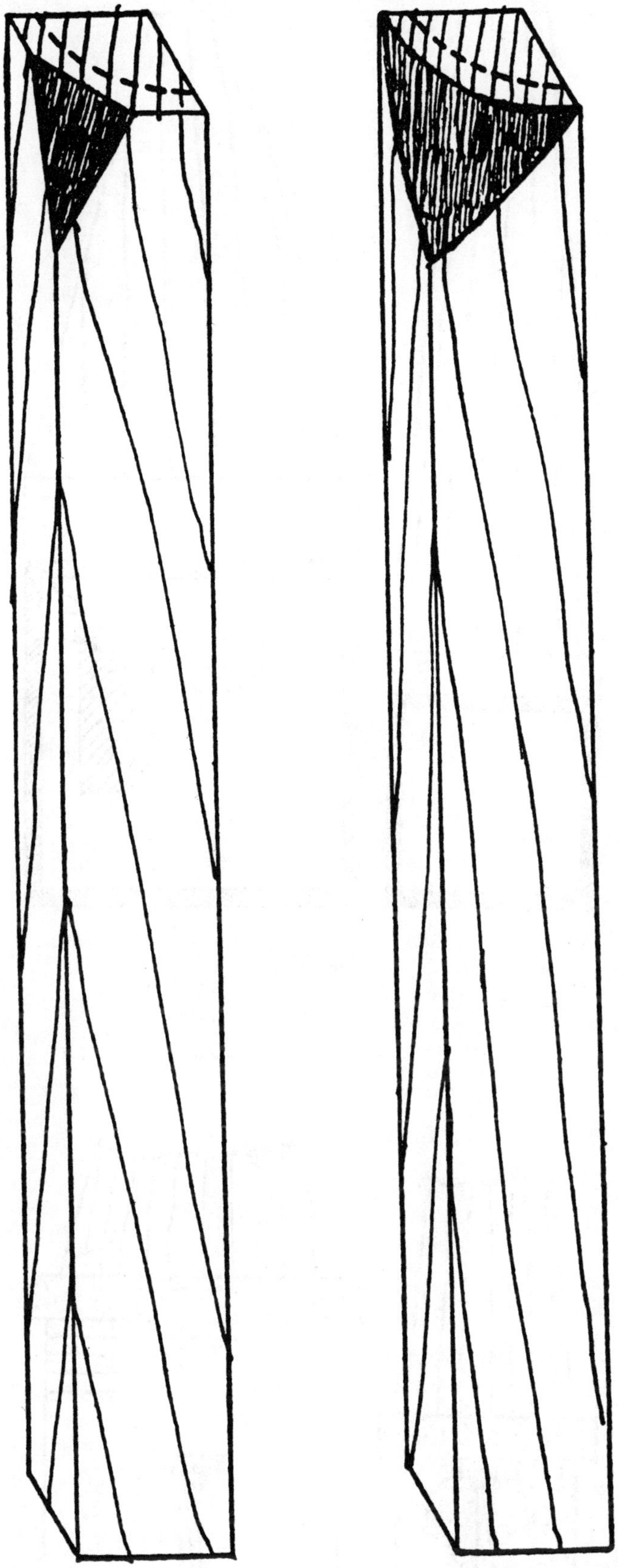

Fig. 253

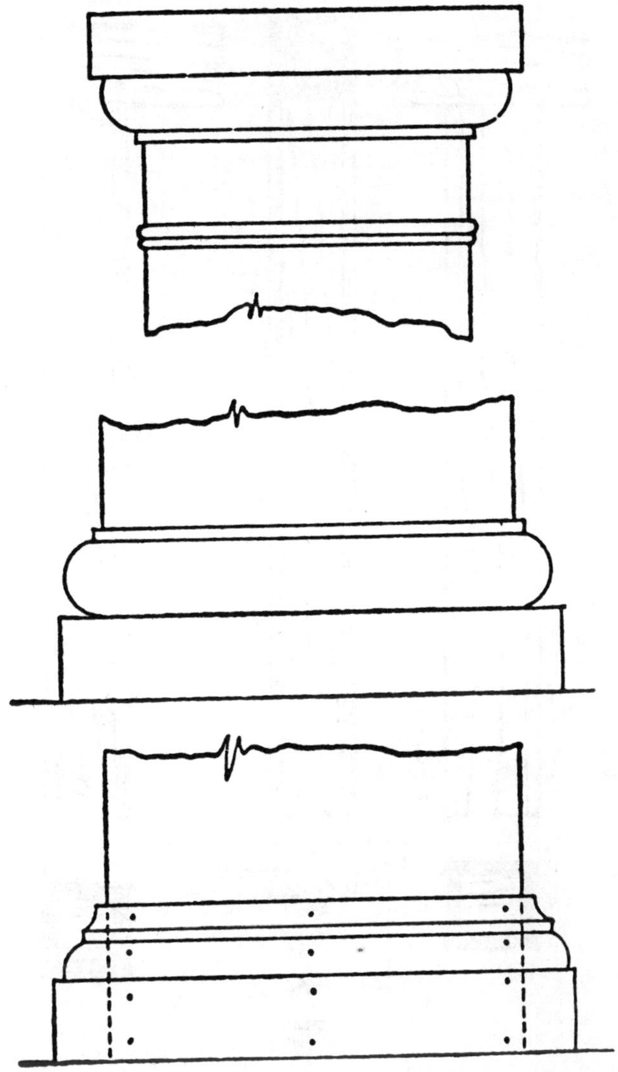

Fig. 254

To the left and center of Fig. 255, are shown two views of a column that goes well with a pedestal similar to that shown at the center. At the bottom, left, is shown a plan of the column with butt joints. A detail of a housed construction of this column is shown to the left, Fig. 256. The column shown to the right, Fig. 255, can be used either with or without cap or base. The plan at the bottom shows a butt-joint. To the right, Fig. 256, is shown a detail of a housed column construction.

Two tapering columns are shown by Fig. 257. At the bottom is shown how the joints are made. It will be noticed that the one to the left is made with miter joints, while the one to the right has joints that are a combination of the miter and butt joints. The inset, as pointed out, shows a large-scale detail of the combination joint used in this column.

The bottoms of the columns shown in Fig. 257, are shown by Fig. 258. They are shown without a base, which is practical, and appears well. A

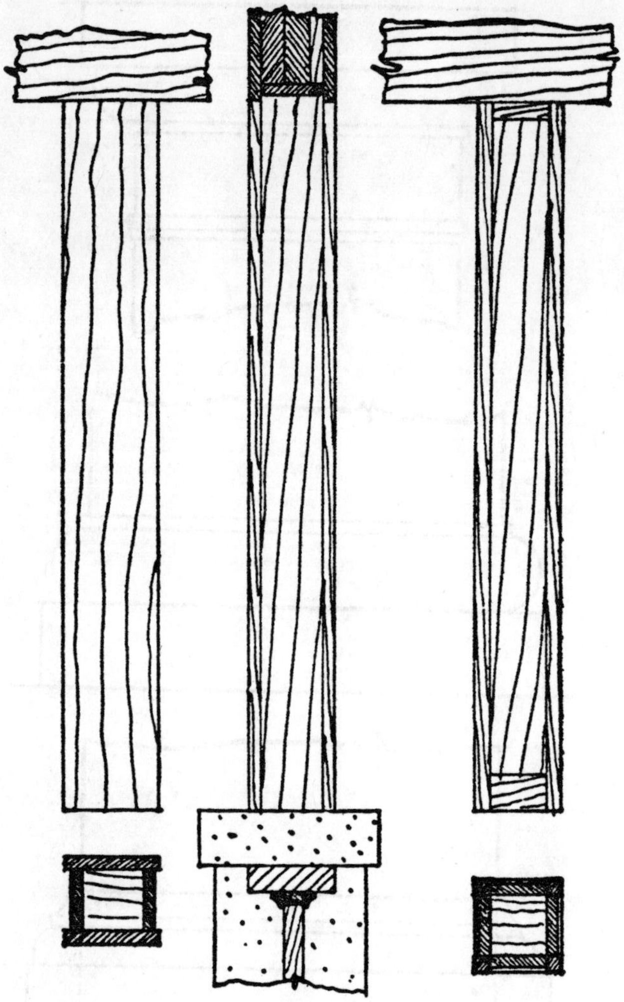

Fig. 255

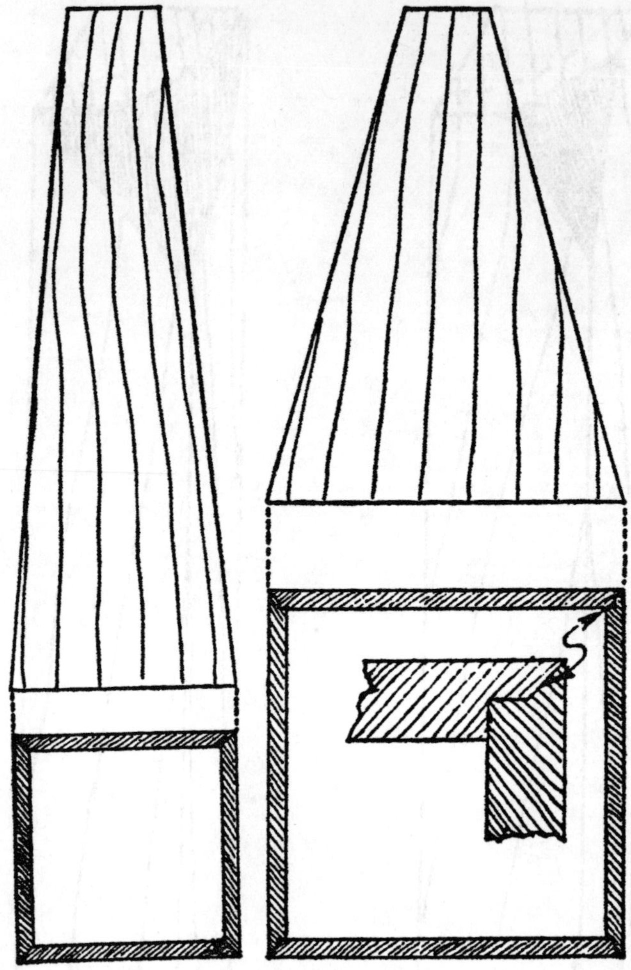

Fig. 257

plain base, such as shown in these figures, dries out more readily than a moulding, or a built-up base. Especially a built-up base holds water, which sooner or later starts rotting. In case a base is desired, it can easily be added.

Altogether too often one sees porch columns that are, in one way or another, out of proportion with the chord. Sometimes the chord is too large for the column, as shown to the left, Fig. 259, and at other times the column is too large for the chord, as can be seen to the right of this figure. The latter is probably the most commonly seen. We work-

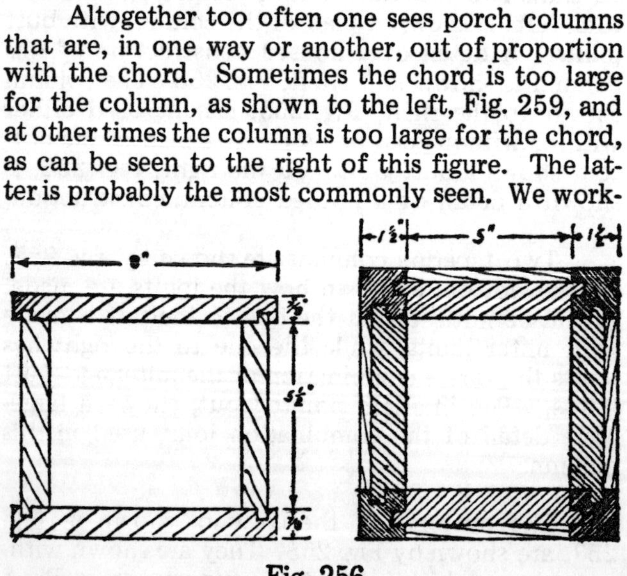

Fig. 256

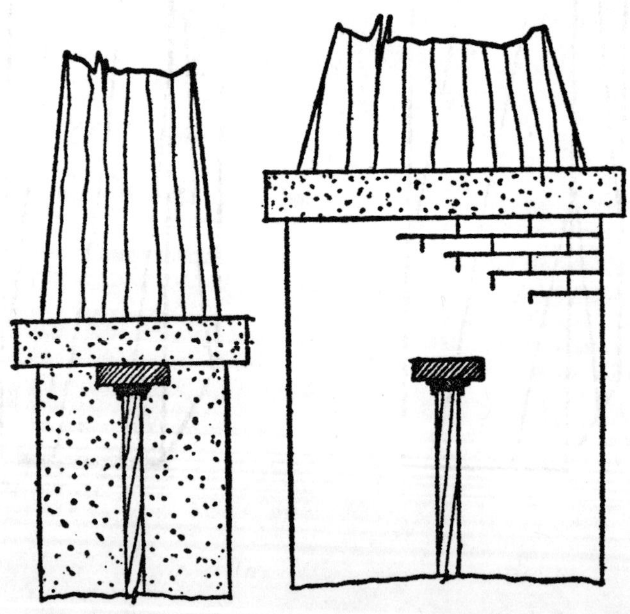

Fig. 258

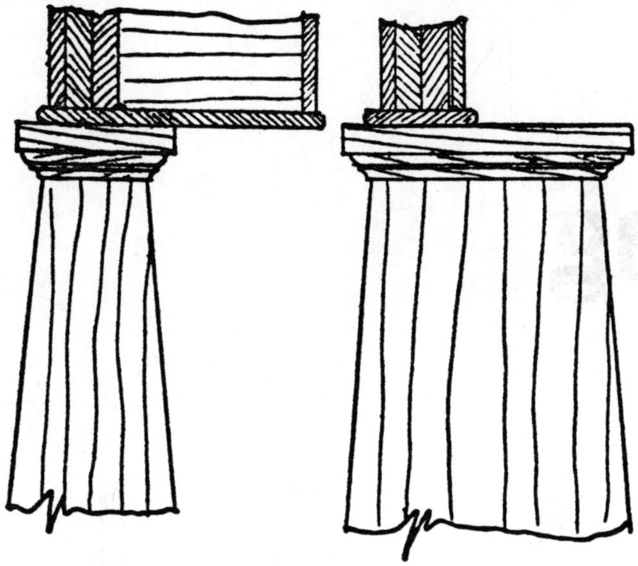

Fig. 259

Fig. 260

ed on a porch some years ago, where this very thing happened. The plan showed a small chord, and extremely large columns.

We show the tops of two columns properly centered with the chord, in Fig. 260. Compare Fig. 259 with Fig. 260 and note the difference. The illustrations in Fig. 259 are exaggerations.

While one plain constructions are shown by the illustrations of large columns, such columns lend themselves readily to paneled work. In this, too, the columns and the chord should be in agreement.

19 Porch Railing

Cheap Porch Railings

This discussion is confined to the simplest of porch railings which will, directly or indirectly, include all such railings that might be found, not only on porches, but on platforms and similar places. A cheap porch in part, is shown by Fig. 261, which is left without the protection of a railing.

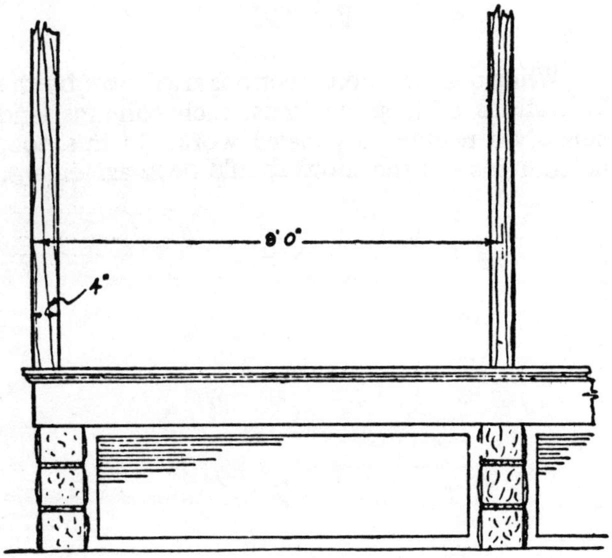

Fig. 261

In Fig. 262 we show the same arrangement, having the protection of one rail placed 2 feet 6-inches above the porch floor. This height is taken for convenience rather than as a standard. Occasionally railings are found to be higher, and quite frequently they are lower. It will be noticed that in both of these instances the porch posts are 4x4's, without base. This arrangement is quite suitable for porches that are to be screened, as many of such porches are.

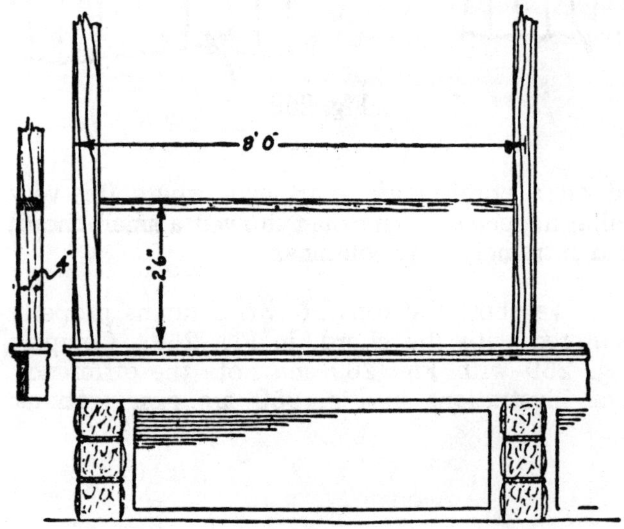

Fig. 262

A two-rail layout is shown by Fig. 263. To the left we show a cross section of it. The top of the upper rail, as in the previous figure, is 2 feet 6-inches above the porch floor. The posts also are 4x4's, but have bases. These bases are usually nailed to the posts, but sometimes the post is set onto the plinth block, and then finished with a quarter-round, cove or some other moulding. Fig. 264 is similar to the layout shown in Fig. 263, excepting that it has three rails and a center support. This is a substantial construction, and suitable for harder usage than the one shown in the preceding figures. A cross section is shown to the left.

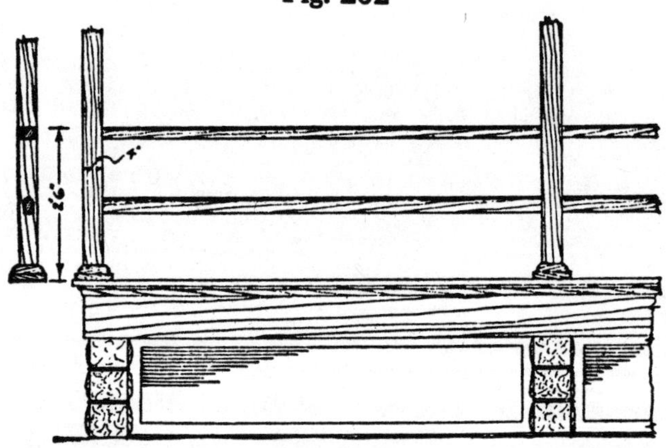

Fig. 263

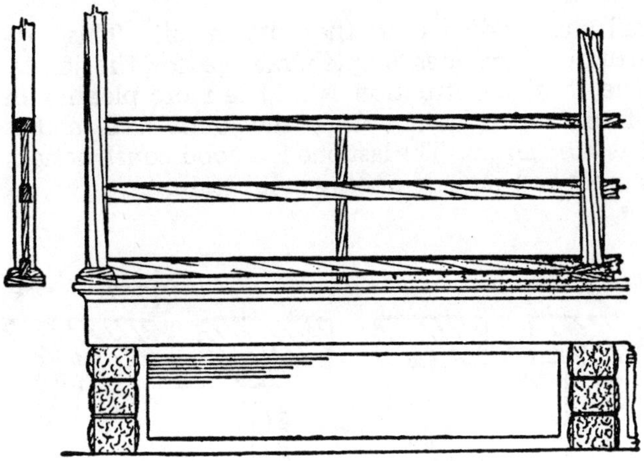

Fig. 264

How to fasten short posts securely to the porch is shown by Fig. 265. To the left we show the post fastened to the porch floor with a two-nut bolt. To the right we have the same post in cross section with the railing and base in place. Depending on toe-nailing alone for holding such posts in position, is not sufficient to make a substantial job -- the bolt reinforcing is of great importance.

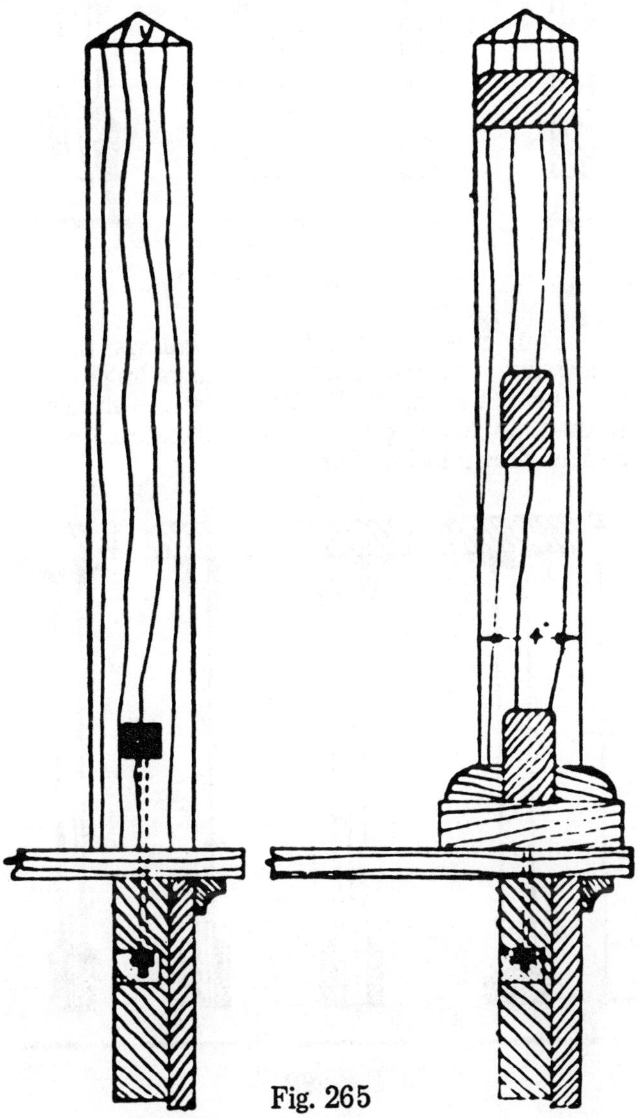

Fig. 265

In Fig. 266 are shown, to the left, two cross sections of a paneled or curtained railing. A side-view of the better one is shown to the right. It will be noticed that the beaded ceiling is shown housed into the rail, but in many instances, where the quarter-round or cove moulding is used, the housing is omitted. The dotted perpendicular lines suggest the appearance of the posts when they extend up to the superstructure of the porch.

Fig. 266

A detail of a curtained railing is shown by Fig. 267 in which the beaded ceiling is not housed into the rail. The dotted lines suggest housing, should it be desired. To the right we have an apron at the top and a base at the bottom. These members lend support to the railing, which, as we indicate at the bottom, is held 1/8 of an inch above the porch floor for drainage purposes. A center support should be provided on this railing construction to prevent sagging.

While the construction just explained, in some form or other, is often used, the 1/8-inch space under the railing is one of the worst dirt-catchers that can be found on any porch.

We are aware that a cheap porch will always be a cheap porch, but that fact only makes it all the more important that the builder endeavor to obtain the very best appearance possible. Homes on which such porches are used, rarely are designed by architects, and for that reason every carpenter should cultivate a taste for harmony and proper proportions.

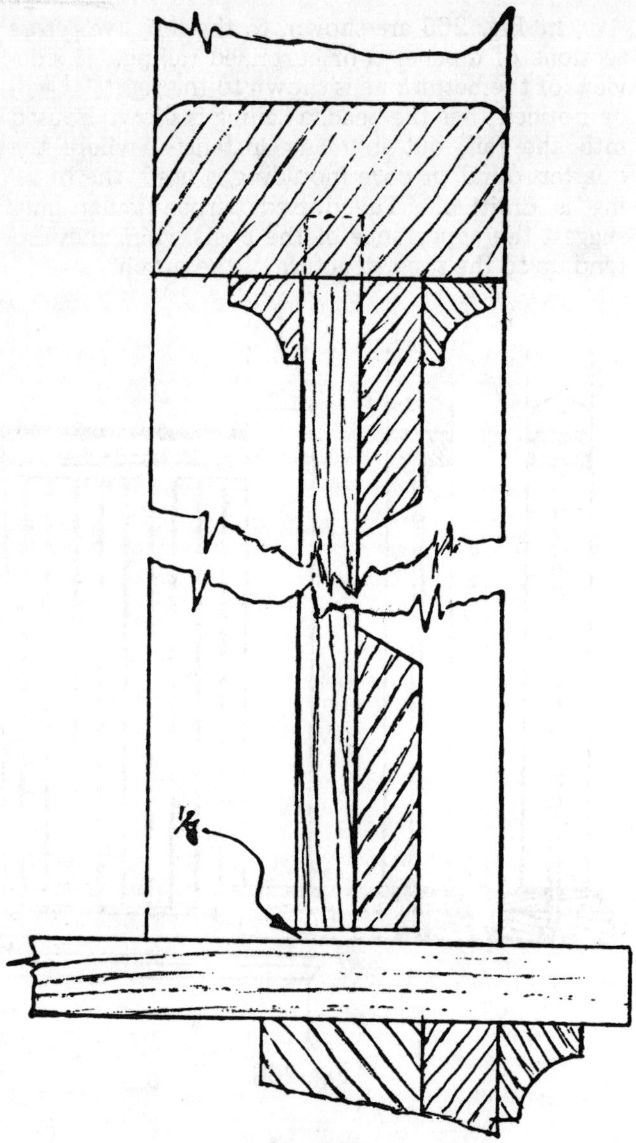

Fig. 267

Painting Joints

If the joints of porch flooring need painting, they need it applied thoroughly on both tongues and grooves alike, otherwise it is only a half-way job. We assume, of course, that the exposed surface will be kept well painted also. All joints, both of the plinth blocks and of the porch columns should be well painted, and the hollow part of the column, speaking of round columns, should be painted as far up as possible. The life of porch railings will be increased by many years, if all joints are thoroughly painted when they are made, and the exposed surface is kept properly painted.

Details of Porch Railings

In Fig. 268 we present four cross sections of porch railings. The first from left to right, is simple, although, if it is used on a house designed in harmony with it, it will have a pleasing effect. The second one is a little more substantially constructed, with the balasters housed into the top rail and saddled onto the bottom rail. This construction provides ample drainage for the joints. The third construction is a little more pleasing in effect, but does not drain the joints well, and is less substantial. The last one is a good construction, simple and capable of easy modification.

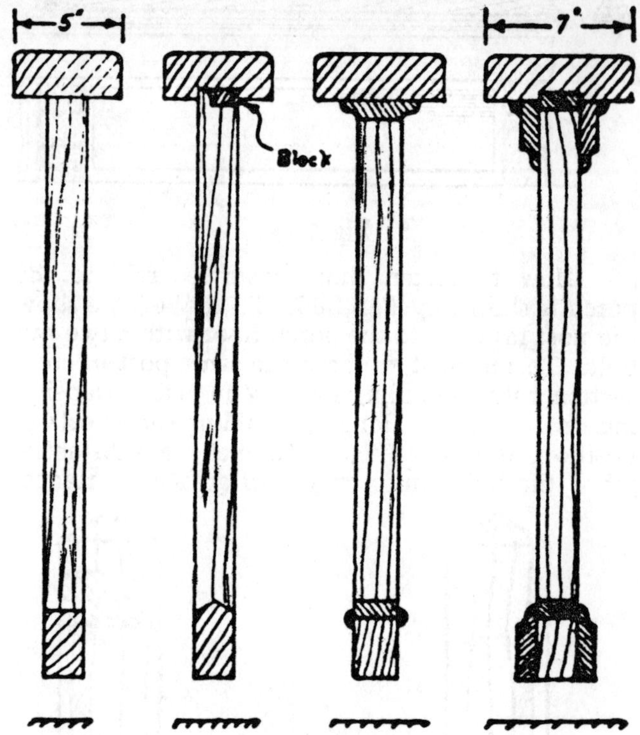

Fig. 268

Four cross sections of porch railings, that are suggestive of various possibilities in railing designs, are shown by Fig. 269. These designs should be studied, in keeping with the explanations of the designs shown in Fig 268.

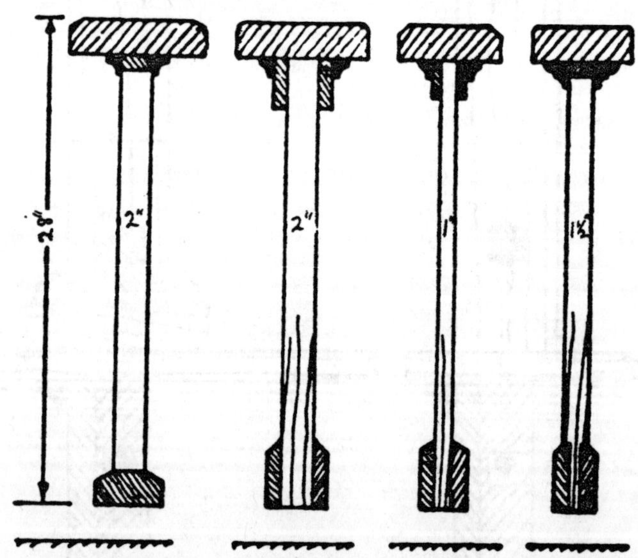

Fig. 269

A commonly used porch railing construction is shown by Fig. 270. To the extreme left is given a cross section, and to the right of this is given a sideview of the railing. At *a* is indicated a center support, which should be provided to prevent sagging of the railing. In case of long railings, these supports should be placed at intervals of 5 or 6-feet, depending on the strength of the railing.

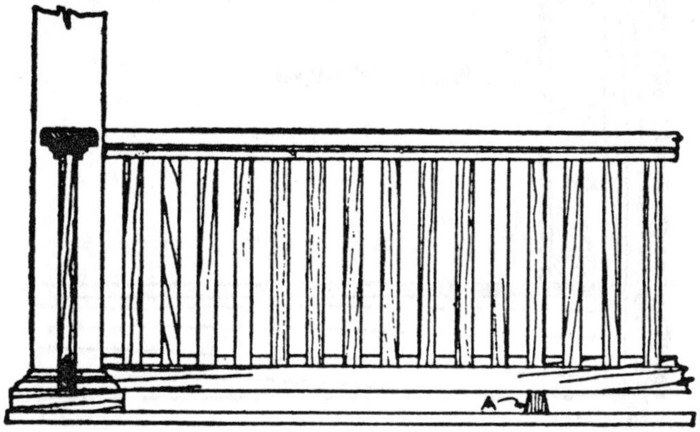

Fig. 270

In Fig. 271 is shown a railing construction, the spindles of which are rather clumsy, but suggestive of variation possibilities. Just the opposite of

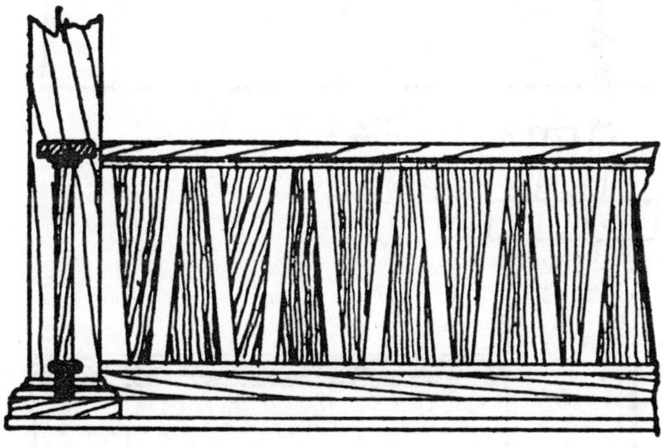

Fig. 271

this is shown by Fig. 272, which is bad, because it is a child-trap, and should never be used. The spindles are set slanting in a sort of cow-stanchion fashion, with the openings ranging from 2-inches on one end and 6-inches on the other, making it possible for small children to get their heads through and hang themselves. The effect of this design is as bad as that of Fig. 271. Bearing in mind the warning just given, both of these designs can be modified so as to be safe and render a pleasing appearance. In the first instance, the size of the spindles should be reduced, and the openings between them increased somewhat -- while in the second instance, the spindles should be set closer together, and more nearly perpendicular, so as to reduce the space between the spindles enough to insure safety.

Fig. 272

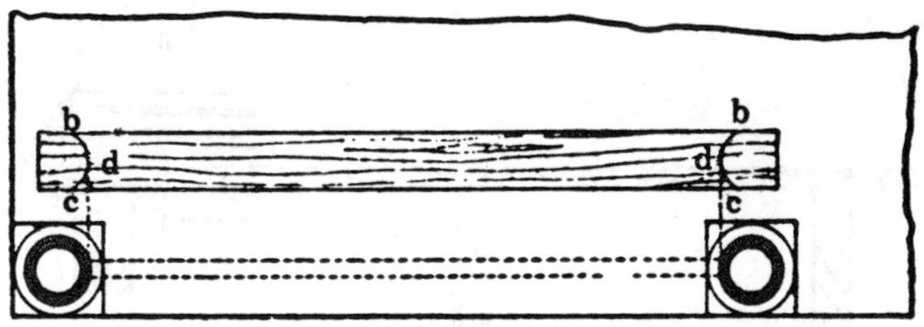

Fig. 273

In Fig. 273 is shown a porch platform in part, with two columns. The dotted lines extending from column to column, represent the railing from the top rail down. This in place, we transfer the distance between the columns to the rail, with a measuring pole. This is indicated by dotted lines extending from the columns to points *d* and *d*, on the un-cut rail. This done, we set the compass to the radius of the columns, and proceed to mark the rail by setting one leg of the compass at point *a* in such a manner that by striking *b-c* the line will contact *d*. Both ends marked, the rail should be cut to the mark accurately, making the rail, if anything, just a trifle strong. If these operations are painstakingly done, the rail should fit snugly onto the columns, somewhat on the order shown by Fig. 274.

How to fasten short posts substantially to the porch, is shown by Fig. 275. The post shown represents a built-up square post, but the method of fastening will answer equally as well on round posts.

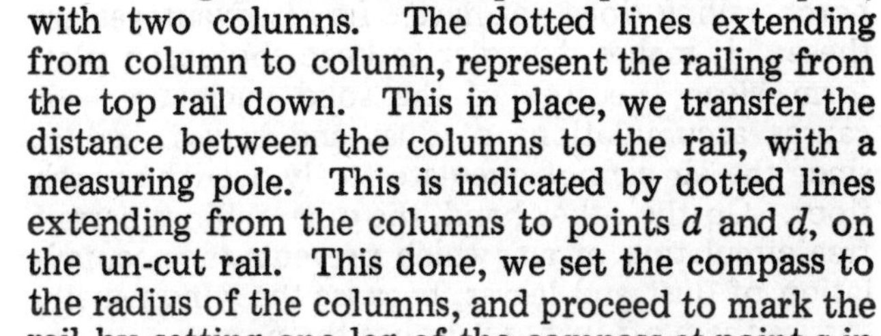

Fig. 274

Fig. 275

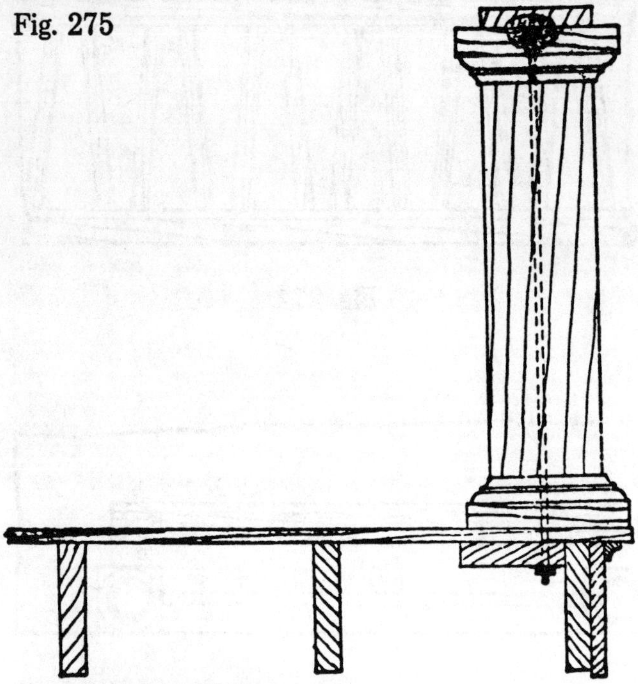

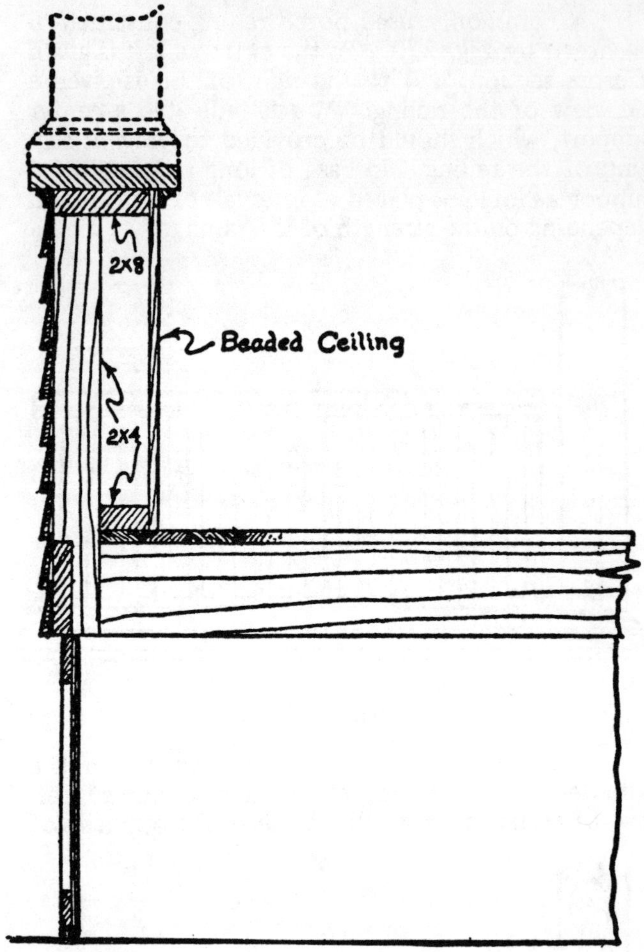

Fig. 277

Closed Porch Railings

Closed porch railings have advantages and disadvantages that open railings do not have. Among the advantages are these: A closed porch railing yields itself readily to screening; it is a little more private, and has protective qualities that the open porch railing does not have. Its disadvantages are these: It makes it harder to keep the porch platform clean, because of the solid enclosure -- it causes accumulations of dust and leaves readily, since the air cannot circulate freely over the porch floor. On the other hand, the open railings permit free circulation of air, which prevents the accumulation of dust and leaves, because the wind usually carries them off just as it brings them on.

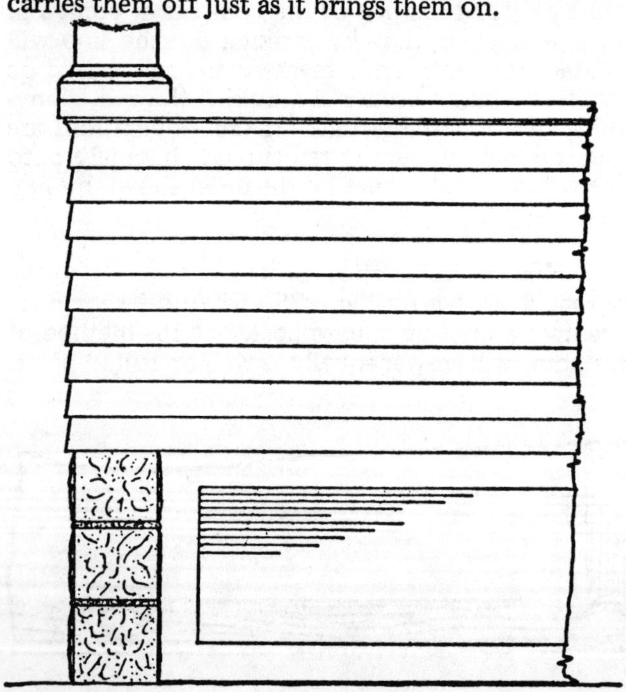

Fig. 276

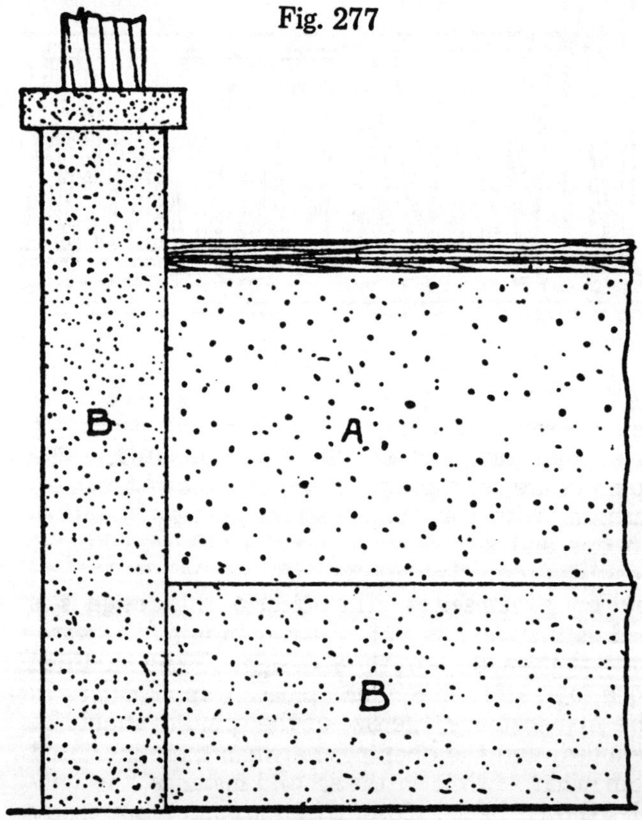

Fig. 278

An exterior corner of a closed porch railing, is shown by Fig. 276. To the upper left, we show the relationship between the railing and the column. Fig. 277 shows a cross section of the railing, looking toward the main building. The column base is shown by dotted lines, and some of the parts of the construction are pointed out.

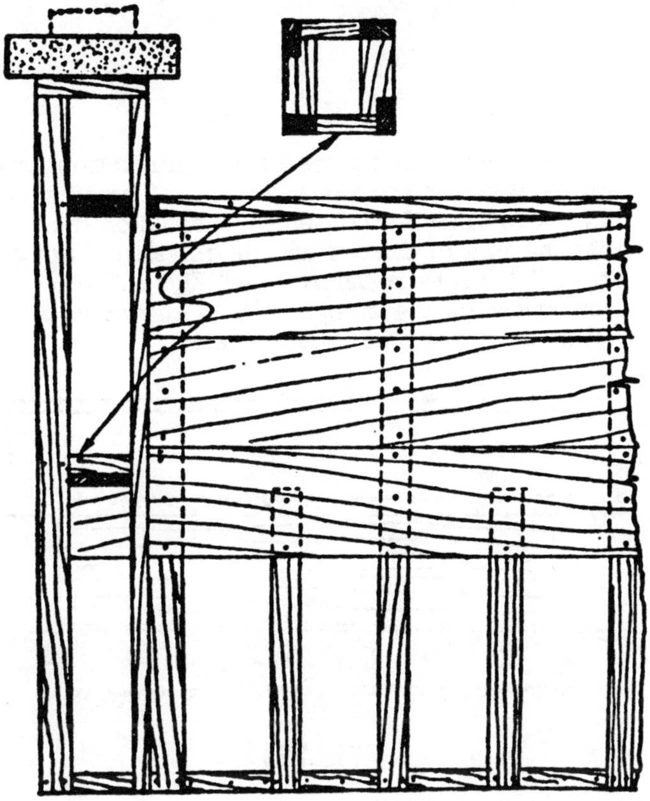

Fig. 279

In Fig. 278 is shown a closed railing which is stuccoed on the outside to correspond with the outside finish of the house. The part marked *A*, shows a pebble dash, while the parts marked *B*, are dashed with cement and sand. The framework of this railing, ready for the metal lath, is shown by Fig. 279. The dotted lines show how far the studding extend up back of the boxing. Every other one of the studding, it will be noticed, goes only part way up. These short studding are necessary to provide ample nailing for the metal lath, and to keep it from bagging. Where the boxing is, the studding are placed approximately two feet on center. A cross section of this construction is shown heavily shaded in Fig. 280. To the right of the cross section is shown the interior or back of the railing, looking from right to left, relative to Fig. 279. The inside of the railing is finished with beaded ceiling -- plywood is being used more and more. This figure does not show the corner support or pier of the porch, which is given by Fig. 281, and is a duplicate, in part, or Fig. 280, showing the pier and footing. A little study of the last four figures, will make the construction clear.

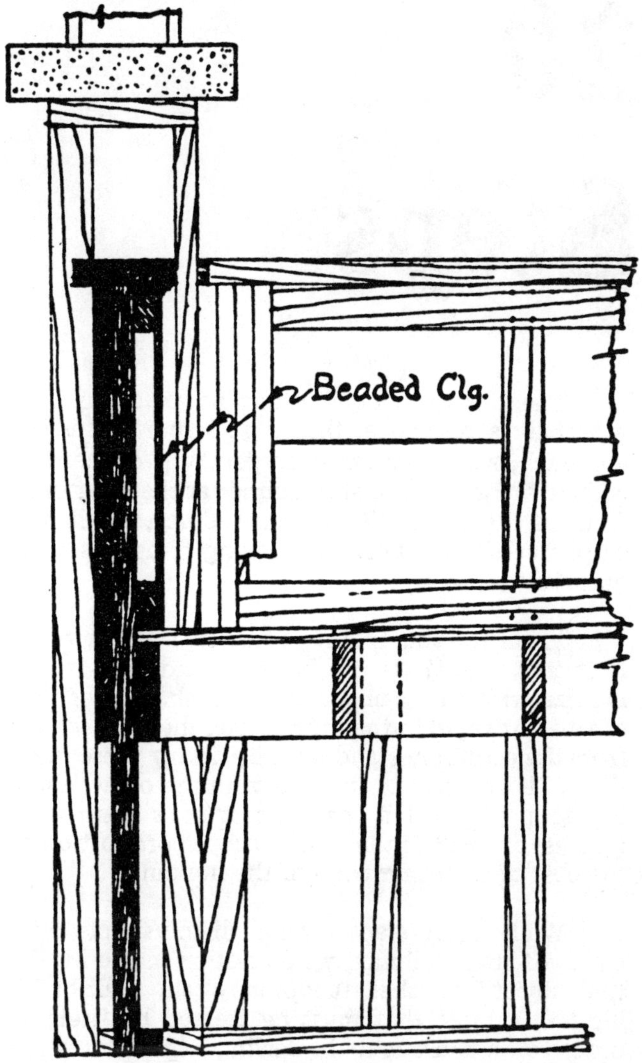

Fig. 280

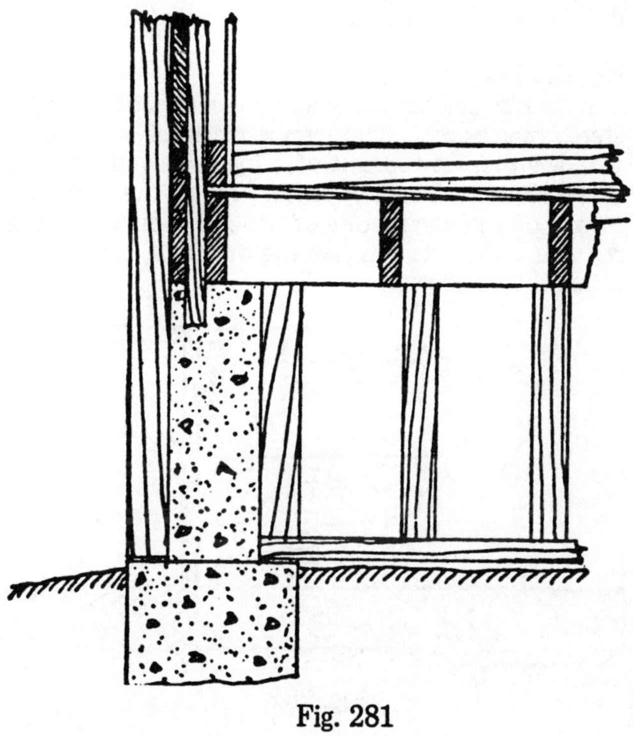

Fig. 281

20 Steps

There is probably nothing about a building, especially a residence, that is so important as the approach; which, so far as carpentry is concerned, is the step or flight of steps, either at the door or in front of the porch. To obtain the best results the carpenter should begin by asking himself several questions:

From the standpoint of service, is it practical? Is it convenient? Is it substantial and does it harmonize with the main building? If these questions can be answered in the affirmative, then consider it from the economy standpoint by asking these questions: Is the cost in keeping with the owner's ability to pay? Will it meet the owner's desires, regardless of cost? Is the cost in proper proportion to the cost of the other parts of the building?

While entry steps involve all the simple principles of stair building, we do not intend to handle the subject from that standpoint at all. Stair building is a subject that must be treated by itself, in order to be thorough. In dealing with entry steps we want to be helpful both to apprentices and to the journeyman carpenters, who often must choose designs for their clients.

Novelty Steps

In Fig. 282 we show a cross section of a simple step for a porch. The step is made of a block of wood, either hewed out of a log or sawed into the shape shown. The step is fastened to the porch by means of the cleats, one of which is pointed out at *A*. The other cleat is like the one at *A*, but on the other end of the step.

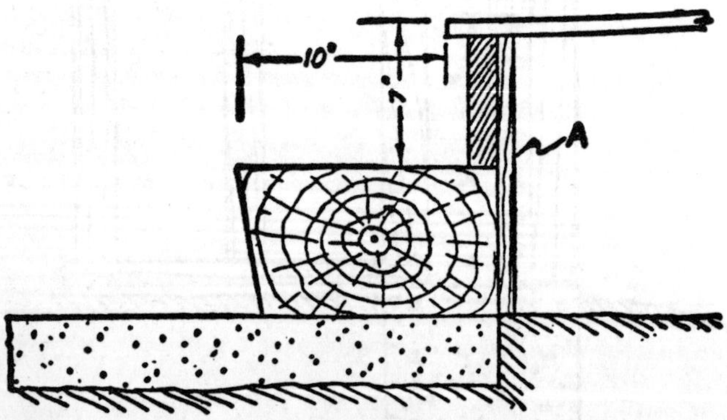

Fig. 282

As shown in figures by the illustration, the run is 10 inches and rise is 7 inches. These figures have been taken for convenience, based on the rule that the sum of the rise and the run should not go above 17 inches or below 16½. Fig. 283 shows a front view of the step, which is shown in both figures as resting on a concrete base.

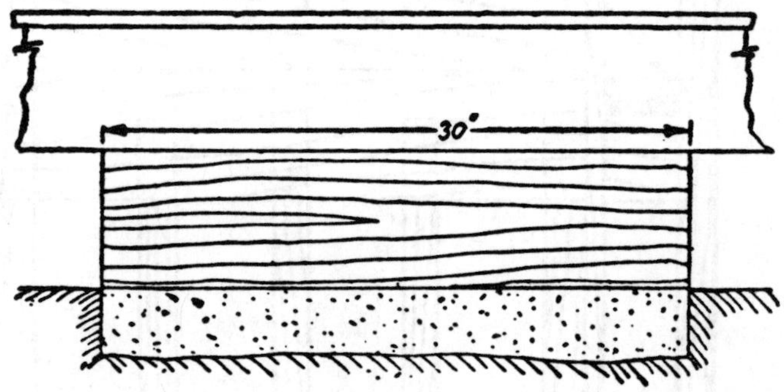

Fig. 283

A novelty in steps is shown by Figs. 284 and 285. The first is a side view and the second is a front view. These steps are made of material, either sawed to the shape shown by the drawing, or hewed out. In cabin construction hewing is still often employed, especially when the hewed effect is desired. And, as is usually the case, such cabins are almost always built near some timberland, which is depended on for the building material. Pieces of slabs are used to hold the steps together, as indicated at *A*. The run and rise are, respectively, 10 inches and 7 inches.

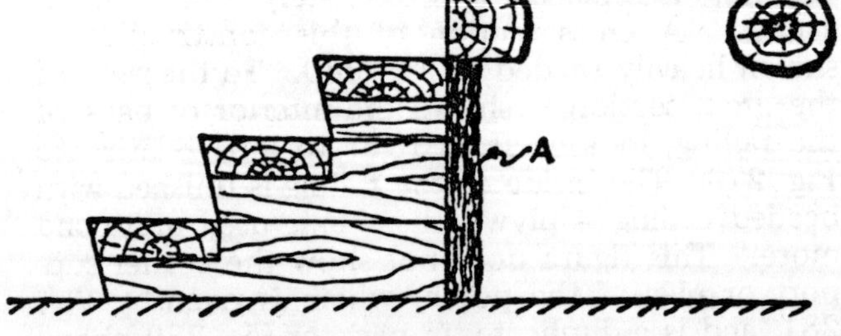

Fig. 284

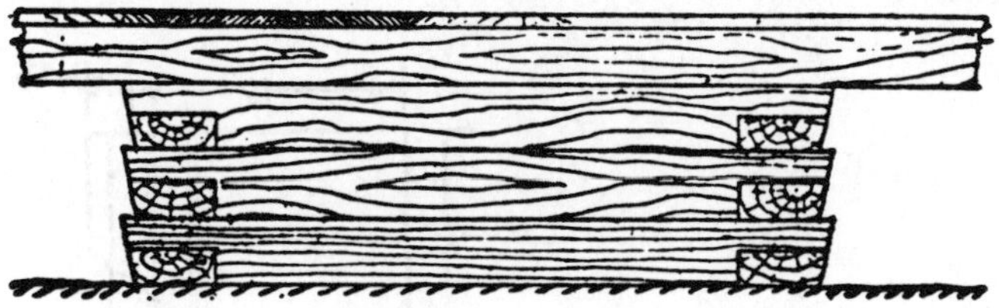

Fig. 285

Rustic Construction

Building material at out-of-the-way summer resorts is more often available in the rough than in the sawed forms. Slabs are frequently used in cabin construction with good results. Fig. 286 shows a flight of porch steps made of slabs. At *A* is shown a cross section of the kind of material from which such steps are made. We show the steps nailed together. This is acceptable for temporary work. If the steps are to be used permanently they should be housed into the stringers.

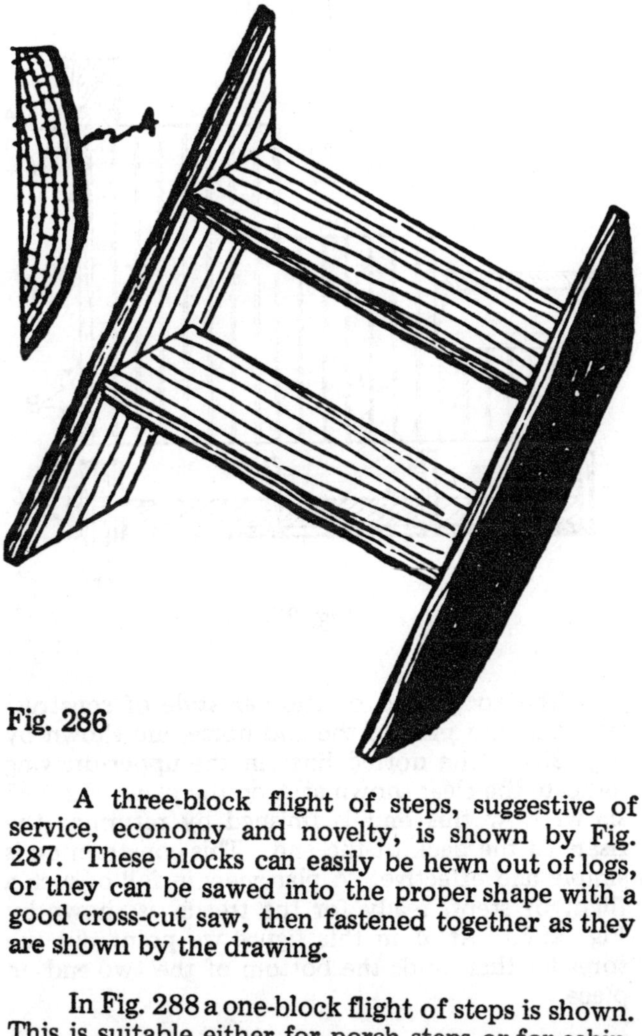

Fig. 286

A three-block flight of steps, suggestive of service, economy and novelty, is shown by Fig. 287. These blocks can easily be hewn out of logs, or they can be sawed into the proper shape with a good cross-cut saw, then fastened together as they are shown by the drawing.

In Fig. 288 a one-block flight of steps is shown. This is suitable either for porch steps or for cabin-door steps. Sawing with a cross-cut saw is the best means of construction for such steps but hewing them out of a log is not impossible. These suggestions are made assuming that the owner wants novelty and is willing to pay for it in wages to his carpenters.

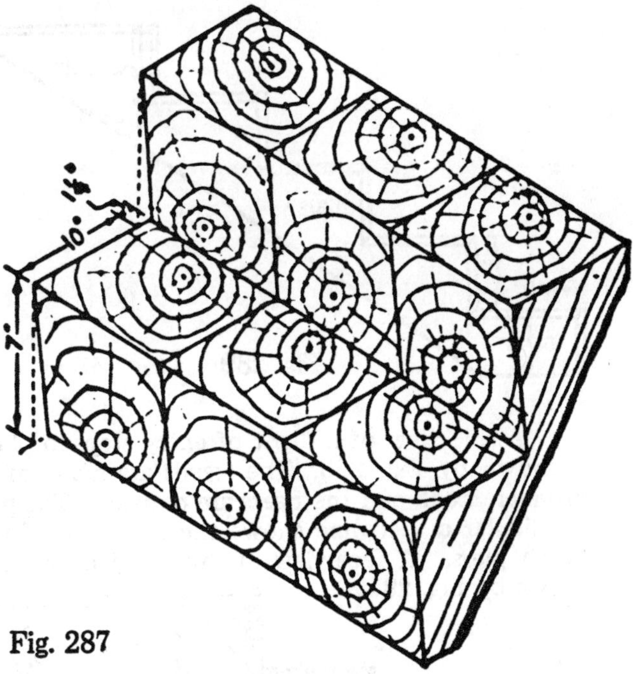

Fig. 287

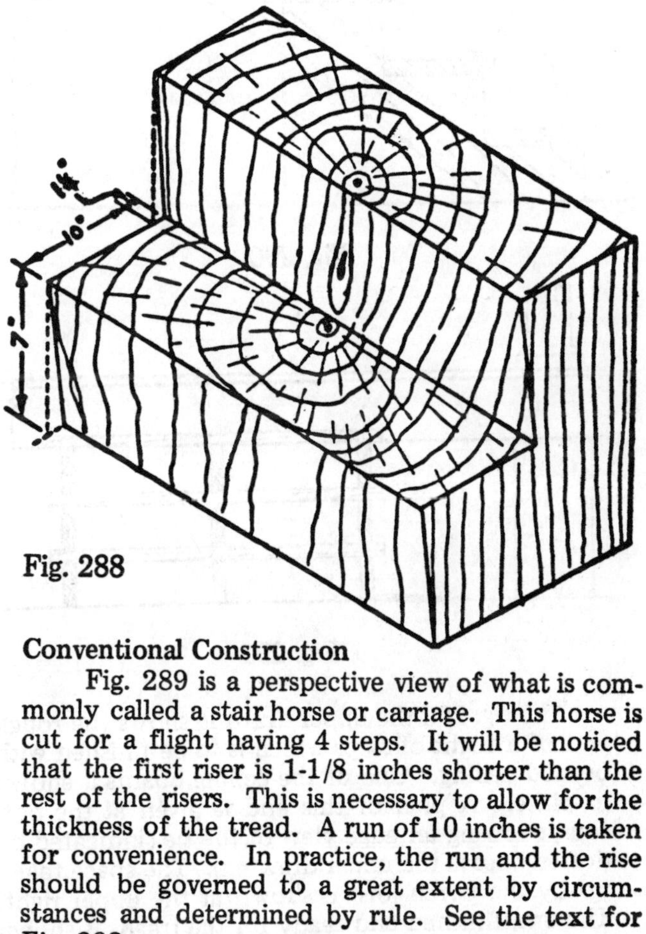

Fig. 288

Conventional Construction

Fig. 289 is a perspective view of what is commonly called a stair horse or carriage. This horse is cut for a flight having 4 steps. It will be noticed that the first riser is 1-1/8 inches shorter than the rest of the risers. This is necessary to allow for the thickness of the tread. A run of 10 inches is taken for convenience. In practice, the run and the rise should be governed to a great extent by circumstances and determined by rule. See the text for Fig. 282.

To the right of Fig. 289 we show by dotted lines how the last step, seemingly, has been made 1-5/8 inches wider than the rest. Actually, this is not the case. The dotted lines indicate the space taken up by the outside porch joist, which in rough work would answer for the last riser.

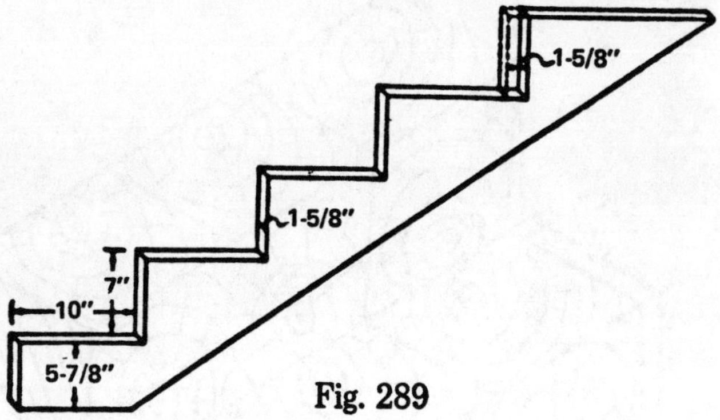

Fig. 289

A side view of a flight of entry steps in place is shown by Fig. 290. The treads are made of 1-5/8-inch stock and the risers are omitted. The perpendicular dotted line shows the face line of the porch piers. A face view of this flight of steps is shown by Fig. 291.

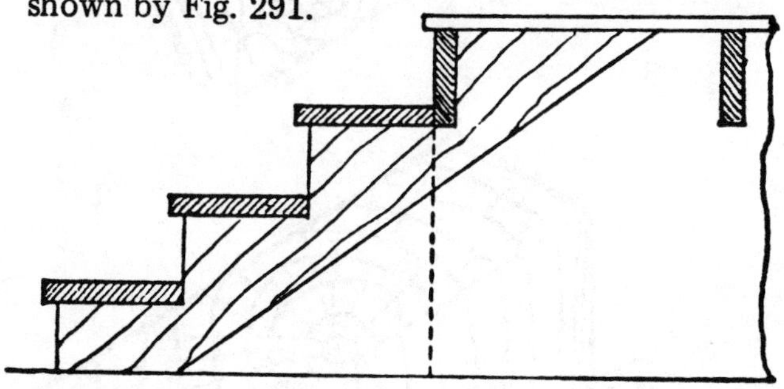

Fig. 290

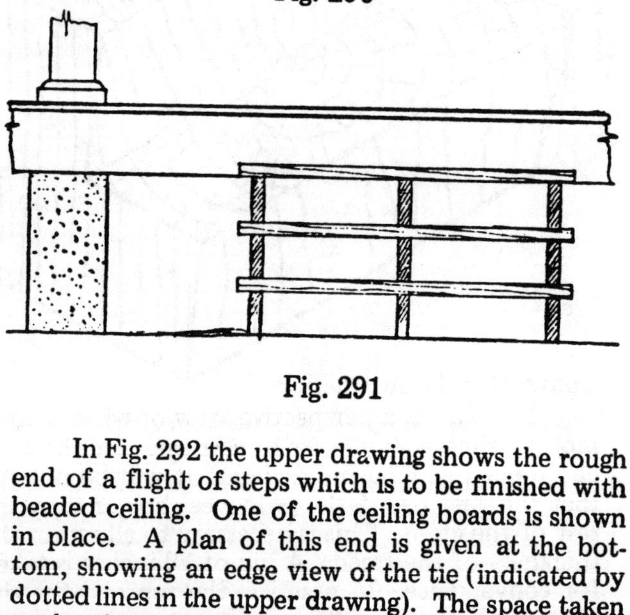

Fig. 291

In Fig. 292 the upper drawing shows the rough end of a flight of steps which is to be finished with beaded ceiling. One of the ceiling boards is shown in place. A plan of this end is given at the bottom, showing an edge view of the tie (indicated by dotted lines in the upper drawing). The space taken up by the porch joist is shown at the upper right. The end finished and ready for the treads is shown in the upper drawing, Fig. 293. To the left is shown a tread and, by dotted lines, where it is to be placed. The cleat at *A* shows the fastening of the flight of steps to the porch. Below is a plan view showing in part the risers and other details. At *B* is indicated the framework for the lattice. At *C* is shown the end view of the tie indicated at *A* in the upper drawing.

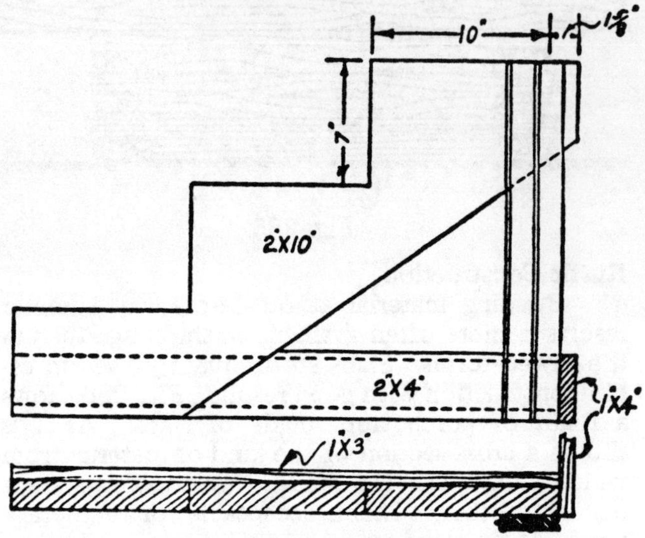

Fig. 292

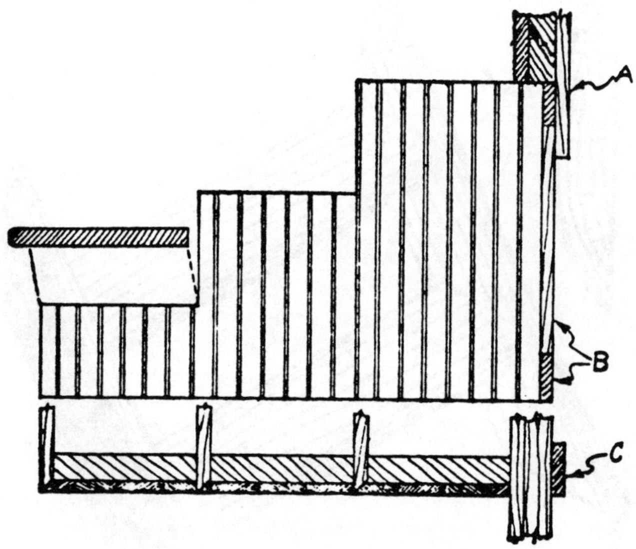

Fig. 293

The rough end of another style of construction and the plan of the end horse, are shown by Fig. 294. The dotted lines in the upper drawing indicate the cleat shown at *A* on the plan. Fig. 295 shows how this end is finished by returning the risers to the back of the end. This construction is simple and effective. A plan view in full of such a flight of steps, ready for the treads, is shown by Fig. 296. At *A* in this figure we point out the spreader that holds the bottom of the two ends in place.

So far as the width of a flight of steps is concerned, no hard-and-fast rule will apply; good taste is the best rule, speaking from the practical standpoint. One finds that steps are more often made too narrow than too wide. Steps that are narrower than the walk leading to them do not appear adequate.

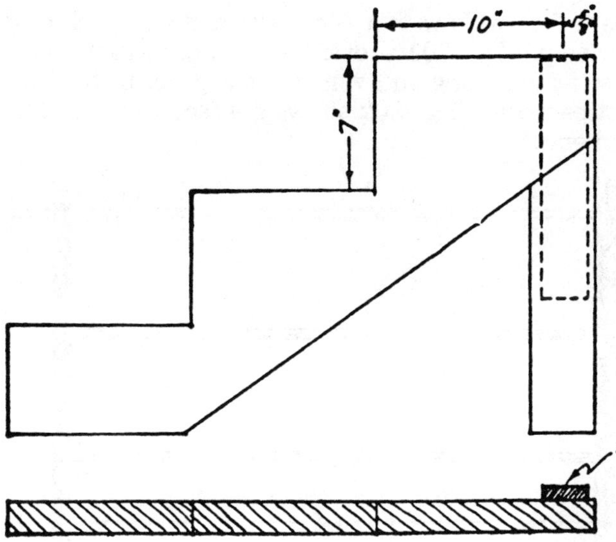

Fig. 294

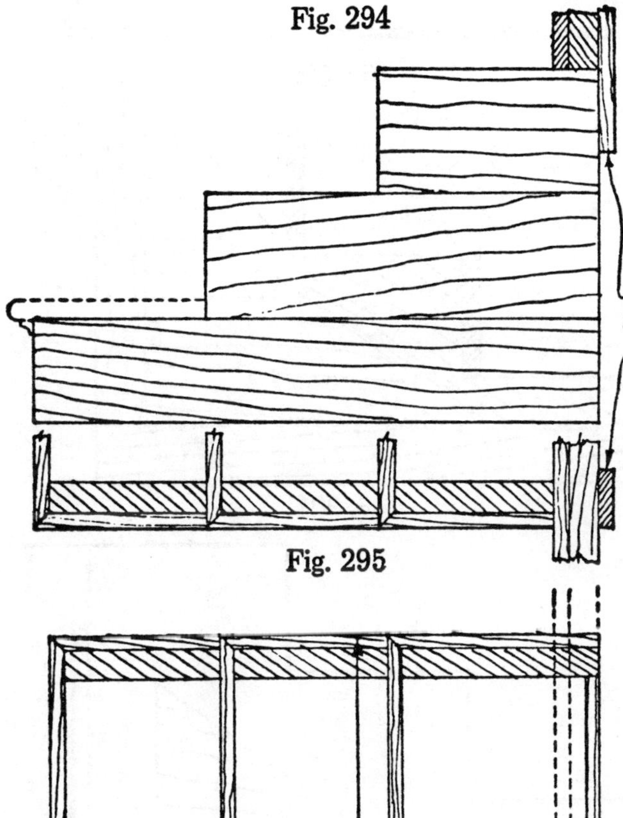

Fig. 295

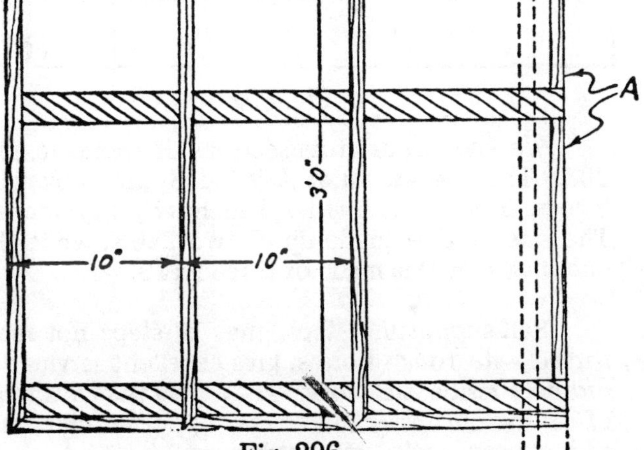

Fig. 296

Rule Explained

Let us suppose we are to build steps for a porch having a total rise from the grade line to the top of the platform nosing of 28 inches and we want steps with about 7-inch risers. In this case we proceed by dividing 28 by 7, which gives us 4, or the number of risers required for the flight of steps. Knowing the number of 7-inch risers we need, which is 4, we need to know what the run is in inches. The rule says that the run in inches, plus the rise in inches should not be more than 17 inches or less than 16½ inches. We will take 17 inches and subtract 7 inches from it, which will give us the run in inches, or 10 inches. Now we have a 10-inch run and a 7-inch rise. To simplify the matter we purposely used figures that would come out in even numbers; in practice, though, fractions are unavoidable.

Other Rules

We list several other rules to show the wide variety of rules that exist on the subject:

"The rise in inches plus the run in inches should equal 17 to 17½ inches." This authority holds out for a little larger base figure. We believe a 16½-inch base figure gives a more comfortable step.

Another authority gives this rule:

"Twice the rise in inches, plus the run in inches should equal 24." Let's see how this will work out, using the first rule. We found there that the width of the riser was 7 inches. Twice 7 equals 14, and 14 subtracted from 24 leaves 10, or the run in inches. Let's try it the other way, with the run established; subtracting 10 from 24 and dividing by 2, gives us 7, or the rise in inches.

Here is another rule: "Let the product of the rise and the run equal 66." This means that the rise in inches times the run in inches should equal 66. Let's try this with the example we have already used. There we had a 7-inch rise and a 10-inch run, which, if multiplied, would equal 70. If we wanted to find the run by dividing 66 by 7, the known rise, we would have 9-3/7, or the run in inches. If we wanted to find the rise by dividing 66 by 10, the known run, it would give us 6-6/10, or the rise in inches. There are other rules, with slight variations, too numerous to bring in here.

And finally this rule in a diagram:

This rule is illustrated by Fig. 297 where we have a triangle, *a, b, c*, with a 24-inch base and a 11-inch altitude. To obtain the run of a step when the rise is known, measure the rise off on the altitude as shown from *b to d*, or 7 inches, and strike the horizontal line *d-e*, which gives the run of the step, or in this case 8-3/11 inches. When

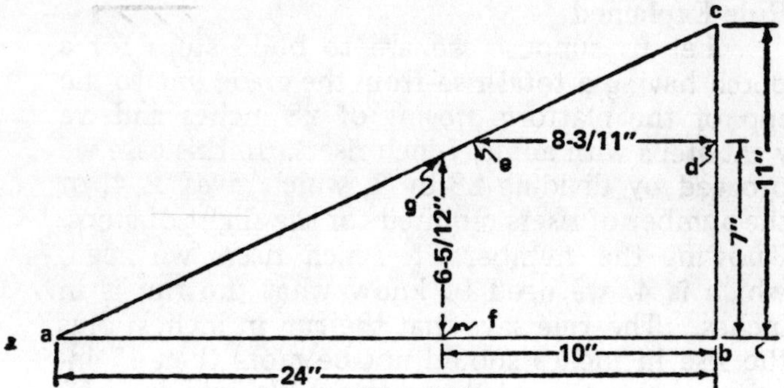

Fig. 297

the run of a step is known, which in this case is 10 inches, measure it off on the base as shown from *b* to *f*, and strike the perpendicular *f-g*, which gives the rise, or 6-5/12 inches. It would be an excellent practice for the student to make a diagram of this rule in a large scale and try it out in comparison with the rules previously given.

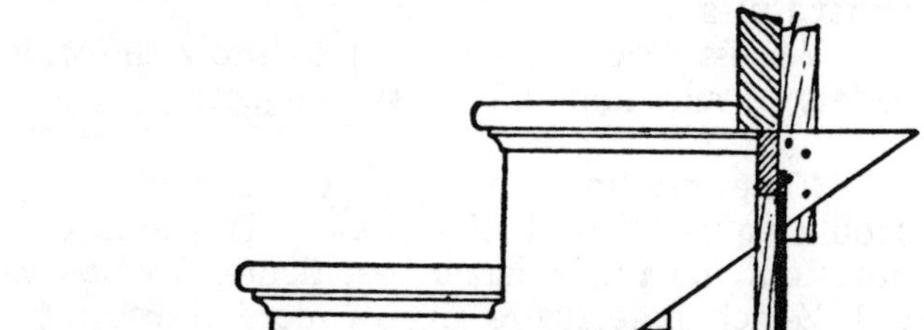

Fig. 298

Self-Supporting Steps

We have an end view of a self-supporting step in Fig. 298, the upper drawing, and just below it is a plan in part. Fig. 299 shows the back construction of the other end of the same step. Here we can see the lattice, the joint ties, the method of fastening the steps to the porch, and how the risers are nailed to the treads to support the back of the tread. Fig. 300 gives a plan of such steps just before the treads have been put in place. The steps are 4 feet wide with a 7-inch rise and a 10-inch run per step. Fig. 301 shows the back construction of a self-supporting end which has a panel instead of lattice-work. Fig. 302 shows the face view of such and end.

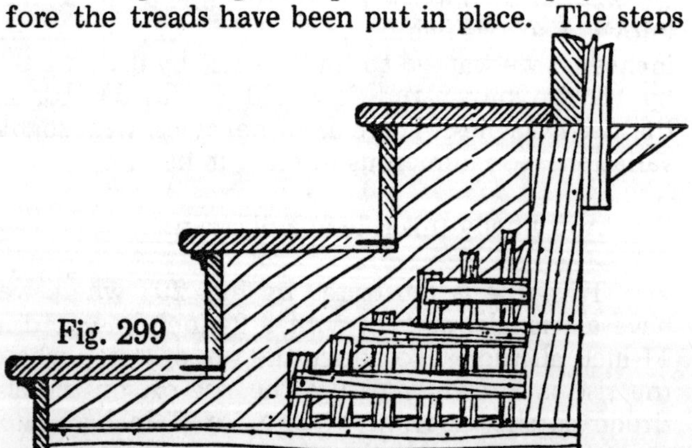

Fig. 299

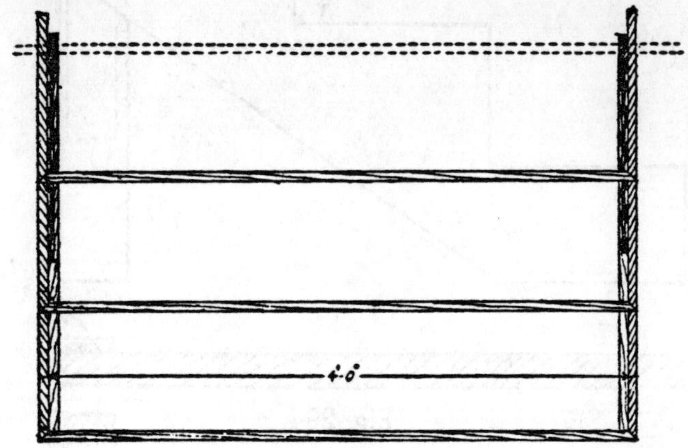

Fig. 300

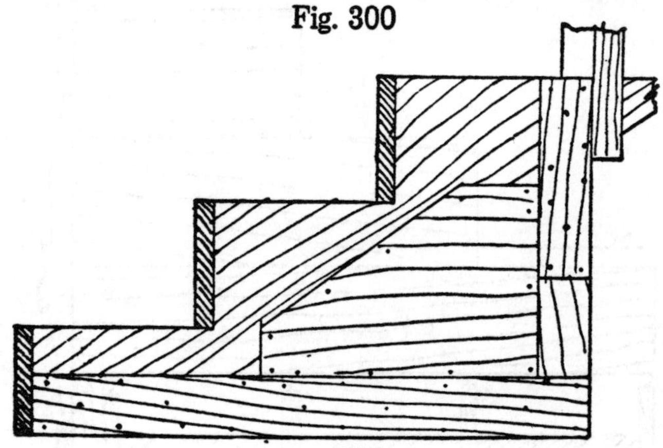

Fig. 301

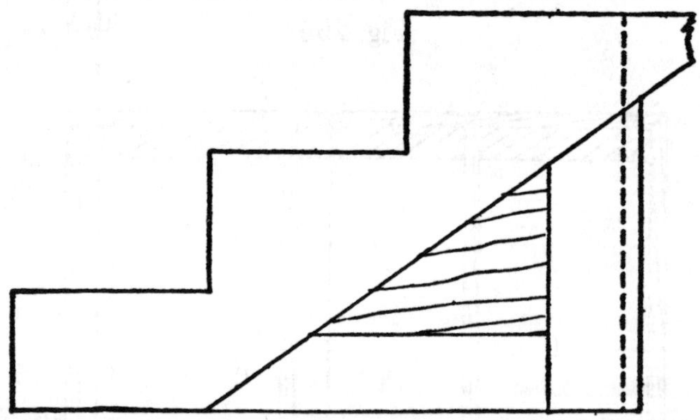

Fig. 302

We show four cross sections of treads in Fig. 303. The one shown at *A* is 1-1/8 inches by 11½ inches. The one at *B* is 1-1/8 inches by 11½ inches. The one at *C* is made up of two 2x6's, while the one shown at *D* is made of three 2x4's.

Self-supporting steps, that is, steps not reinforced with rough horses, give excellent service for ordinary residential work. They have the advantage of drying out faster than steps supported by horses and, consequently, will last longer.

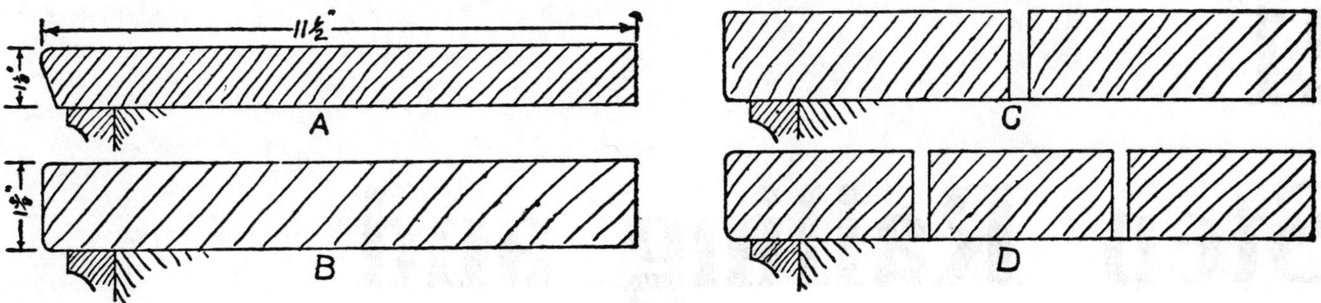

Fig. 303

21

Step Railing and Cheap Lattice

There are perhaps two reasons for the omission of step railings. One is that it reduces the cost, and the other is that it eliminates a seemingly useless part of a porch that is overly exposed to the elements. No other part of a building, as a rule, will take and hold water like step railings. Even if kept well painted, steps with railings will have a shorter life than those left without railings. Still, there are two important reasons why porch steps should have railings. One is that step railings improve the appearance of both the porch and the steps. The other is that such railings are safety necessities.

Step Railings

A simple step railing, put up primarily for service, is shown by Fig. 304. The main drawing shows the height of the railing relative to the treads. The bottom drawing gives the plan view in part, showing the fastening of the lower support to the bottom step. Fig. 305 gives a little more complete construction, showing two 4x4 posts, a handrail and spindles. A plan, in part, of the layout is

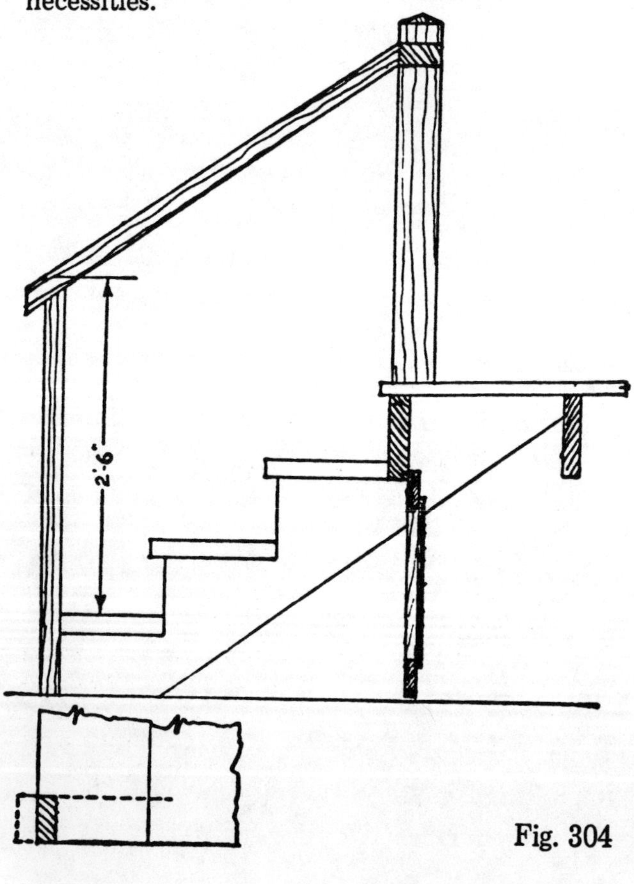

Fig. 304

Fig. 305

shown just below the main drawing. In both of these examples we are holding to the practical rule of service, but if a more elaborate railing is desired, a slight modification of Fig. 305 is all that is needed. We do not, as stated in a previous chapter, intend to go into the stair building field. Our purpose here is more nearly suggestive than technical.

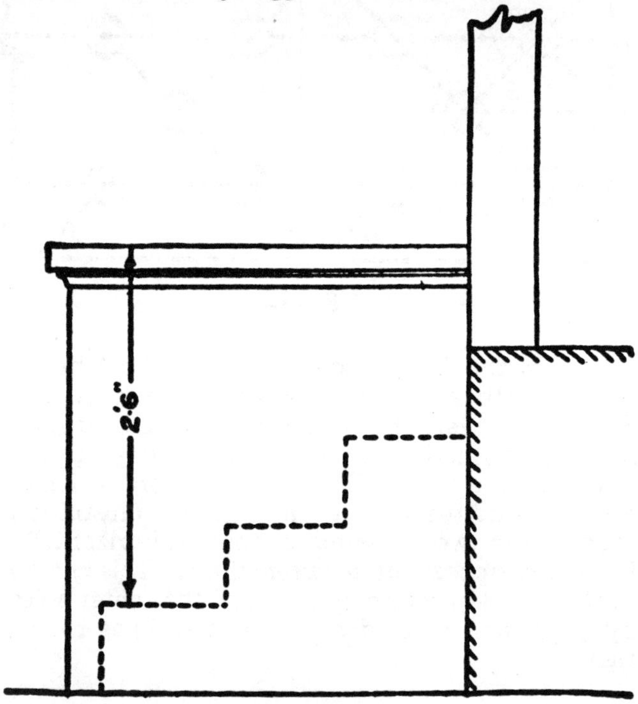

Fig. 306

A sort of diagram of a step railing that is frequently used is shown by Fig. 306. The distance from the top of the bottom tread to the top of the rail is 2 feet 6 inches. The rail is level, making it suitable for holding flower pots or boxes. A cross section of the steps and the inside finish of this railing is shown by Fig. 307. We suggest beaded ceiling, which is often used, but plywood or some other material can be used for finishing the inside of such railings. Whatever is used, it should harmonize either with the inside or the outside finish of the porch.

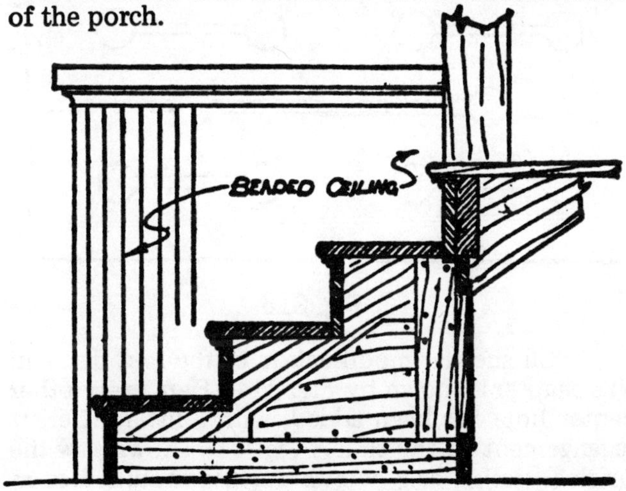

Fig. 307

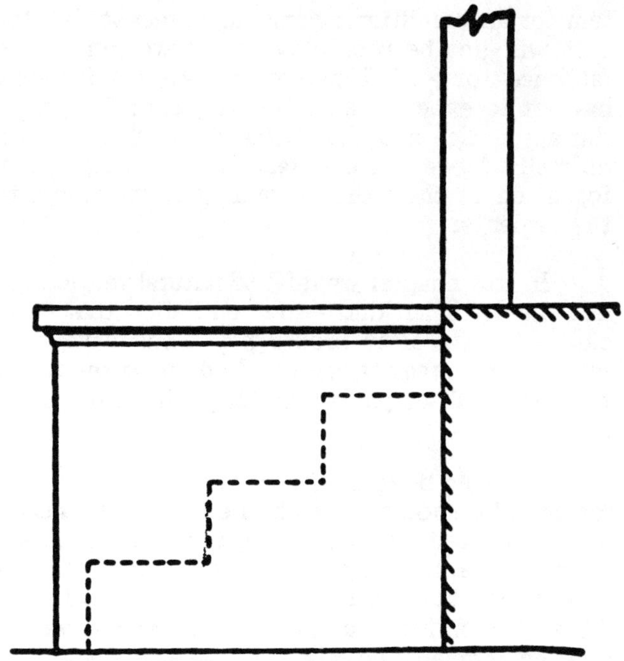

Fig. 308

A simple railing for steps is shown by Fig. 308. The rail is kept level and intersects with the top of the porch floor. Where there are more than three steps, a railing similar to what is shown by Fig. 309 gives quite satisfactory results.

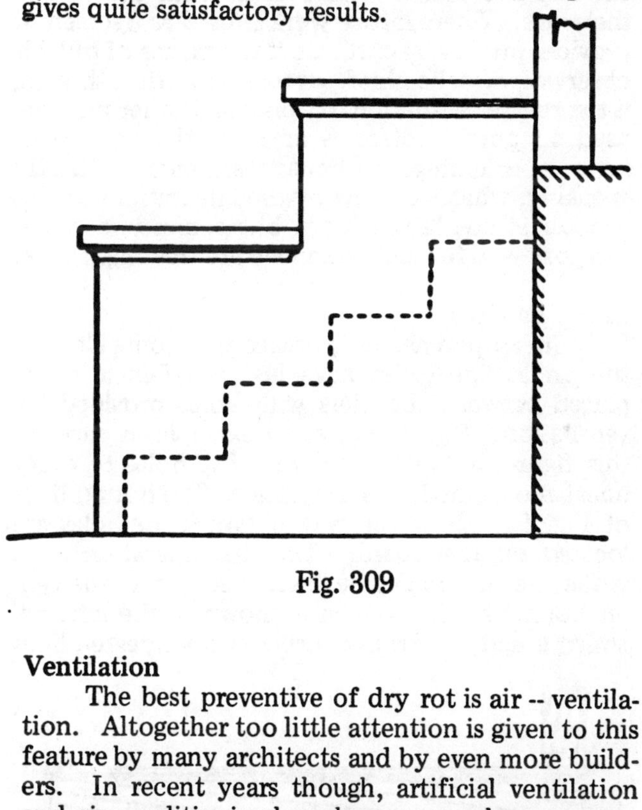

Fig. 309

Ventilation

The best preventive of dry rot is air -- ventilation. Altogether too little attention is given to this feature by many architects and by even more builders. In recent years though, artificial ventilation and air-conditioning have become much more common in residential construction. This is especially true in larger towns and cities. The apprentice who is looking around for something definite to specialize in will find this branch of the industry a fertile field. The young fellow who can work out a system for air conditioning existing homes so that the cost will not be prohibitive will have himself established for a life-time career. Air conditioning has yet to experience its climax. Therefore, both the apprentice and the journeyman carpenter will do well to give this new feature a thorough study for much of the work pertaining to it belongs to the carpenter.

In this chapter we refer to natural ventilation, not the artificial kind - the ventilation that takes care of itself after proper provisions have been made. For example, attics of buildings should be provided with ample ventilation. This is too often neglected.

It should be remembered here that ventilation for an attic should be with the outside air rather than with the inside. If the air can freely circulate between the studding from the basement to the attic, or the reverse, it becomes a fire-hazard. Under such conditions a fire in the basement would in a few moments be carried to the attic -- while an attic fire would feed on air from the basement and every part of the building.

Lattice Work

Lattice work, in addition to providing for ventilation, aids in the appearance of a building — if the designer makes it harmonize with the rest of the work. Where lattice work is used as a screen, it provides privacy by cutting off the means of outside observation. The chief purpose of lattice, though, is to provide ventilation for joists and other material used on porch platforms or under the superstructures of buildings without basements. Strictly speaking, what we show by the illustrations of this chapter is not lattice work but comes under that category, and for that reason we are treating it here.

Cheap Lattice

Cheap porches, or porches that come close to the grade line often have just a 7/8-inch board placed between the piers with holes provided for ventilation. Fig. 310 gives an example in part. In this figure, a 1x12 board with five holes to every lineal foot constitutes the lattice. The dotted lines of Fig. 311 show the way in which the holes are located on the board. The first operation is to strike the full 45-degree lines. Then strike the center horizontal line, which is shown to the left, between *a* and *b*. From *a* strike two 45-degree lines

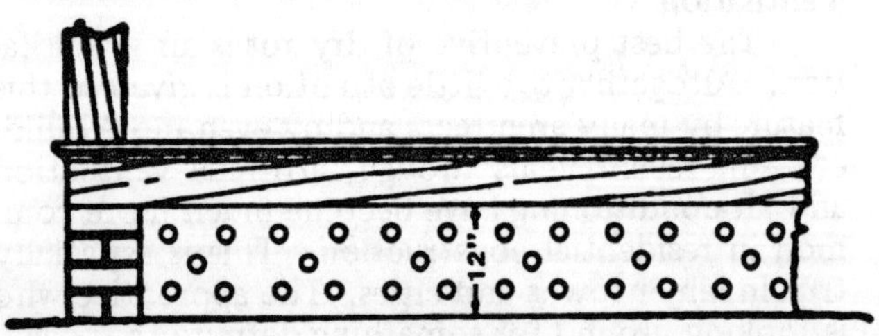

Fig. 310

that end in *c*, *c*. Where these lines cross the full 45-degree lines draw the upper and the lower horizontal lines, locating the centers of the upper and the lower holes.

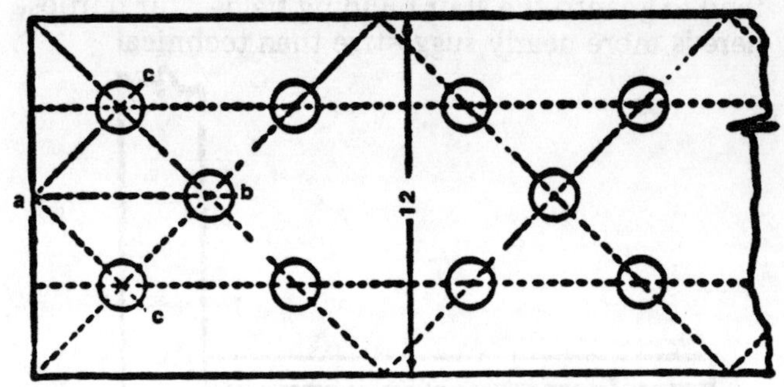

Fig. 311

How the design shown in Fig. 310 can be developed into a different design is shown by Fig. 312. To the left of this drawing (shown by dotted lines) is the marking necessary for cutting out the design to the right. Fig. 313 shows another modification of the first design. In this illustration every other two holes are connected horizontally by a slot ripped out between them. This can be further developed by connecting all the center holes by slots, as indicated by the dotted lines to the right.

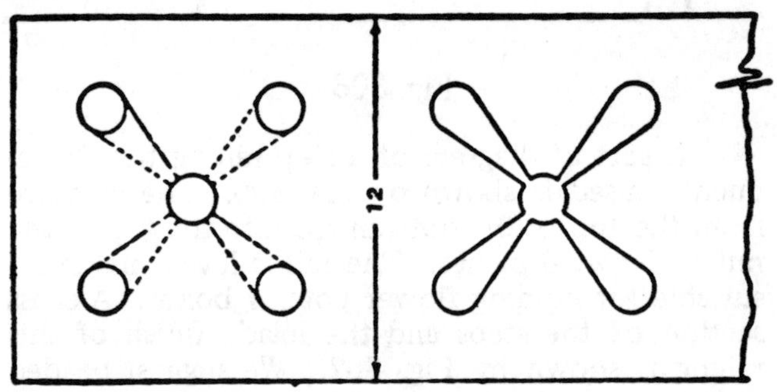

Fig. 312

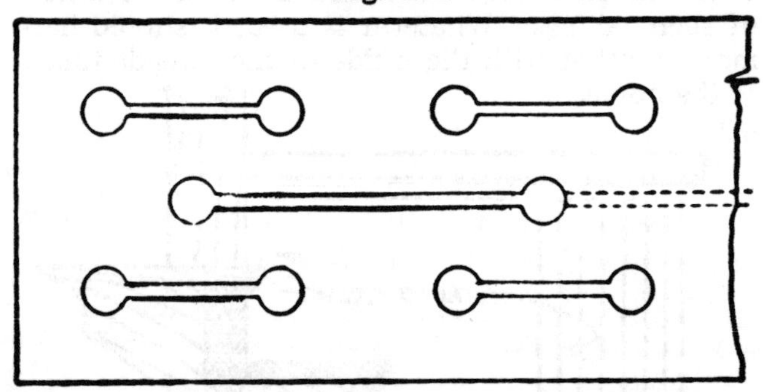

Fig. 313

Still another modification of the first figure in this chapter is shown by Fig. 314. Here every other center hole has been added, with a little different arrangement of the slots. Fig. 315 shows how the original design can be still further modified so as to increase the capacity for ventilation.

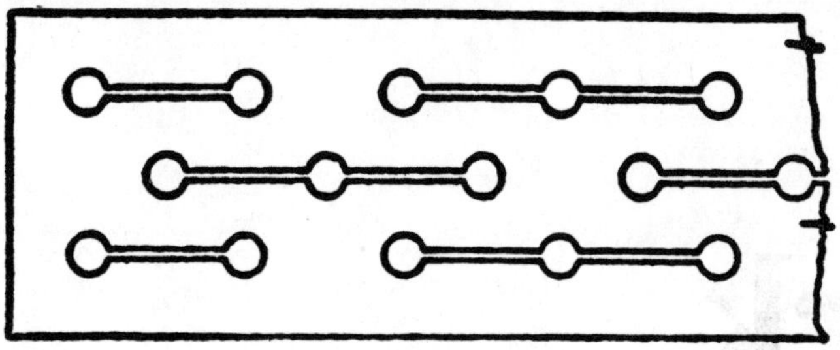

Fig. 314

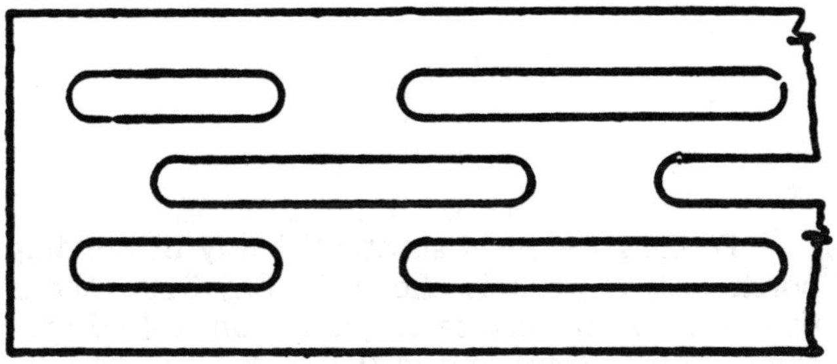

Fig. 315

Fastening Cheap Lattice

Fig. 316 shows cross sections of four methods for fastening the lattice work illustrated in this chapter. The example to the extreme left is a good one. In this the joist answers for the outside finish of the porch platform and the lattice-board is fastened to it by means of cleats nailed to the back. How the lattice-board should be braced is shown in part at the bottom and to the right of the drawing. The second drawing from the left shows how a lattice-board should be fastened where the outside joist is concealed with a finishing board.

In this method a 2x2 is notched over the joist in such a way that when the lattice-board is fastened to it the outside finishing board will lap over the upper edge, as shown by the drawing. The third drawing from the left shows a method that is often used though it is not adequate except for temporary porches. The lattice-board is toe-nailed to the outside joist and held in position at the bottom by means of a stake driven into the ground. The method shown to the extreme right is not good because the board is hung to the joist by toe-nails without bottom stays.

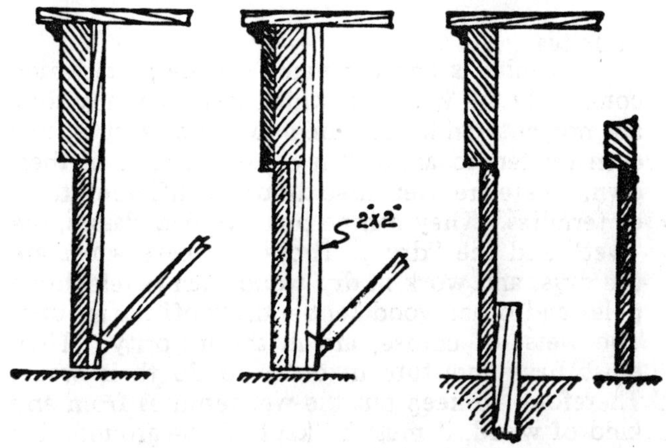

Fig. 316

22 Lattice Work

Termites

Termite is another name for the pests which commonly are known as white ants. Actually they are not ants and, according to the scientists, not even related to ants. They are in a class of their own. There are over three thousand different kinds of termites. They can be put into two classes, the "wet" and the "dry". Those that have wings are the drys, and work in dry wood such as telephone poles and other wood kept entirely off the ground. The wets, of course, are in the majority. They must have moisture in order to do their work. Therefore, to keep out the wet termites from any kind of wood, it must be kept off the ground. No wood is entirely termite-proof but some woods have a greater termite-resistance than others. Oak, cedar, hedge and hackberry fall in this category. A treatment of creosote for wood that must contact the earth will keep the pests from doing their destructive work.

Lattice Work

Fig. 317 shows a whole section of porch lattice work in place. The front pieces of lattice should always be placed as they are in this illustration, perpendicularly, while those in the back should run horizontally. They should be fastened together with nails on every fourth lattice. There is, however, an objection to this way of fastening because it puts all the nails in one of the front pieces. A better way is indicated to the right where the nail line is on a 45-degree angle, placing only one nail in each lattice at the fastening.

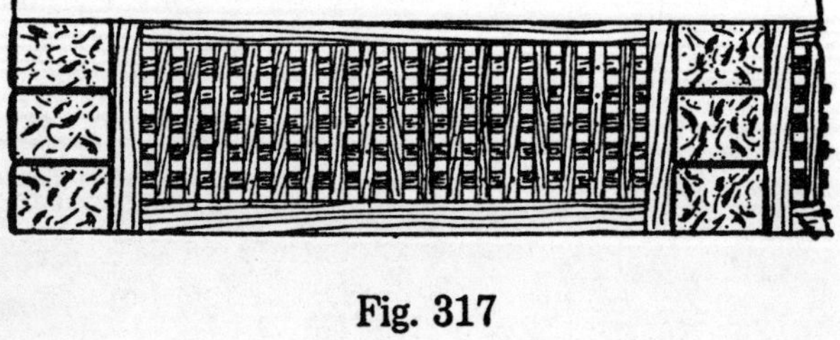

Fig. 317

Perhaps the most substantial way of building porch lattice work is shown in Fig. 318. When this is used the lattice is placed on a 45-degree angle both in front and in the back. The dots indicate how the front and back should be fastened together with flat-head nails.

Fig. 318

Another substantial way of building lattice work for porches is shown by Fig. 319. This method places the lattice at a 60-degree angle which makes the lattice in every way as substantial as the lattice shown in Fig. 318. Another way of fastening the front and the back lattice together is by placing the nails on a horizontal line through the center.

Fig. 319

The porch lattice shown in Fig. 320 is used less than the lattice in the previous illustrations but is still very acceptable. It has the advantage of being easily made and, if anything, a little less expensive than the others. It provides more room for ventilation, looks as good as the others and is much easier to paint. It keeps dogs and cats out from under the porch about as well as the others and by making two or three of the upright pieces removable, will provide a way to get under the porch.

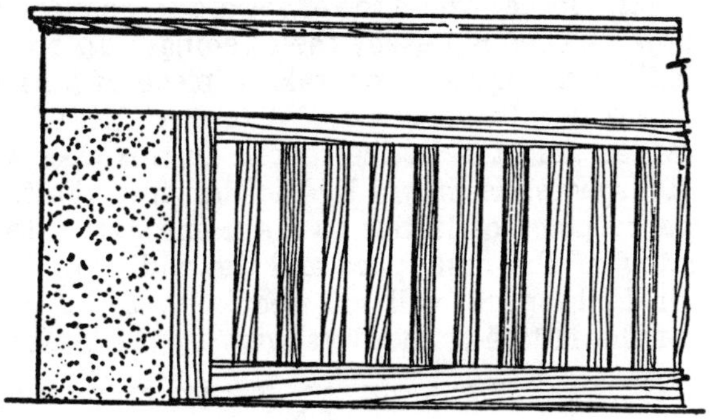

Fig. 320

A part of the back of this section of lattice and a cross section are shown to the left of Fig. 321. By these drawings we show how the corners of the frame are held together with cleats and how the lattice work is fastened to the joints. To the right of this figure is shown the back construction and a cross section of the lattice shown in Fig. 319. Also shown is another method of fastening the lattice to the porch. The cross section and the drawing to the right may require study and comparison to make the construction clear.

Fig. 321

Fastening Lattice

Fig. 322 shows by cross section four fastening methods that are very much like those given in the previous chapter. The one to the extreme left is a good one. The brace that holds the bottom of the lattice is shown in part. The part of the brace not shown, naturally, must be fastened to the bottom of the porch platform. The second drawing from the left shows another good method of fastening lattice. The 2x2's, heavily shaded, are put into place and then the lattice is fastened to the upright pieces either with nails or with screws. At least one section of porch lattice should be made removable by fastening it with screws so access is possible for repairs or for storage.

The third drawing from the left shows a faulty but frequently used method of fastening lattice. The lattice is toe-nailed to the joist at the top and held in place at the bottom by a stake driven into the ground. The drawing to the extreme right shows the poorest way of fastening lattice. At *A* we point out how the framework is fastened to the porch with nails without anything at the bottom to hold it in place except the friction at the piers.

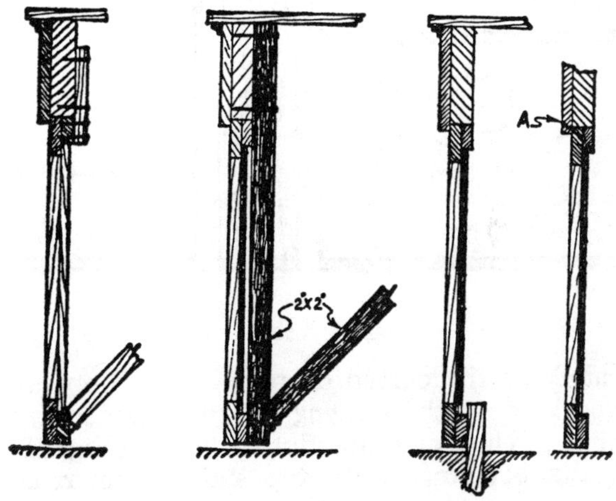

Fig. 322

Wood

While wood deteriorates, is destroyed by fire, and eaten by termites, it is still the most satisfactory building material in existence. There is no substitute for wood. Cement has displaced wood in many instances, but cement is not a wood substitute. Concrete is only a different kind of building material which gives results different from those obtained with wood. Reinforced concrete structures, while fireproof, are also more expensive. For the superstructure of many buildings, concrete is not practical. Cement floors have their uses, and, under certain conditions, are superior to wood floors, but cement floors do not take the place of wood floors. The various composition flooring materials have their own places in the building world, but when the qualities of wood are desired, only wood will give satisfactory results.

Lattice Problems

In Fig. 323 we show how to scribe lattice to stone or cement-block porch piers. One method is to take the end pieces of a lattice frame before the frame is put together and place them in a plumb position to the pier (somewhat as shown in the figure) and scribe them with compasses or scribers as

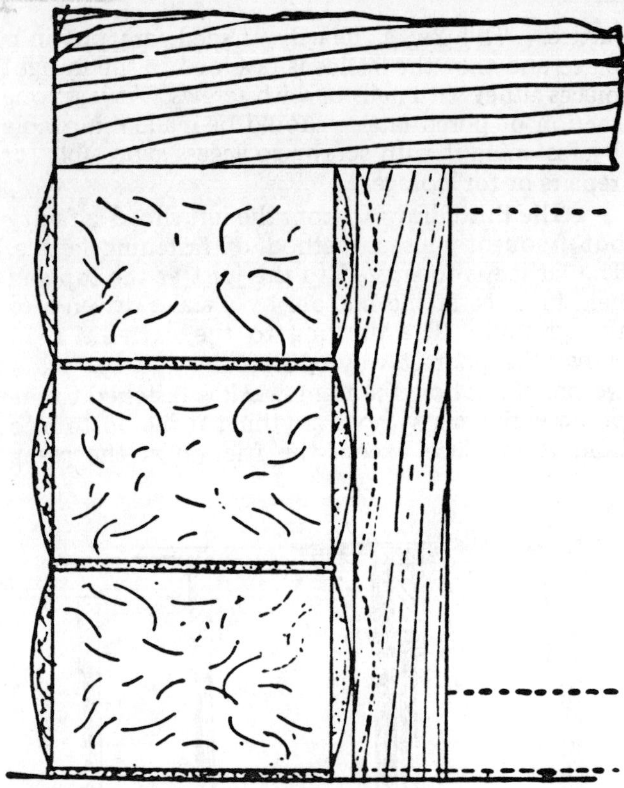

Fig. 323

indicated by the curved dotted lines. The other method is to do the scribing after the frame is put together. (The horizontal dotted lines in Fig. 323 show the position of the top and bottom rails.)

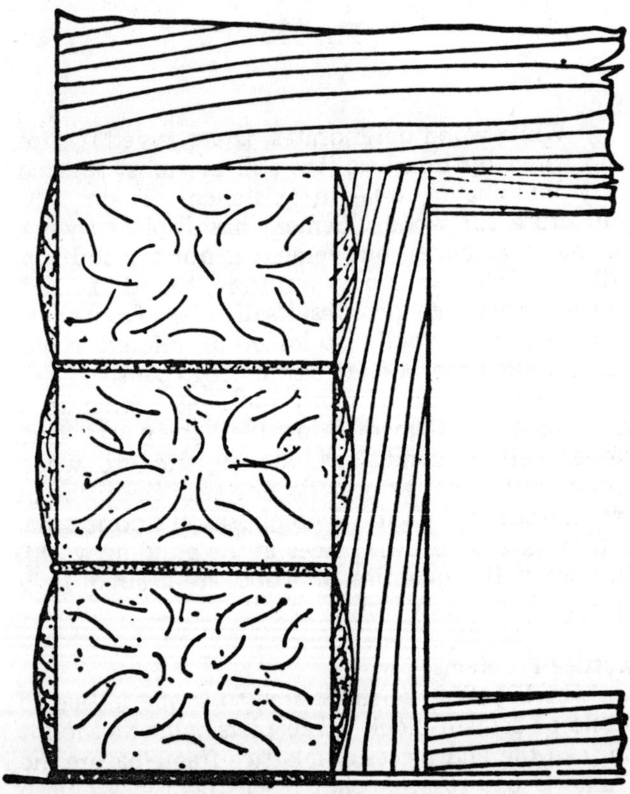

Fig. 324

After placing the lattice frame in the position shown, scribe one end in the way explained before. Then transfer the distance between the piers to the lattice frame, and scribe the other end of the frame in such a manner that it will fit neatly between the piers. One end of a lattice frame is shown in place by Fig. 324.

We show in Fig. 325 a method of spacing for horizontal lattice strips, assuming that the strips and the openings between the strips are of the same width, in this case 1-3/8 inches. The *X's* to the left indicate the spaces where the strips are to be placed. The other spaces represent the openings. In order to simplify the matter, we take a piece of lattice and divide it into spaces of 1½ inches, 1/8 of an inch wider than the lattice strips. In this case we have 17 spaces, as shown by the drawing. Fasten the spaced piece of lattice to the frame so that the beginning of the first space and the ending of the last, or 17th space, will intersect with the inside edge of the lattice frame, as shown by the illustra-

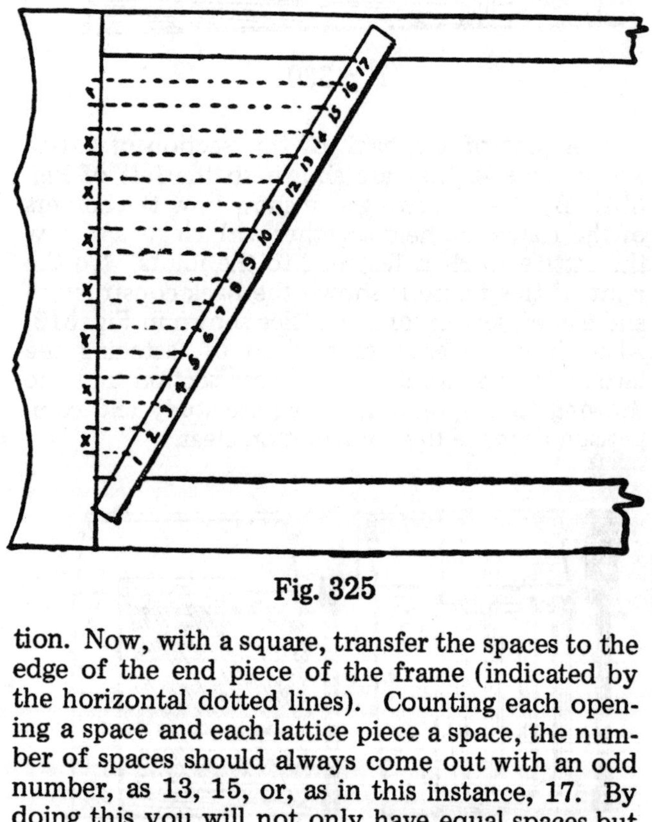

Fig. 325

tion. Now, with a square, transfer the spaces to the edge of the end piece of the frame (indicated by the horizontal dotted lines). Counting each opening a space and each lattice piece a space, the number of spaces should always come out with an odd number, as 13, 15, or, as in this instance, 17. By doing this you will not only have equal spaces but the first and the last spaces will represent openings, as they should. In order to space for the upright lattice pieces we must first find out what is the most nearly right spacing for them. Then we proceed to space the frame, say, up to within about 17 single spaces, as we used in the previous figure, or 8½ double spaces, as shown by Fig. 326. The rest of the spacing, in order to overcome any variation, is done by placing the spaced lattice piece in the position shown and with a square transferring the spaces to the bottom rail, which is indicated with the dotted perpendicular lines. The *X's* represent the single spaces where the lattice pieces are to be placed. The bottom rail spaced in this way, the spacing can easily be transferred to the top rail.

About as good a way to space cross-lattice as any, especially if the lattice material is all of the

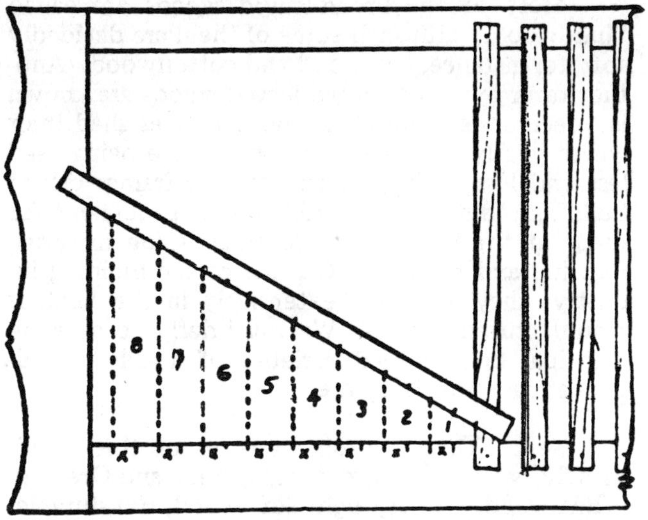

Fig. 326

same width, as it should be, is shown by Fig. 327. This shows the back of the lattice frame almost completely covered with strips of lattice. These strips have been placed on a 45-degree angle, and every other one is nailed, which is indicated by the dots at the ends and the shading. When the nailing is completed, the loose pieces are removed, leaving

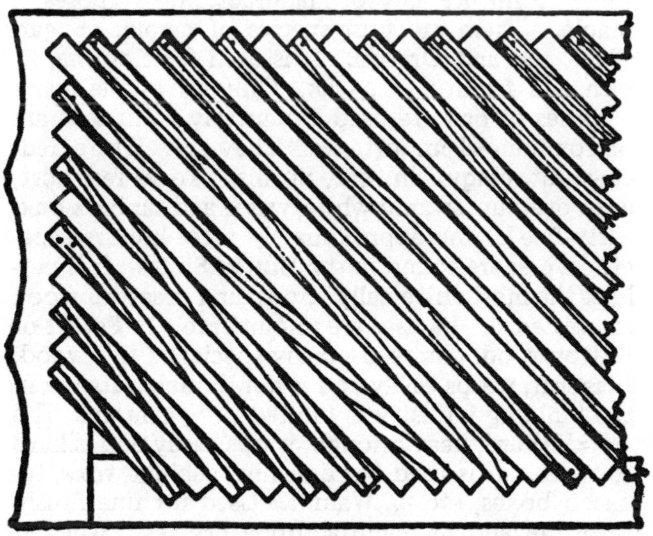

Fig. 327

the results shown by Fig. 328. This done, proceed in the same way to put on the back pieces, making them cross the front pieces at the same angle, as we show by the dotted lines.

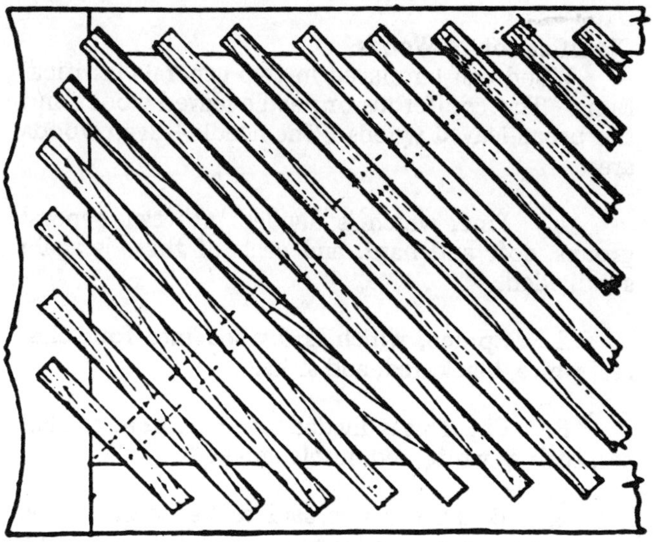

Fig. 328

Four ways of fastening the corners of lattice frames are given in Fig. 329. The one shown at number 1 is the same as what was shown in Fig. 321, to the left. It is made by means of nailing a cleat over the back of the joint. Number 2 is the same kind of a joint, but held together by means of fastening a piece of sheet-iron on the back with screws. This is indicated by shading. Number 3 shows a mortise and tenon joint. This joint is not often used for lattice because it will hold moisture and consequently cause rotting. However, if both the mortise and the tenon are well painted before the joint is put together, it makes a good construction. Number 4 shows by the shaded part how a butt joint can be held together by means of nailing heavy galvanized iron on the back . . . Miter joints for lattice frames, in our opinion, do not look good. They are sometimes used.

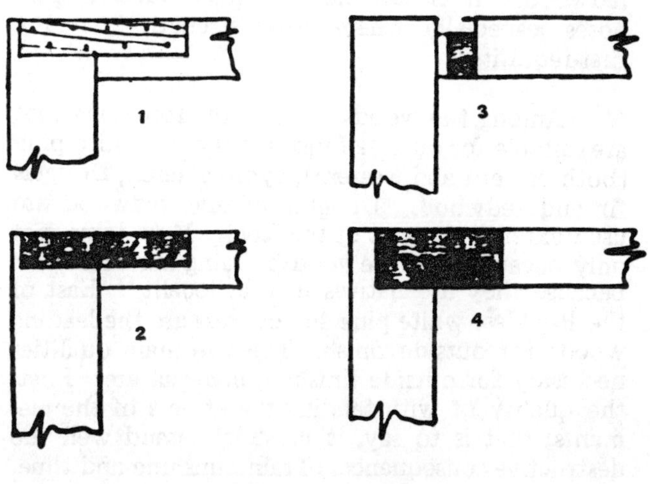

Fig. 329

Classification of Woods

Wood can be placed in two general classifications: The coniferous or needle-leaved woods and the broad-leaved woods. The needle-leaved woods are:

1. Pine, which is divided into two general groups, soft and hard, and each of these is again subdivided.

2. Spruce, which is divided into four classes, white, black, red, and sitka.

3. Fir. This is divided into a number of classes, such as white, red, noble and so forth.

4. Douglas Fir is said to be neither pine, spruce nor fir, but more nearly a bastard hemlock.

5. Hemlock, which is divided into two classes, the eastern and the western. Of these two the western is claimed to be the better. It has many legitimate uses, especially in light construction. The eastern variety is used for studding and framing on cheap work, and sometimes for finish work.

6. Larch or Tamarack. The western species of this wood is of a better quality, as a rule, than the eastern. It is used for posts, sills, railroad-ties, etc.

7. Cedar, of which there are white, red, yellow and several other species. This wood has many different uses and is a valuable building material.

8. Cypress, a durable building material especially for outside finish. It is also suitable for inside finish, works easily, and has a very luxurious grain if given a natural finish.

9. Redwood, which is divided into two classes, the common redwood and the giant or big-tree redwood. It is suitable for many building purposes, especially where great strength is not a first-requisite.

Among the woods listed ir the foregoing that are suitable for outside finish are these: White pine (both eastern and western), cypress, larch, Douglas fir and redwood. Douglas fir and redwood are used extensively west of the Rocky Mountains, not only because they are good building materials, but because they are natives in that locality. East of the Rockies, white pine and cypress are the leading woods for outside finish. The two main qualities necessary for outside finishing material are: First, the quality of withstanding the effects of the elements; that is to say, it must withstand well the destructive consequences of rain, sunshine and time. And second, it must work well and permit nailing without splitting. When these two qualities exist in wood it can be used for outside finish. Wood that does not work well and will warp easily, even though it might be durable enough otherwise for outside use, should not be used for outside finish, especially on the better class of buildings.

Most of the broad-leaved woods are called "hardwood," although some of them are decidedly soft, for instance, basswood and cottonwood. Another term by which broad-leaved woods are known is "deciduous." meaning that the trees shed their leaves yearly. Broad-leaved woods are being used less and less in this country for the framework of buildings, but there is still a strong demand for them in the building world as finishing material, flooring and so forth. Outside of the building industry, they are used extensively in the field of manufactured articles. We will briefly go over the principal broad-leaved varieties of woods. In alphabetical order, they are:

Ash. There are six varieties of this wood, namely, white, black, red, blue, green and Oregon . .. **Basswood.** A very light soft wood, not durable, used in the field of manufacturing rather than as a building material ... **Beech.** A hard wood, used for furniture, flooring and inside finish. Not suitable for outside work ... **Birch.** A hard wood used for inside finish. There are six species, white, paper, red, yellow, sweet and western ... **Cherry.** A hard wood used for furniture and fine finish ... **Chestnut.** Used for cabinet work, posts and cooperage ... **Elm.** More or less used as a building material. During the broad-axe days it was used in heavy timber framing. It is increasingly finding favor as a finishing material and for furniture ... **Gum.** Used for finish, furniture and for manufacturing purposes ... **Maple.** There are four species of maple: Sugar, silver, red and Oregon. It is used for flooring, finish and furniture ... **Oak.** Only six of the many varieties of oak are used extensively. Thirteen are of commercial importance. They are: White, red, chestnut, cinquapin, bur, Spanish, Texas red, post, cow, overcup, swamp white, water and tanbark oak. Fifty-five of the approximately three-hundred species of oak are found in the United States. It is perhaps the most universally known and practical wood in existence. Its uses are unnumerable ... **Poplar or Cottonwood.** There are many species of this wood. It is soft, warps easily and is not durable. Its use in the building industry is limited ... **Sycamore.** The wood is hard, heavy, tough, warps easily and is hard to work. It is used for finishing, cabinet work, tobacco boxes, etc. ... **Walnut.** Used for fine finish, furniture and in manufacturing articles. When it was more plentiful it was used for fence posts and for many building-purposes, but it is becoming more and more one of our rare woods ... **Whitewood.** This is also known as yellow poplar, tulip poplar, tulip tree, and hickory poplar. It has many uses, among them are interior finish, furniture and manufactured toys.

Other than the woods we have enumerated in the foregoing, there are many varieties of woods which are of little importance so far as the building industry is concerned. They are such woods as apple, hackberry, hedge, hickory, mulberry, willow, and others.

23 Corner Boards and Outside Base

Corner Boards

A simple design of a corner board is shown by Fig. 330. The dotted lines indicate the relationship of the siding to the board. The upper drawing gives the elevation, in part, of this corner board. The board is made of 7/8-inch material and is suitable for cheap work, such as garages, two-room houses and other buildings of that kind where only one side or end of a building is exposed to the public eye. This construction of a corner board can be made to look reasonably professional. It is often used where a row of small buildings are placed rather close together. A similar corner board is shown by Fig. 331. Here, it will be noticed, the building is boxed and the board is made of 1-1/8-inch material with a beaded corner. The dotted lines shown on the plan, or the bottom drawing, indicate the siding. The top drawing shows the elevation in part. This kind of board can be used successfully on houses that, as it were, hug other buildings on both sides. On the other hand, it will give fairly good results on a building standing out by itself.

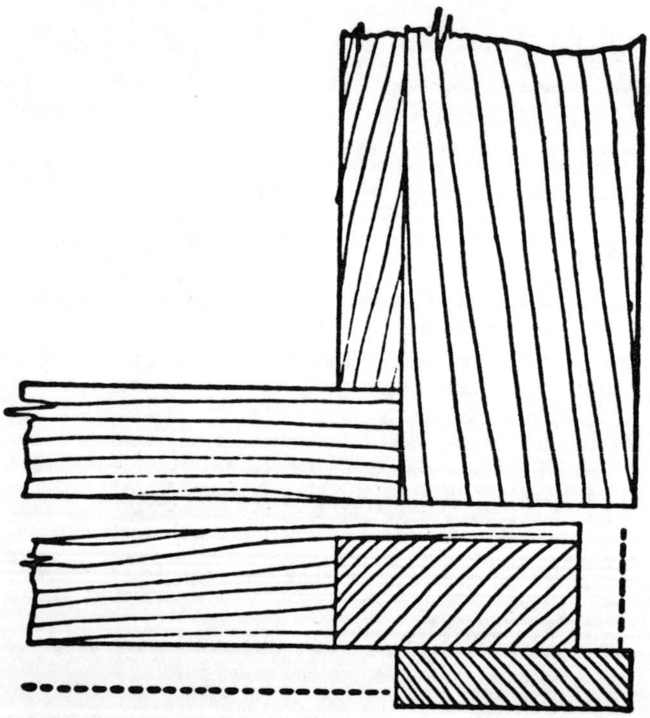

Fig. 330

Fig. 331

Details of two corner boards, each one evolved one step higher than the two just explained, is shown by Fig. 332. The one shown to the left is made of 7/8-inch material and is plain. The dotted lines indicate its relationship to the building. The one shown to the right is made of 1-1/8-inch material and has two beads which serve as ornaments and conceal the joint.

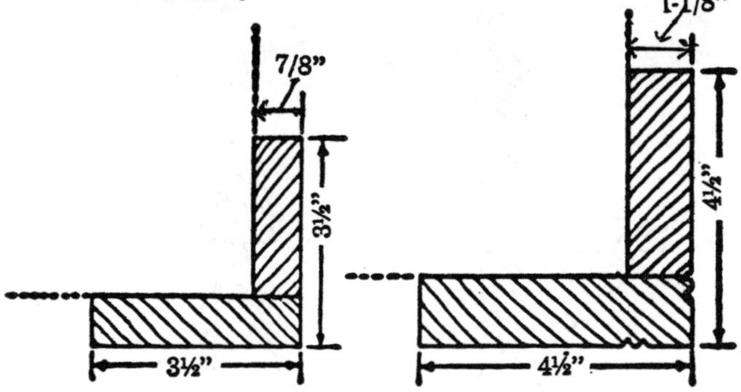

Fig. 332

Two details of built-up corner boards are shown by Fig. 333. The one to the left shows a 7/8x7/8-inch fillet in the corner, resting on a base block.

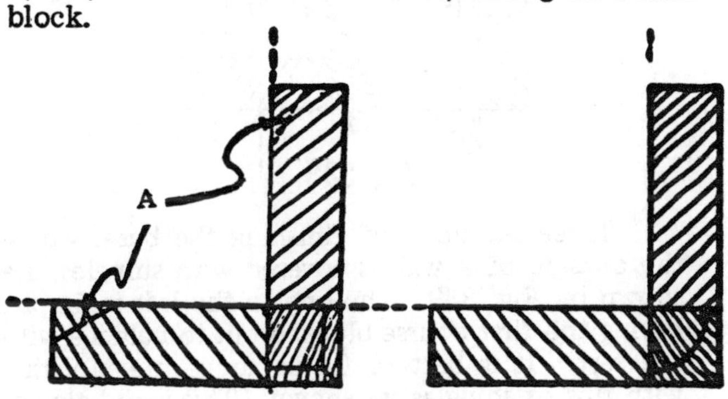

Fig. 333

An elevation of this is shown to the left of Fig. 334. The chamfers indicated by dotted lines and pointed out with the indicators at *A* show how the corner board should be back-chamfered when the outside of the building is to be finished with stucco. The detail to the right shows a built-up corner board with a quarter-round in the corner, resting on a base block, as shown, by the elevation to the right of Fig. 334. The two edges should be beveled when siding is to be joined to the board. In case shingles are used instead of siding the edges should be kept square.

A cap of a corner board is shown by the left drawing of Fig. 335. The perpendicular dotted line indicates the joint. A detail of this cap is shown at number 4 of Fig. 336. There is one difference; the frieze in the detail does not fit the cap as it is shown by the drawing in Fig. 335. The open feature, we think, is an advantage insofar as it provides the joint with ample drainage. To the

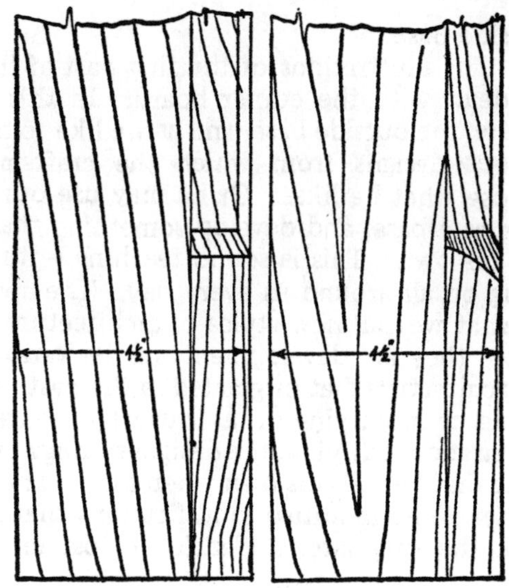

Fig. 334

right of this figure we show how joints, if joints are necessary, should be made. That is to say, the boards should be cut on an angle and put together so as to drain the joint outward. The shaded parts show two nails driven with an upward slant, for the purpose of not only keeping the water out, but of draining the nail holes, should water get into them. The nails should be set, as shown, and the holes puttied.

Fig. 335

In Fig. 336 is shown: 1, a plain joint between the corner board and the frieze; 2, a simple cap for the corner board; 3, a somewhat better cap, and 4, a cap that is more or less standard.

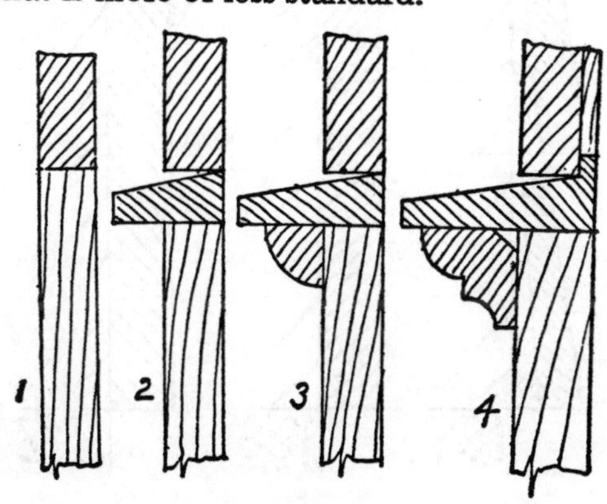

Fig. 336

Outside Base

The illustrations of the first part of this chapter dealt with the corner board. In this part we take up the outside base, and as we like to do, show various designs from which the craftsman may choose what he likes. Or he may use our suggestions as a base, and develop something that he can call his own. This is sound teaching -- it is what is going on all around us every day. The new styles, or what we call new styles of architecture, are just that -- they are developments of the various styles of architecture that originated in the past. By combining or modifying or, as is usually the case, combining and modifying, the designer brings out something that he calls his own whereas, the truth of the matter is, he modified modifications that go so far back the originator's identity is lost in oblivion.

Three methods of finishing the base of the outside wall, when siding is used as a covering, is shown by Fig. 337. The one to the left is a sort of baseless base. That is, the base is made of siding with the first board blocked out at the bottom, as shown, to keep it balanced with the other members. This method is extensively used, even on important buildings. The base shown at the center is plain, but can be made to give quite satisfactory results. It should be noted that the first siding board is placed right against the boxing and fits the top of the baseboard. The one shown to the right is more nearly standard, having a watertable supported by a cove moulding. The first siding board is shown blocked out.

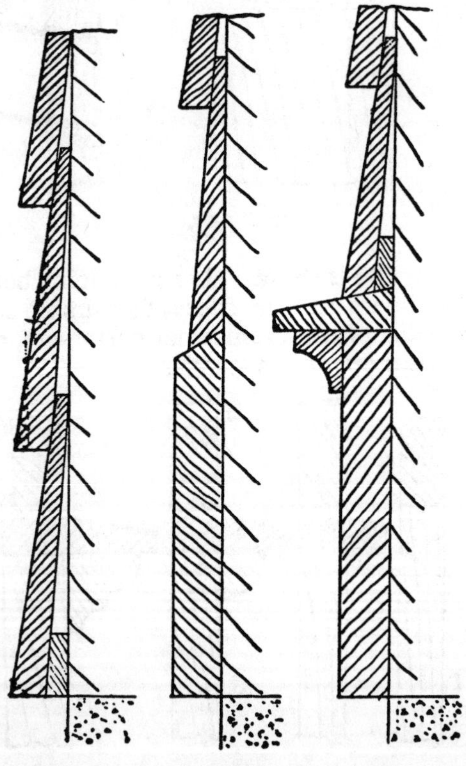

Fig. 337

To the left of Fig. 338 is shown a standard base. It has a regular watertable supported by a bed moulding. The baseboard is blocked out. To the right we show a sway-base construction. This base provides plenty of drip. The blocking shown under the third board should be noted. Its purpose is to prevent splitting when the nailing is done.

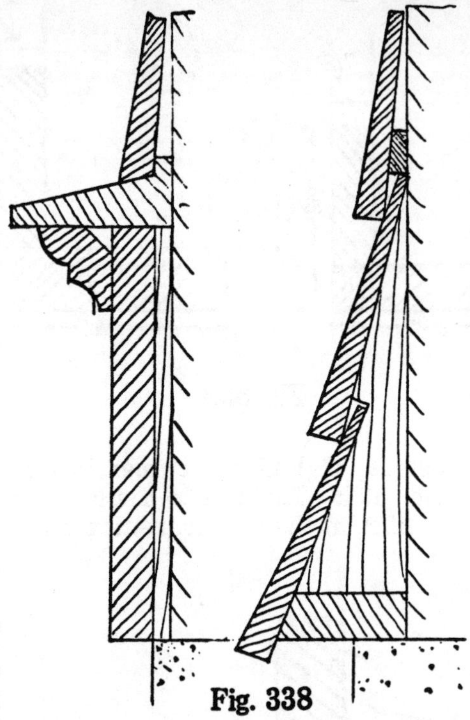

Fig. 338

Three methods of finishing the base, where the outside of a wall is covered with shingles, are shown by Fig. 339. The one to the left is simple, having the first course blocked out to balance with the rest. It is best to fill in above the blocking with tips of shingles, as shown. This is not always done, but it makes a better job if it is. The method shown at the center is a good one when a little sway is desired for the purpose of providing drip. To the right we show a graceful sway, with a full measure of drip. Baseboards, such as are shown in Figs. 337 and 338 are often used with a shingle finish.

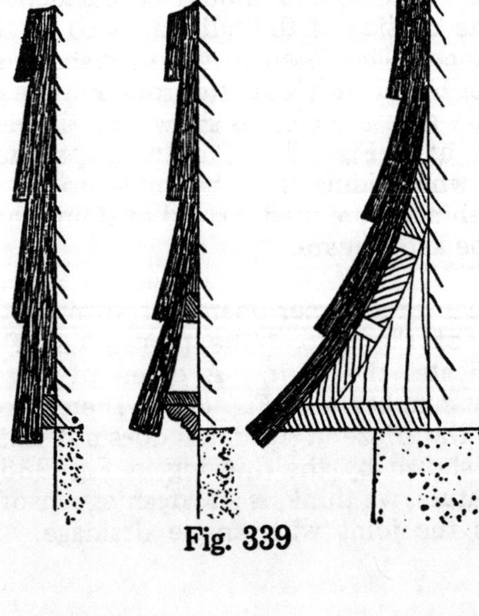

Fig. 339

In Fig. 340 three ways of finishing the base, when the outside of a wall is covered with stucco, are shown. To the left is shown a plain finish, which is extensively used, but it is not altogether satisfactory. The edge should have some means of reinforcement -- something similar to what is shown at the center. Here the heavy loop, shown at the bottom, indicates a rust-proof metal bead which protects the edge of the stucco base. The irregular lines in all three designs represent metal lath. The center design is not so extensively used and is more expensive. The method of finishing the base, shown to the right, seals up the joint between the foundation and the superstructure and the stucco extends down to the grade line.

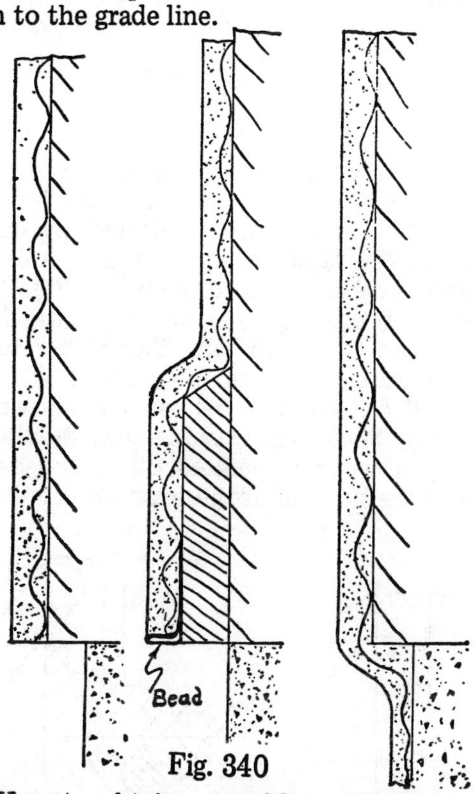

Fig. 340

How to obtain a moulding effect on a stucco base, is shown by Fig. 341. The inverted moulding at *A* is used as a form, which should be covered with a thin coat of, preferably, hard oil, to keep it from sticking. Fig. 342 shows the base after the moulding form has been removed. The irregular line in the body represents metal lath.

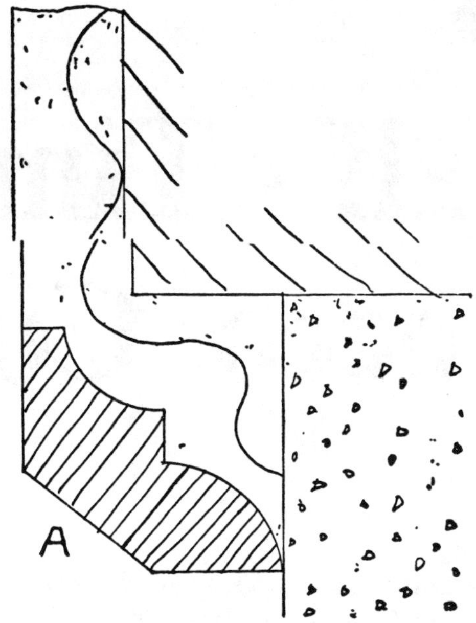

Fig. 341

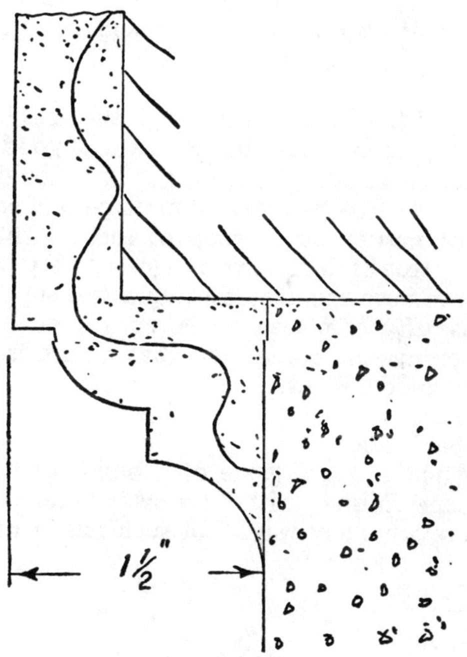

Fig. 342

24

Half - Timbers and Stucco Corners

Good Design

The best critic, speaking of designs, is the public. If the ordinary observers are pleased with a design it must have merit, although it might not be perfect. But if the people, with one accord, like a style of architecture, it must have something in it that is fundamentally right.

Most of the people up through the ages have liked nature, so it is logical that the basis of good design should be nature. That is, a design should represent a sort of conventional summing-up of some product of nature, or a group of them. This is particularly true of decorative designs. A very plain building can have the decorative qualities supplied by means of well-placed vines, shrubs or trees, which supports the contention that nature is the basis of all good designs.

Half-Timber Work

Fig. 343 shows a gable of a building that is stuccoed and finished with false half-timber work. While there are many ways that such false work is used, we show two commonly used designs, one to the right and one to the left. Shown to the left are belt courses with watertables indicated by extra lines, which to the right are shown plain. A detail of the plain course is shown to the left in Fig. 344. To the right of this figure is shown a standard watertable added to a plain belt. Two rather novel but practical watertables are shown by Fig. 345. These are built onto a plain belt, such as shown to the left in Fig. 344. Tin flashing that should be used over such work is pointed out. The back-chamfering at the bottom of these details should be noted.

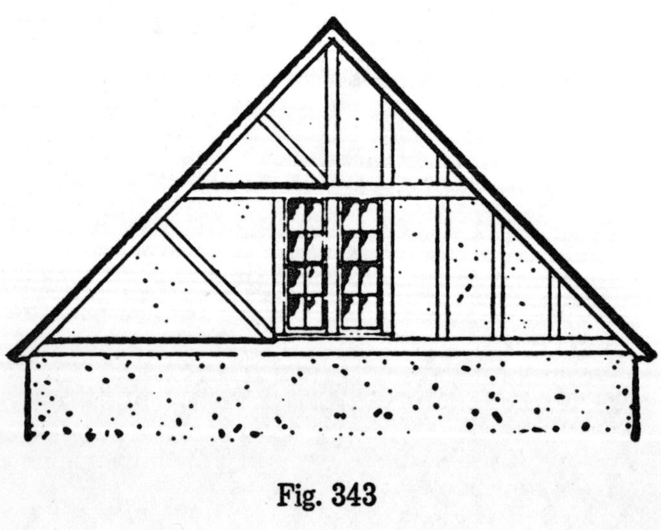

Fig. 343

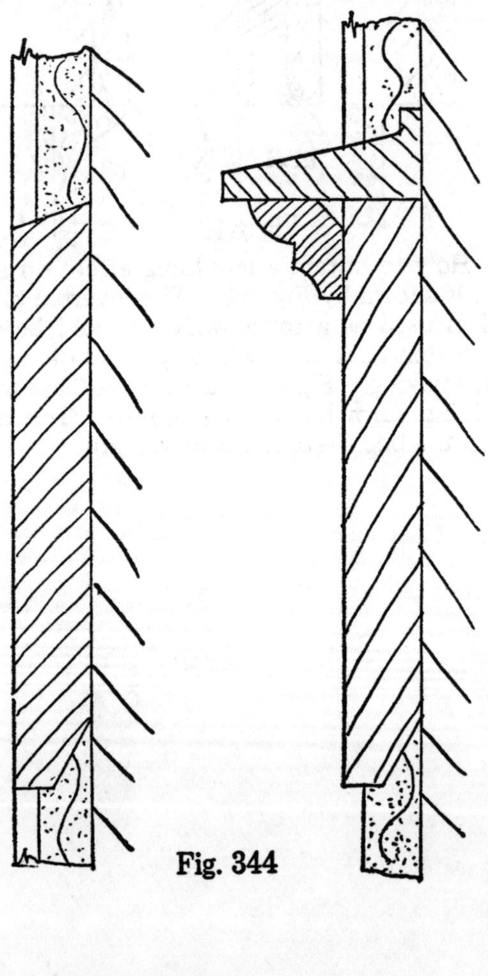

Fig. 344

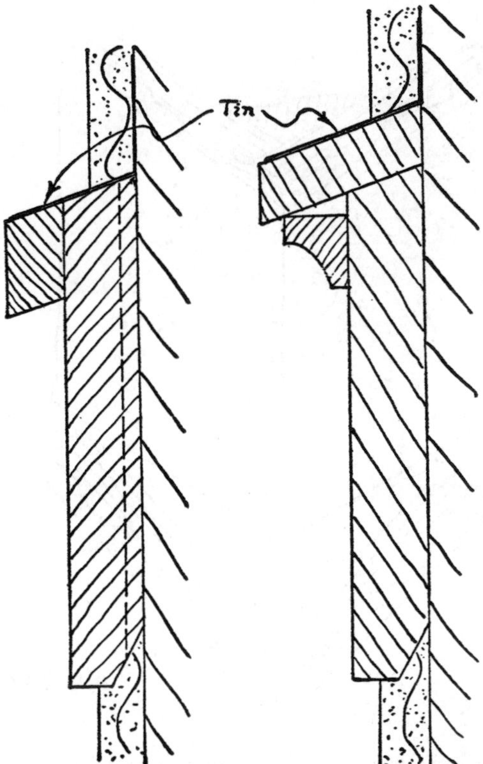

Fig. 345

Two details of half-timber work placed on an angle, similar to what is shown to the left of Fig. 343 are given by Fig. 346. The one to the left is made of 7/8-inch material and blocked out as shown, while the one to the right is made of 1¼-inch material and put on without blocking. The irregular lines shown in the stucco of these illustrations represent metal lath. Fig. 347 shows two details of false half-timber work placed in a perpendicular position. The top one is made of 7/8-inch stock and blocked out, while the one at the bottom is 1¼-inch stock and put on without blocking. Fig. 348 shows another method of backing out the edges of false half-timber work.

Fig. 346

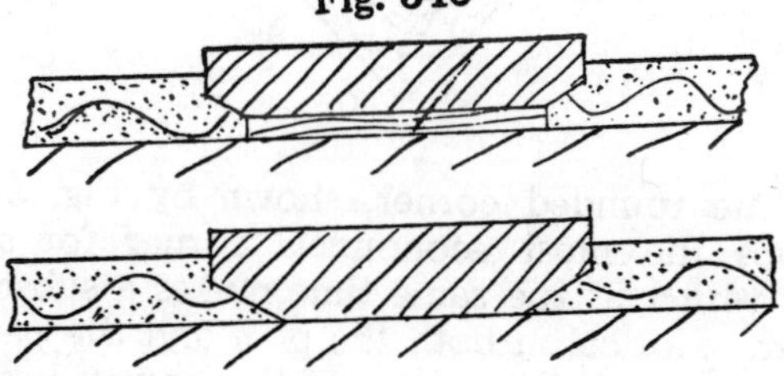

Fig. 347

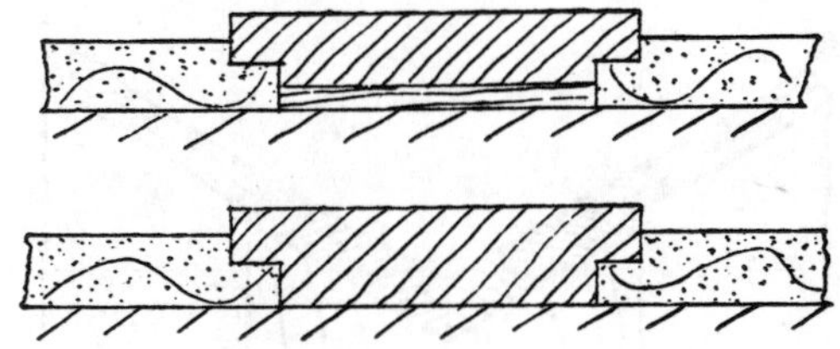

Fig. 348

In Fig. 349 are shown two methods of using false half-timbers. The top one is put into place after the first coat of plastering has been applied, while the bottom one is made of heavier material and fastened without blocking.

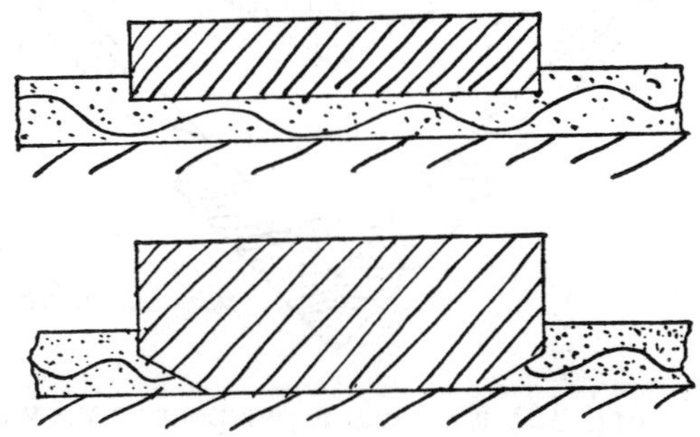

Fig. 349

False, or as it is sometimes called, sham, half-timber work varies in design. The widths of such work run from 4 inches to 12 inches, and the thicknesses of the material vary from 7/8 to 2 inches or even more. The projections beyond the surface of the stucco vary from 1/8 to 1 inch or over, depending on the taste or the notion of the designer.

Stucco with Portland cement is probably the most durable material for this kind of work. There are other stucco materials on the market that give satisfactory results. But any material that will not endure the test of time should never be used on permanent work.

Stucco Corners

Many of the principles involved in outside finish hold good on inside work. Imitation beams are often used on the interior, but in this chapter we are dealing with the exterior problems, and the illustrations will deal with corners for stucco work. In Fig. 350 we show a corner board used with stucco work. The upper drawing shows a perspective view, while a plan is shown at the bottom. This corner board has a mitered corner, and the edges are back-chamfered to receive the stucco. The irregular lines shown in all of the illustrations represent metal lath. The upper drawing of Fig. 350 is a sort of perspective view, while the bottom one is a plan of the corner.

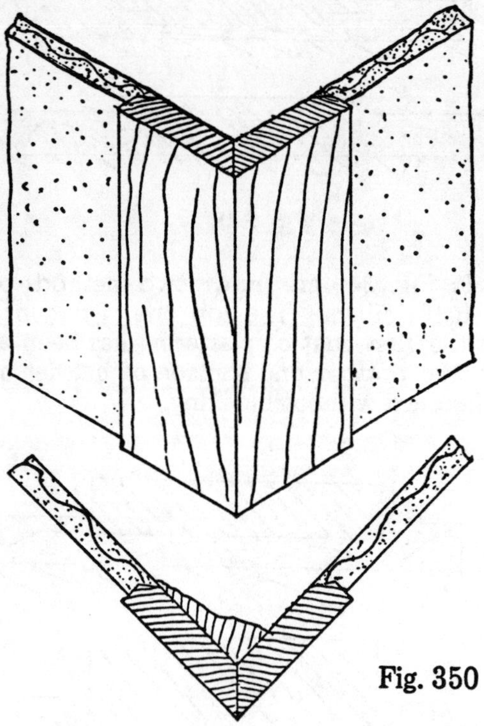

Fig. 350

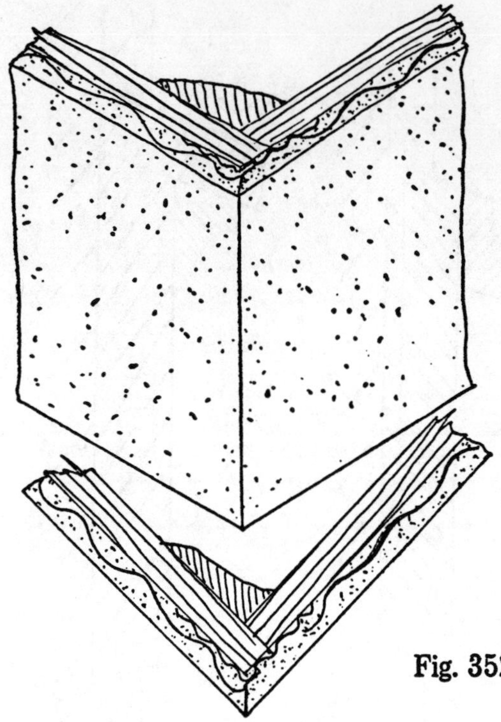

Fig. 352

In Fig. 351 is shown a perspective view and a plan of a corner that is in the making. That is to say, a ground with one edge beveled to a 45-degree angle has been set to establish the corner. The beveled edge of the ground should be given a thin coat of hard oil to prevent the stucco plastering from sticking to it. After one side of the plastering has been applied and is set, the ground should be removed and the other side plastered, leaving the results somewhat as shown by Fig. 352. It will be noticed that we show an extra irregular line at the corner which represents metal lath as reinforcing.

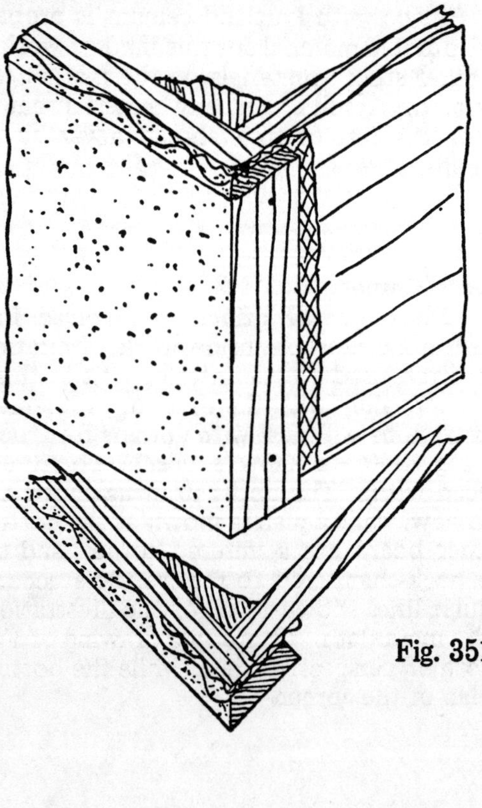

Fig. 351

Another good construction for a stucco corner is shown by Fig. 353. In this the corner is established by setting a rust-proof metal corner bead at the corner and plastering to it. This is one of the best corners for stucco work.

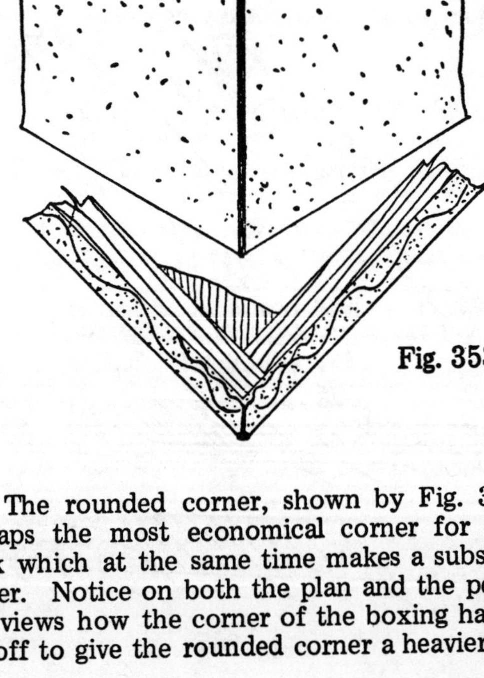

Fig. 353

The rounded corner, shown by Fig. 354, is perhaps the most economical corner for stucco work which at the same time makes a substantial corner. Notice on both the plan and the perspective views how the corner of the boxing has been cut off to give the rounded corner a heavier body.

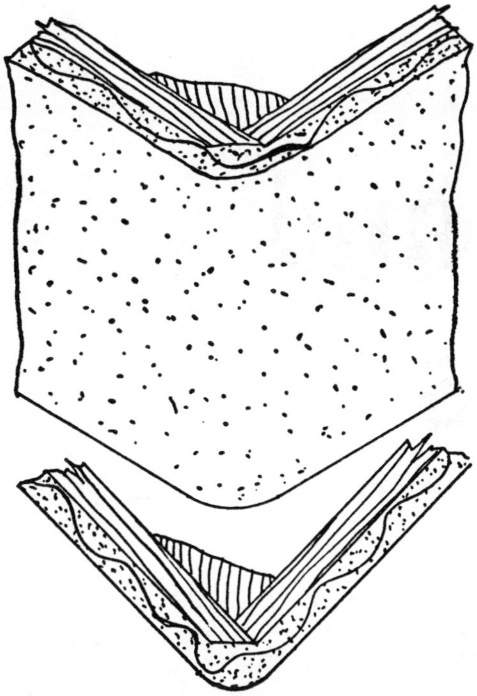

Fig. 354

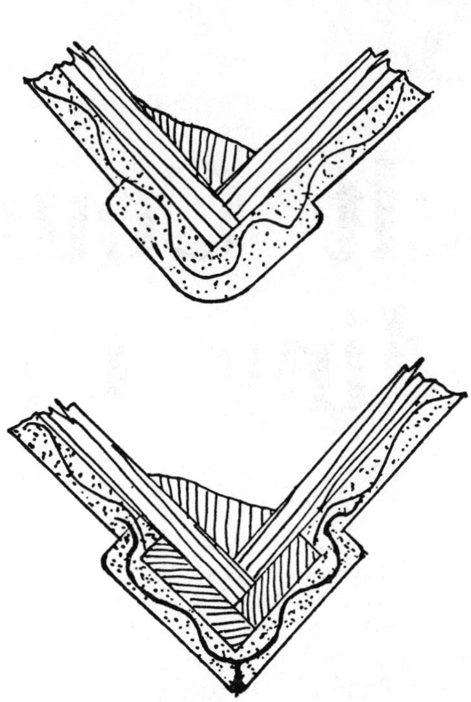

Fig. 355

Two corners that have decorative value are shown by Fig. 355. The upper one is simple, having the corners rounded. The main corner is rounded with a much greater radius than the two edge corners. The design shown by the bottom drawing gives a more expensive construction. For here a blind corner is nailed to the building as indicated by the drawing and is covered with metal lath that is on the corner bead. The main corner as well as the edge corners are all slightly rounded, as will be observed by studying the drawing.

The corners of stuccoed buildings are of great importance, for the corners carry with them a great deal of the decorative value. More than that, there are the corners of piers, pilasters, columns, lintels, doorposts and so forth. Originally all of these were principally constructed of masonry, but with the advent of stucco they are mere make-believe.

During the construction of a building the corners that are likely to be damaged should be carefully protected with shields or with barricades. But after a building is completed and is in use there is still need for such protection. Driveways, by all means, should be so located that there will be no danger of vehicles bumping or rubbing against any of the corners.

25

Belt Courses and Siding Corners

The purposes of belt courses are manyfold. Perhaps the principal use is to form junctions for different sections or for different kinds of wall finishings, both for the interiors and for the exteriors of buildings. The designs usually are in keeping with the rest of the work. The sizes of courses vary from little wooden bands to rather bulky bands of masonry. When a belt course is used between two kinds of wall finishes, its purpose is to conceal the joint and at the same time provide a decorative element. The decorative value is often great enough to justify the use of blind belt courses in order to break the monotony of plainness. Belt courses also form drips and in this respect prevent leaks.

What has just been said about belt courses is in many ways true of corners. The principal difference is in their position; one runs horizontally while the other is in a perpendicular position. They both form junctions and they both carry with them decorative qualities.

A gable in which the upper part represents shingles, the middle part siding and the bottom stucco, is shown by Fig. 356. The following illustrations will show details of the different joining constructions:

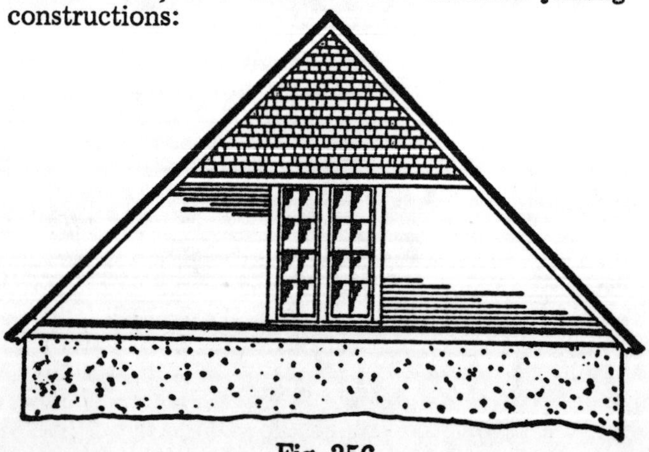

Fig. 356

Belt Courses

Fig. 357 shows by the main drawing a joining of shingles to siding by means of a sort of belt course with a wide drip. The construction as shown does not necessarily need the bed moulding and the the joint is simpler without it. When the bed moulding is used as shown, the frieze and the plancher can be reduced in width somewhat as shown by the dotted lines. The inset drawing represents

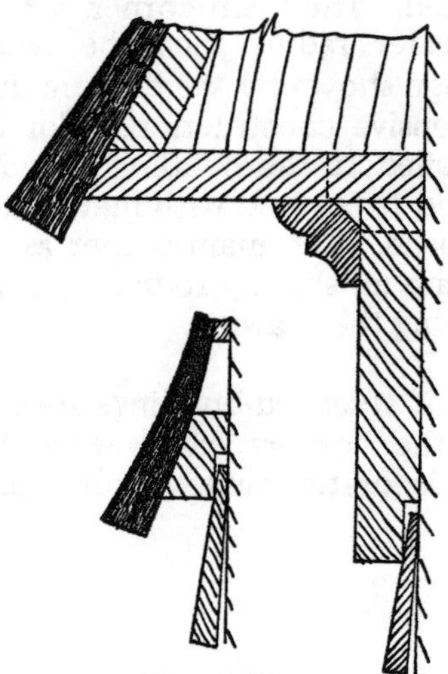

Fig. 357

a method of joining shingles over siding without a frieze and without a moulding. Fig. 358 shows to the left a detail of a belt-course joint between shingles and siding with a blocked out frieze but no moulding. The detail shown to the right is similar in construction except that the drip is wider and a bed moulding is shown, although the moulding can be omitted and yet leave a good piece of workmanship.

Shown to the left of Fig. 359 is a simple joint with the shingles lapping over the siding. This is not a bad joint provided it harmonizes with the rest of the outside work. To the right is shown a watertable belt between stucco and siding.

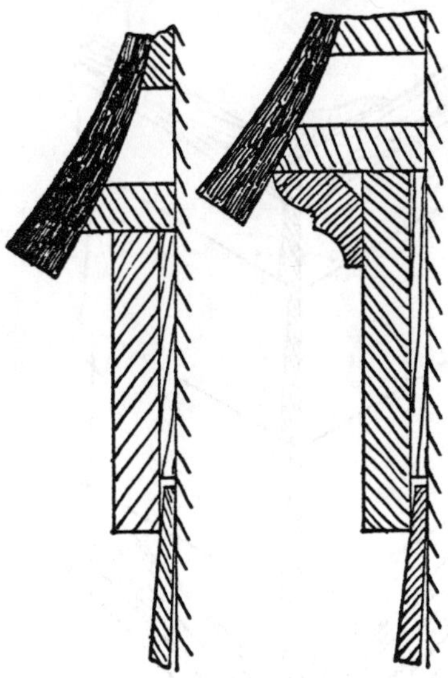

Fig. 358

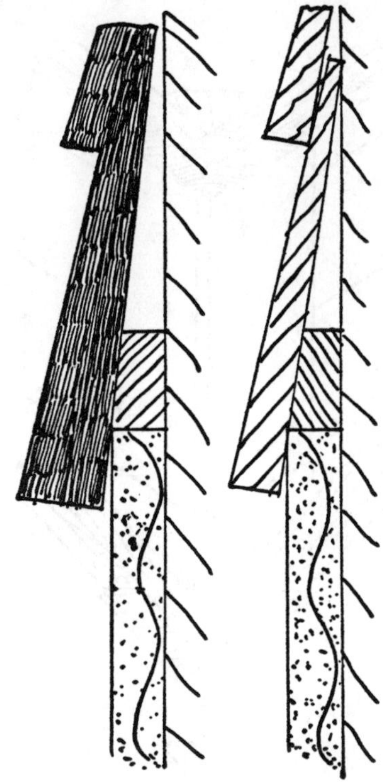

Fig. 360

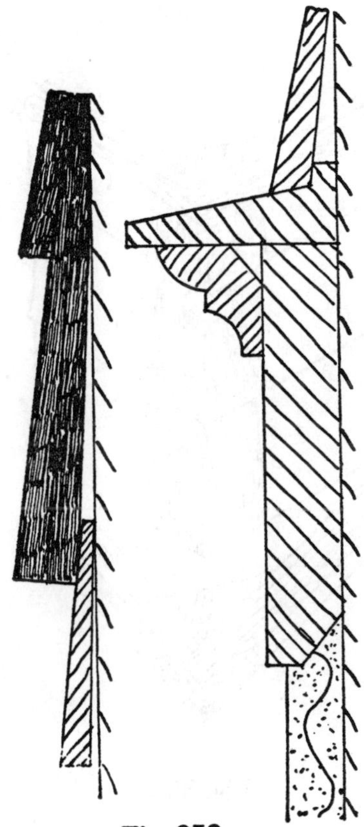

Fig. 359

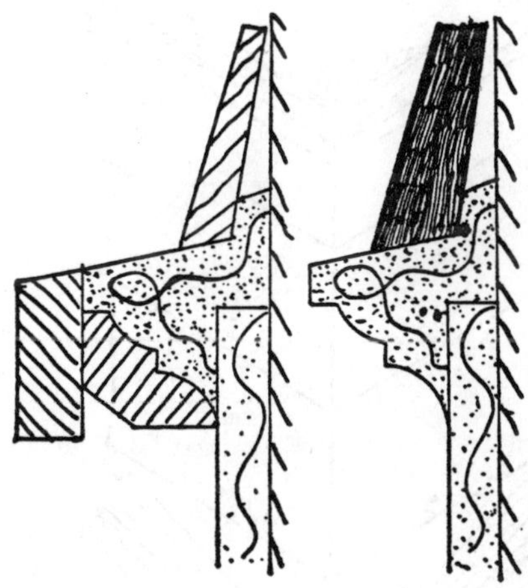

Fig. 361

A simple but substantial method of joining shingles over stucco is shown to the left in Fig. 360. To the right the same method is shown for joining siding and stucco. Fig. 361 shows to the left how to form for a cement watertable coming between stucco and siding. The irregular lines represent metal lath. To the right the watertable is shown with the form removed, but shingles are shown instead of siding.

Corners for Siding

Fig. 362 shows at the bottom a plan of a corner, the upper drawing shows the same corner in perspective. To the upper left at *a* is a nail that has been driven too close to the corner joint. The trouble with this kind of nailing is that if it does not split the board when the nail is driven, it is almost sure to cause damage sooner or lager. Nails should be driven far enough (approximately 2 inches) from the joint to prevent cracking.

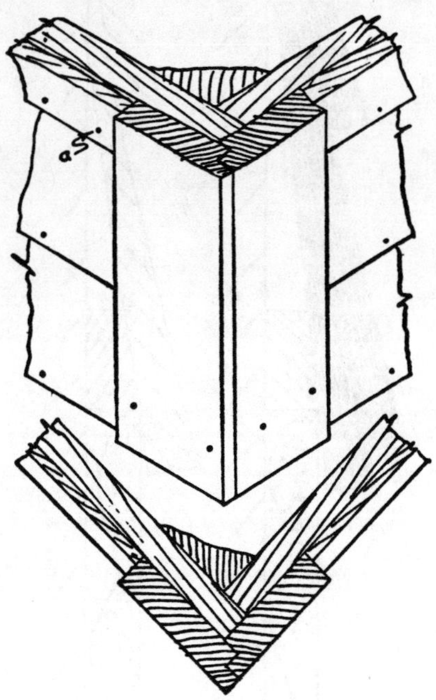

Fig. 362

In Fig. 363 we show a plan and a perspective view of what is called a mitered corner for siding. This corner, if made in a workmanlike manner, is neat and gives good service. The heavy dots indicate the nailing. A plan and a perspective view of a

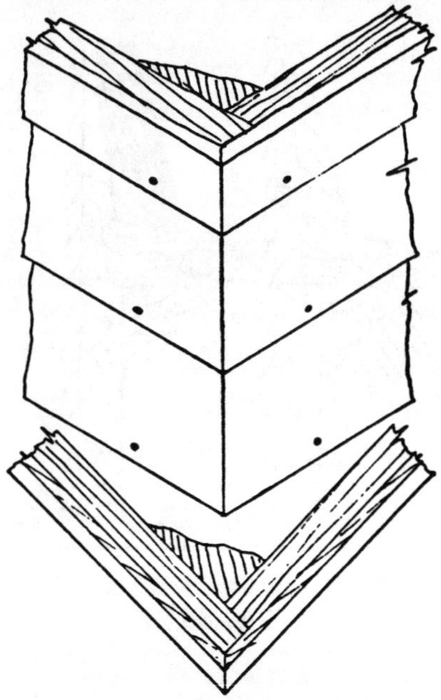

Fig. 363

corner which is to be covered with metal corners is shown by Fig. 364. The metal corners in place are shown shaded in Fig. 365.

In using metal for corners care must be taken to keep them in perfect perpendicular alignment. The nails should not be too large and can be either

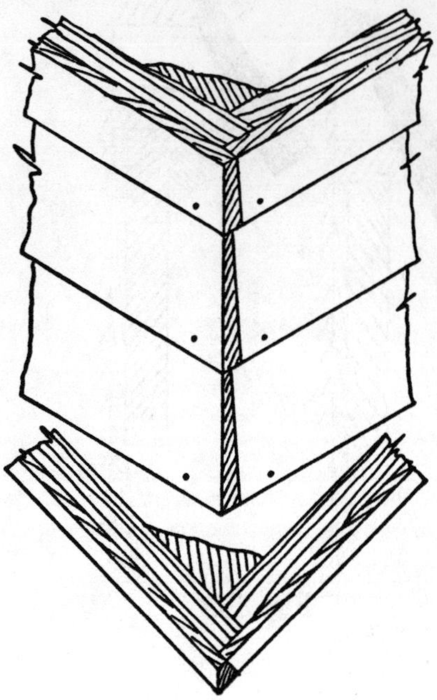

Fig. 364

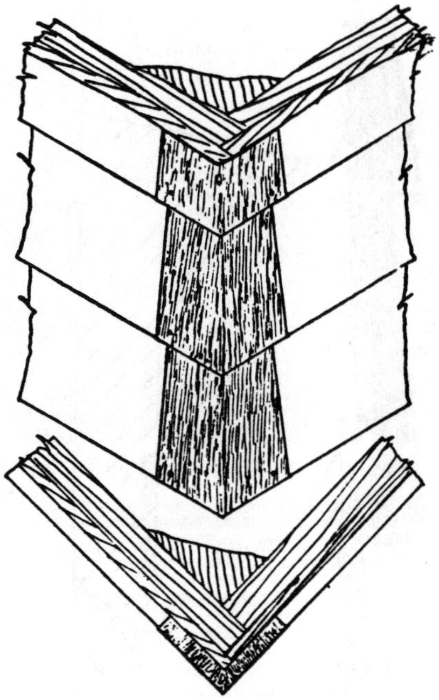

Fig. 365

flat-head or finish nails. The metal corners should slip up under the siding at the top and may be slightly bent under at the bottom edge of the siding.

A plan and a perspective view of a corner in reverse is shown by Fig. 366. In this the siding of one side is scribed onto the siding of the other side. This makes a satisfactory job, especially if the work is carefully done. Another corner in reverse is shown by the drawings of Fig. 367. Here the siding is fitted to a square angle piece corresponding in thickness with the casings or corner boards.

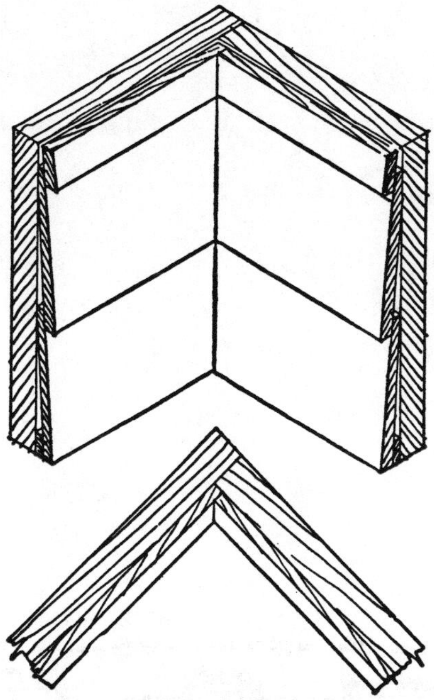

Fig. 366

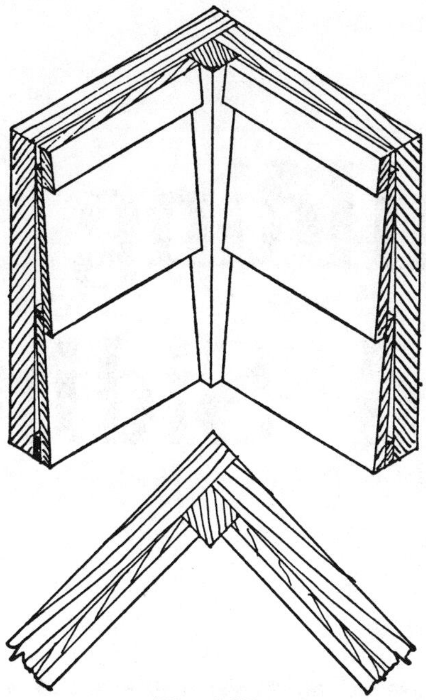

Fig. 367

Corners are always more conspicuous than any other part of the exterior of a building and, for that reason, they should be constructed, not only to give service, but to carry with them a certain degree of ornamentation. This element, as a rule, costs almost nothing in time or money so there is no reason why it should not be used in every corner construction of at least the better classes of buildings.

26

Building Paper and Siding

Insulation

The use of adequate insulation is becoming more and more standard, not only because it insures a greater amount of comfort, but because it saves on the cost of heating homes in cold weather and helps in keeping them cool during the hot summer months. There are many kinds of insulating materials on the market: Mineral wool, quilting, insulating boards (known by various trade names), and, of course, building paper.

Applying Building Paper

In Fig. 368 is shown a side of a house, in part, where we take up the application of building paper. The shaded parts of the drawing represent paper already in place. The dotted lines indicate the joints of paper that is yet to be put on. It will be noticed that at either end a strip of paper is shown extending out from the corner boards. These strips of paper are either fastened to the corner of the building before the corner boards are in place or else they are fastened to the corner boards and put in place with them.

The strips of paper shown at either side of the windows and over the top are usually fastened to the building before the frames are set. Sometimes the paper is put on the window and door frames before the frames are set. But the way paper is put on is not important. The important thing is to get the paper strips on so they will have plenty of lap, both for the frame and for the other paper when it is applied. The purpose of paper on the corners and around openings is twofold: First it stops the air from passing through at the corners, and window and door frames, and second, it prevents leaks around the frames.

At the bottom of Fig. 368 is shown a full-width strip of paper in place ready to receive either siding or shingles. To the left is shown a method of fastening the paper with laths. This is shown further in Fig. 371 and 372. To the right the paper is shown fastened to the wall with roofing nails.

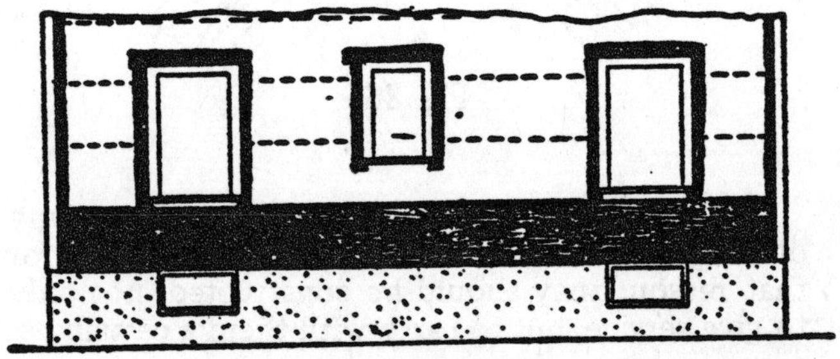

Fig. 368

Fig. 369 shows an old method of saw-cutting rolls of paper into widths that will be suitable for use at the corners and around the window and door frames. Another method of cutting paper strips, especially for windows and doors, is shown by Fig. 370. In this method a full-width strip of paper is cut to the proper length and then rolled up again into a rather small roll, somewhat as shown by the drawing. Then it is placed on a block, and

Fig. 369

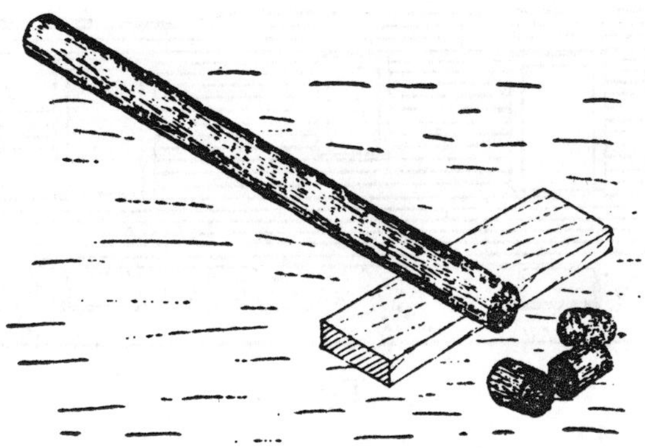

Fig. 370

with a hand-axe cut into small rolls which will be suitable for use around window and door openings.

A detail of the fastening of paper (shown to the left in Fig. 368) is shown by Fig. 371. It will be noticed that the laths are held up from the bottom edge of the paper and that they are nailed only at the ends. The reason for this is shown by Fig. 372 where a siding board is shown in place and the laths have been loosened at the bottom and turned to the position shown. When the siding is finished up to these laths they are removed and the same process is repeated with the laths on the next strip of paper, and so on until the siding is on.

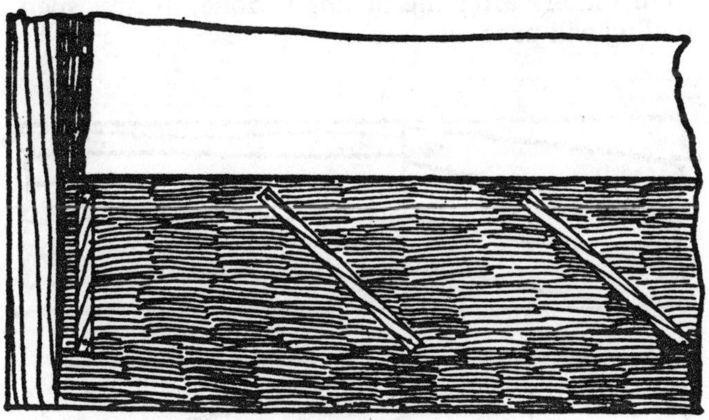

Fig. 371

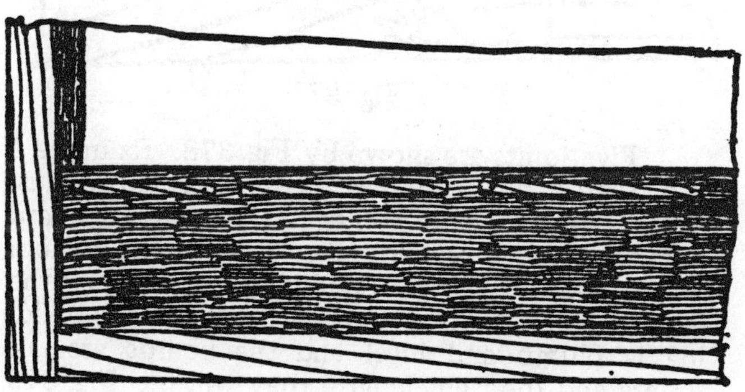

Fig. 372

The right way to cut full-length strips of paper to a certain length (for example, to be used between window and door frames) is shown by Fig. 373. The end of the paper is held to the floor with a block of wood, as indicated by the drawing, and then enough paper is unrolled to make the strip. This done, a short straightedge is placed on the paper, as shown, where it is to be cut. With a foot on the straightedge to hold it in place, take the roll of paper with a hand on either end and tear the paper along the straightedge from one edge to the other, leaving it somewhat as shown by Fig. 374. If several of the same length strips of paper are needed, cut them in the same way, one on top of the other.

Fig. 373

Fig. 374

There are many other ways of cutting paper. Some carpenters fasten the paper to the wall just as it is unrolled, and when they come to a window or door or corner, they cut the paper using the casing or corner for a guide.

Outside-Wall Coverings

Of the four commonly used coverings for outside walls, siding is still the most economical. Brick veneer is probably the most expensive, although it produces a pleasing appearance. Stucco is not so expensive as brick veneer, gives a good appearance, and if it is of good quality, it will give excellent service. Both of these coverings for outside walls have a greater initial cost than either shingles or siding.

Siding

Fig. 375 shows part of a side of a house that is to be covered with siding. The first thing to be done, after the corner boards and frames are set, is to establish the spacing. This is usually done with a compass.

After the various spaces are established, spacing poles are made and marked. Sometimes, though, only one pole is used for a whole story, but in most cases, and for most practical purposes, at least three poles are used. One is used from the foundation up to the window sills, carrying the spacing all around the building. Another pole is made for the window frame. A third pole is used from the window frame up to the belt course, or to the second story window, if it is a two-story house.

The drawing in Fig. 375 shows the corner boards and window frames marked for the siding. It also shows four courses of siding in place. The dotted line on this board and the one to the right of it represent chalk lines. These lines are necessary to guide the workman in keeping the courses straight when he puts on the siding. The joints should be broken somewhat on the order shown.

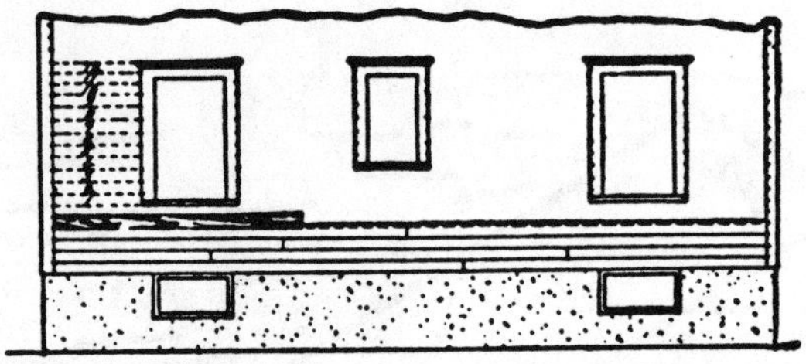

Fig. 375

To the left of Fig. 375 are 12 spaces with dotted lines, numbered 1, 2, 3, 4 and so on. These spaces indicate the 12 siding boards needed for the section between the window frame and the corner board. These 12 boards should be cut square on the end that joins the corner and about one-half to three-fourths of an inch longer than needed to allow for marking and cutting. Put a board in space number *1* and mark it for length and lay it aside. Similarly, mark boards for space number *2*, number *3*, number *4*, and so on, until all are marked, being careful that they are kept in the right order so that number *12* will be on top of the pile. Now start to cut the boards: First cut number *12*, then number *11*, then *10* and so on until number *1* is cut, again being careful that the boards are kept in the right order. Now board number *1* will be on top of the pile. Then nail board number *1* into place, number *2*, number *3*, and so forth, until all 12 boards are in place. Leave the boards as shown by Fig. 376 where the 12 numbered boards are in place.

Fig. 376

This method of marking, cutting and putting the siding into place between window frames or between frames and corner boards, if it is painstakingly done, gives excellent results. The same method, modified somewhat, can be used in siding gables and other small spaces. The common practice, though, is to square one end of the siding boards, then mark and cut the boards one by one as they are nailed into place.

Two siding joints and the nailings are shown in Fig. 377. The upper joint shows the nails approximately 2 inches from the joint and ½ inch from the bottom edge of the board. We recommend this method of nailing joints because it keeps the nails far enough from the end to prevent the siding from splitting. The nailing shown in the bottom drawing is too close to the end and the result is cracks. Even if the cracks do not appear immediately after the nailing is done, in time such nailing will produce cracks.

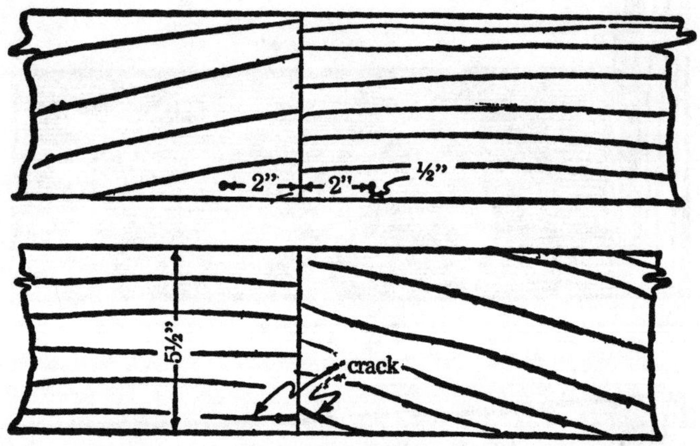

Fig. 377

Five joints are shown by Fig. 378. Example *1* and *5* are good. Number *2* is open, probably caused by cutting board number *1* too long or board number *2* too short. Number *3* hits at the bottom but is open at the top, and number *4* is just the reverse. Such joints are not necessary. If the building is plumb and level and the boards are cut square and the right length, they will fit. Fig. 379 shows three other joints. The joint numbered *1* is as it should be; number *2* is open at the top and number *3* is open at the bottom. If the courses are kept straight and the boards are cut square, joints such as shown at numbers *2* and *3* will not occur.

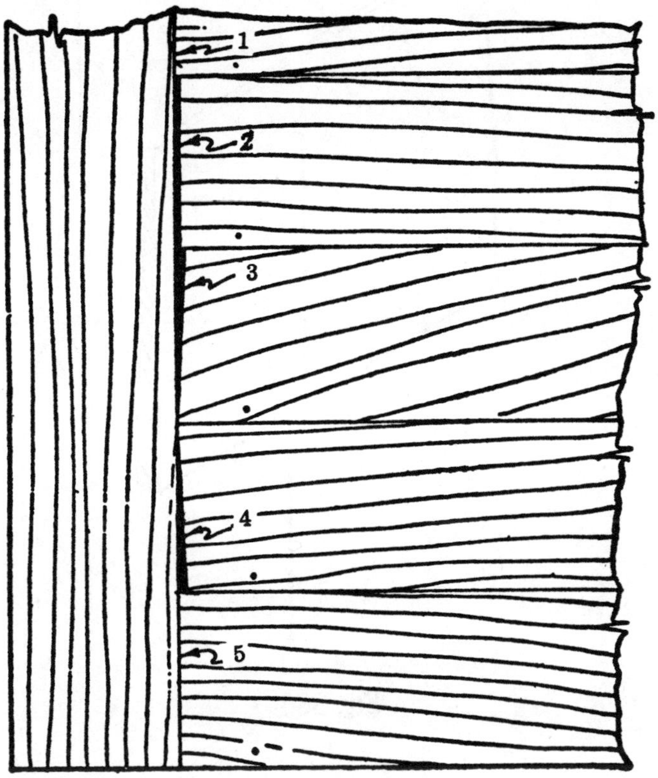

Fig. 378

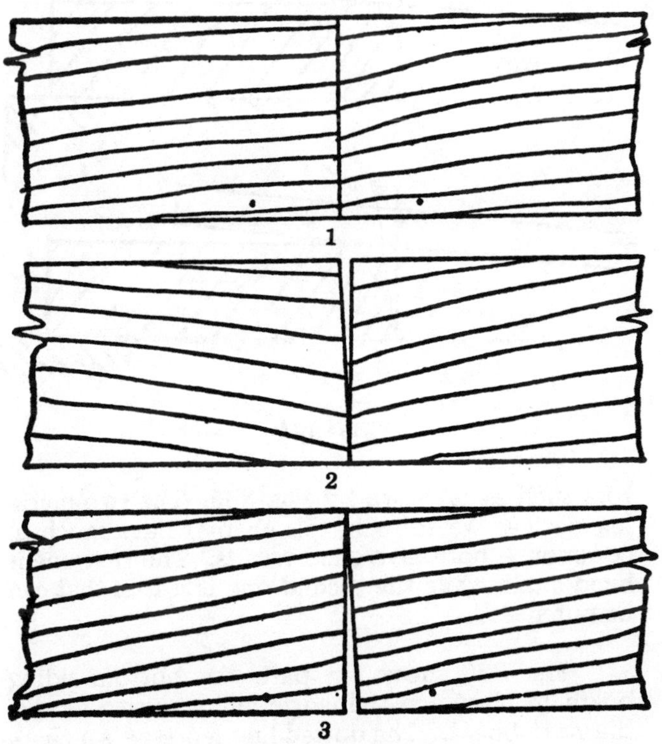

Fig. 379

Fig. 380 at *A* shows how the joints should be beveled as viewed from an edge. At *B* the joint is completed. *C* is an edge view of a defective joint. The joint shown at *D* is very bad, though exaggerated.

The size of nails used for nailing siding must be governed largely by the kind of siding used. For ½-inch siding, 6d box or 6d casing nails are often used. The nails should be placed directly over bearings and should be in perpendicular alignment. If this can't be done, no attempt should be made to make the nails line-up on boxed buildings. Curved or twisted lines of nails look bad.

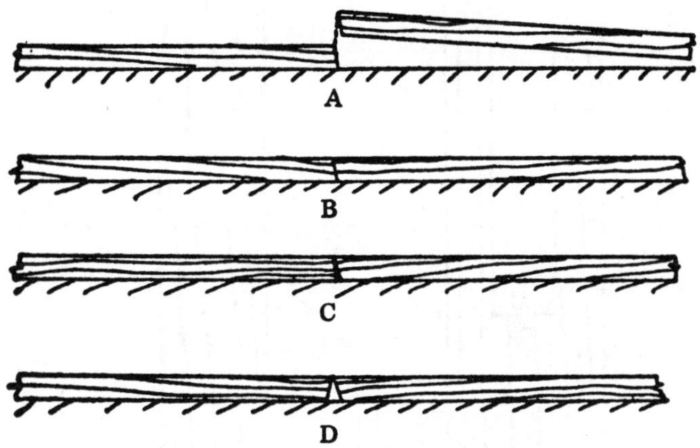

Fig. 380

Siding Hooks

We have in Fig. 381 two views of what is called a siding hook, or commonly known as a hand-made siding marker. Fig. 382 shows the application of the hook. The siding hook must be held firmly against the casing while the marking is done along the edge of the outside line. Fig. 383 shows, by dotted lines, how the mark will appear on the siding after the hook has been removed.

The importance of the bevel on the edge of casings or corner boards is shown by Fig. 384. At *A* we show, viewed from the bottom, the layout shown in Fig. 383. The indicator points to the difference between the marking of a siding board for a square-edged casing and a casing which is properly beveled. Here it will be noticed that the siding for the beveled casing will be marked just a trifle longer, which is a desirable feature because it will keep the joints tight without crowding the casing. At *B* the indicator points to the difference in the marking of a siding board for a square-edged casing and a casing that has been beveled too much. The dotted line to the right shows where the siding would be marked with the siding hook. This would make the board too long. The bevel shown at *A* is right, while the bevel shown at *B* is wrong.

What we have said about Fig. 384 will apply to cases where the marking is done with a marking

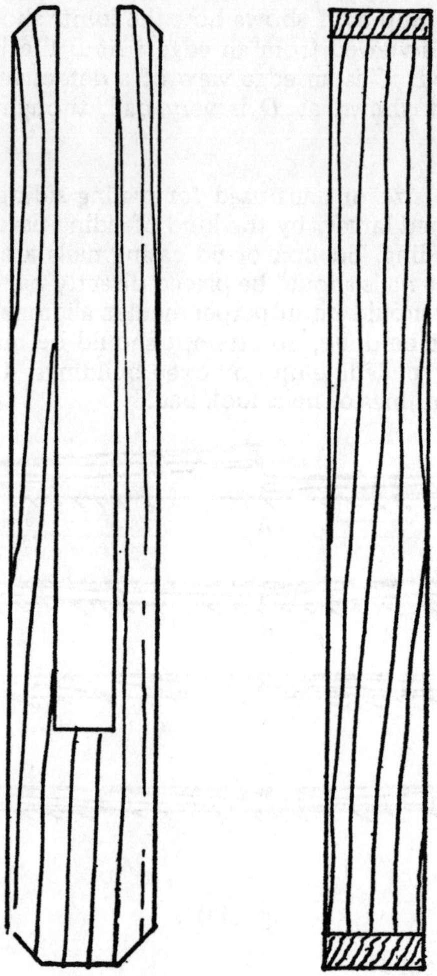

Fig. 381

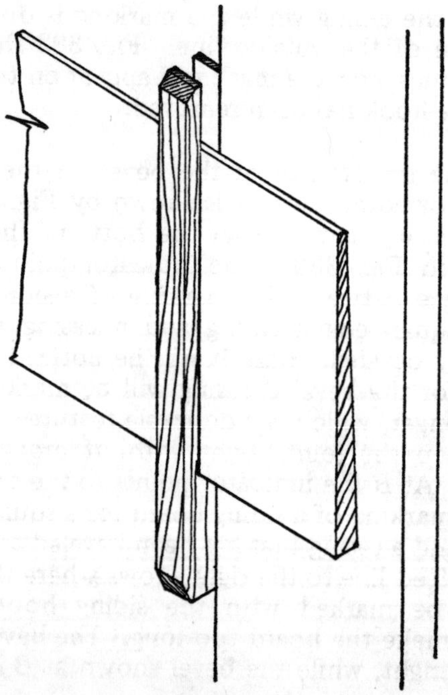

Fig. 382

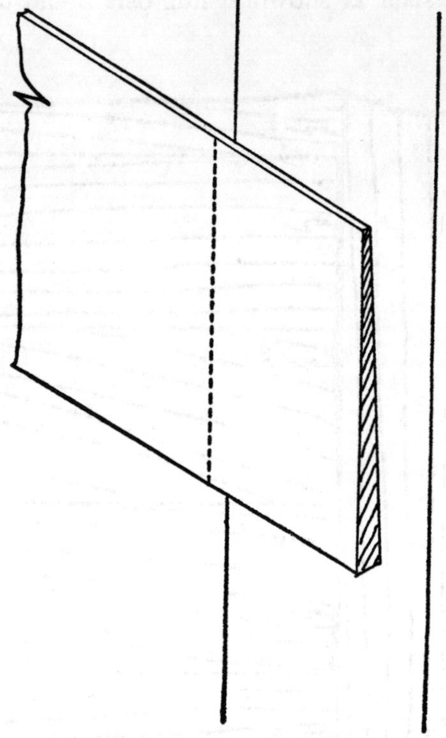

Fig. 383

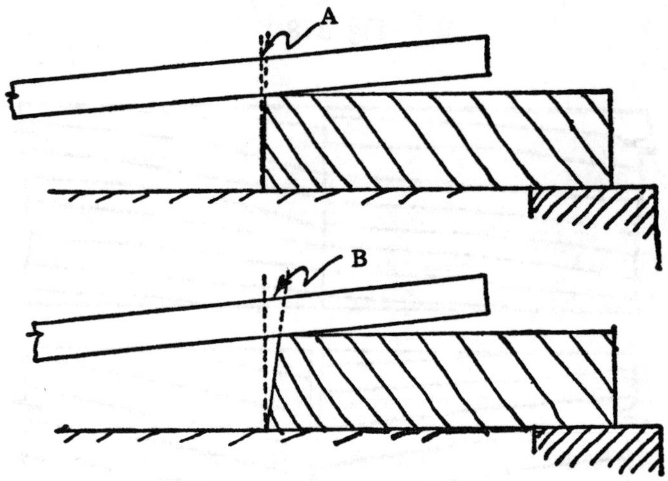

Fig. 384

tool such as is shown by Fig. 385. The two marking devices we show in this chapter have much in common -- both give good results. The first one is hand-made while the second one is a tool that can be purchased.

Fig. 386 shows, in part, the bottom siding board in place and the ledger-nail stuck to receive the next board. The dotted line represents a chalk line struck to guide the workman in keeping the

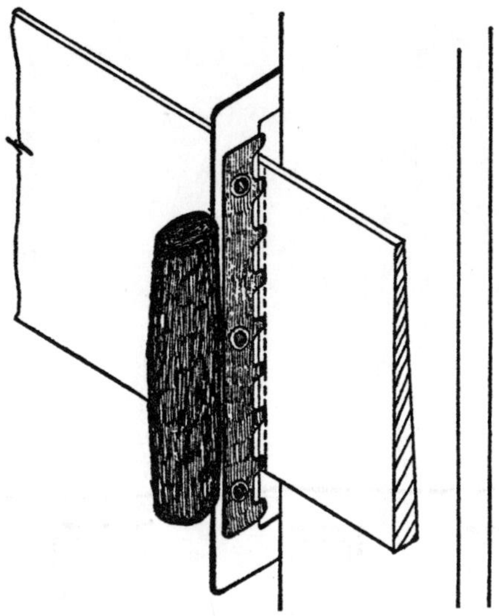

Fig. 385

bottom edge of the siding straight. Short runs of siding do not require the chalk line. Fig. 387 shows the board, which is shaded, resting on a ledger-nail. When the board is nailed, the ledger-nail is removed and stuck for the next board and then for the next and so on until the siding is completed.

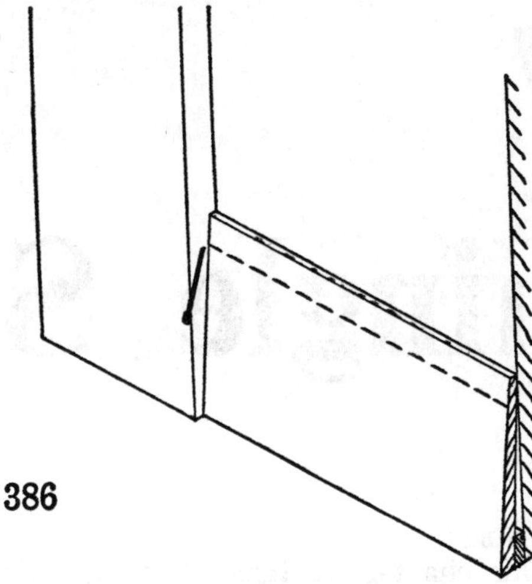

Fig. 386

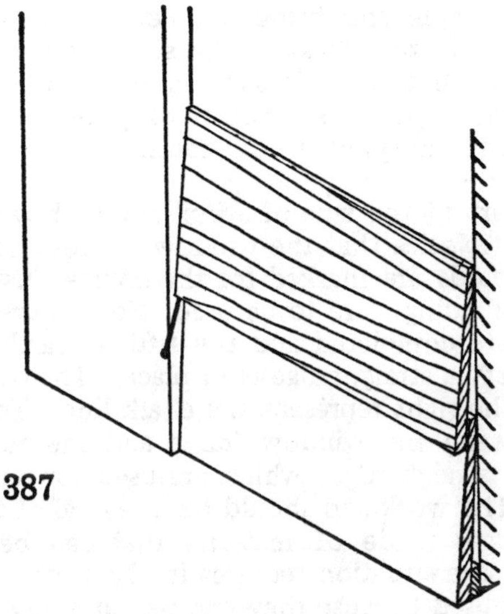

Fig. 387

27

Shingle Siding

Side Shingling

In this chapter we take up the question of shingling the outside of buildings. The illustrations here are simple and bring out certain principles that are valid regardless of the style of shingling that might be used. We show narrow courses although wide courses are becoming popular. The illustrations will apply in both cases.

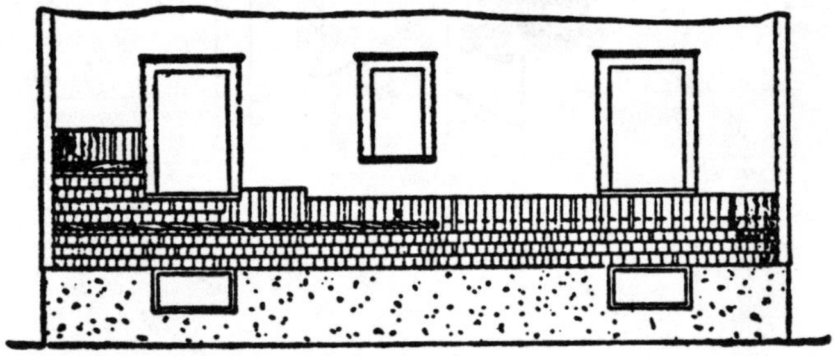

Fig. 388

A part of the side of a building is shown by Fig. 388. Notice that the window frames and the corner boards are marked for the courses just the same as if siding were to be used. Four courses of shingles are completed and the fifth is partly on. To the left a straightedge is in place. The dotted lines to the right represent the chalk line. To the left, between the window frame and the corner, are short straightedges which are used for narrow spaces. The workman should have several of these straightedges made of material that can be cut whenever the situation requires it. Lath or lattice are often used because they can be cut easily with a hatchet.

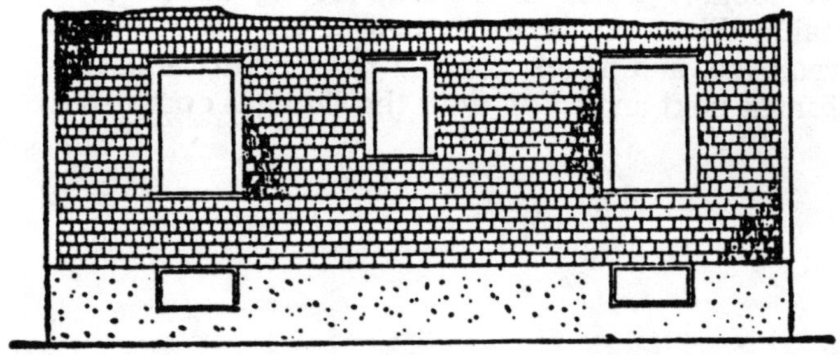

Fig. 389

As in roof shingling, the first course must be doubled. The first course is usually held to a line stretched from one corner to the other and fastened at various intervals to keep it from sagging. After the first course is on, a chalk line is struck as a guide for the straightedge, which holds the shingles in line when they are put on. This process is repeated for each long course until they are all in place. The short courses are put on without the use of a chalk line. Whenever the straightedge reaches across the space to be shingled, no chalk line is necessary. Fig. 388 should be compared with Fig. 389, which shows the shingling of the side completed. It will be noticed that the course joining the window sills line up with the bottom of the sills, and the course that joins the window cap lines up with the top of the caps.

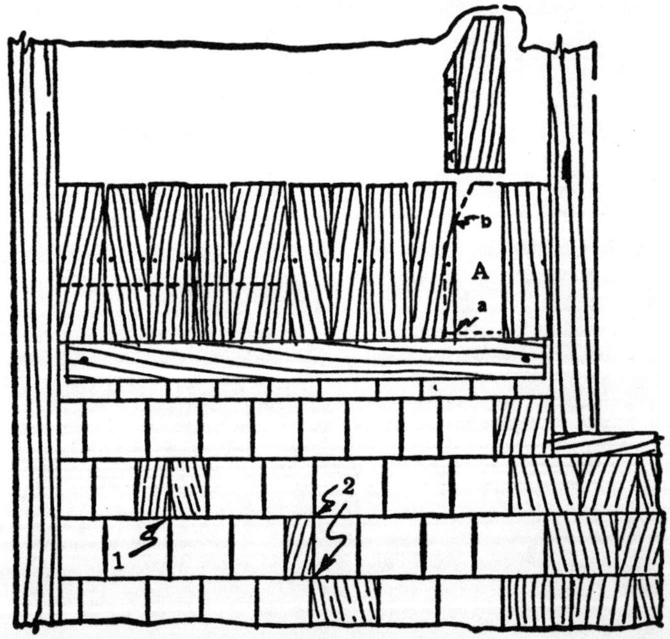

Fig. 390

The heavy dots on Fig. 390 indicate nails, which should be kept about one-half inch from the edge of the shingle. Use only two nails to a shingle. Some carpenters use three nails on extra wide shingles. Shingles over 4 inches wide are in reality more than one shingle.

The horizontal dotted line, shown by Fig. 390, indicates where the next course is to be applied and shows approximately how much it should lap over the nails. At *A* is shown by dotted lines how by holding the finishing shingle away from the straightedge (point *a*), the shingle can be marked. To mark the top of this shingle, clip off the corner of the shingle, somewhat as shown, and mark it where it intersects with the other shingle, or point *b*. The shingle shown above the space marked *A* shows by the X's the part of the shingle that must be trimmed off. At *1* and at *2* are joints that are bad and should be avoided.

Fig. 391 shows by dotted lines and light shading how to cut shingles to be used for finishing under window frames and frieze. First square and straighten the butts in the bunch of shingles. Then mark it according to the length needed and cut as

Fig. 391

many of each length as required. Fig. 392 shows the bunch of shingles after one length of shingles has been cut from it. Each length, of course, must be marked and cut separately.

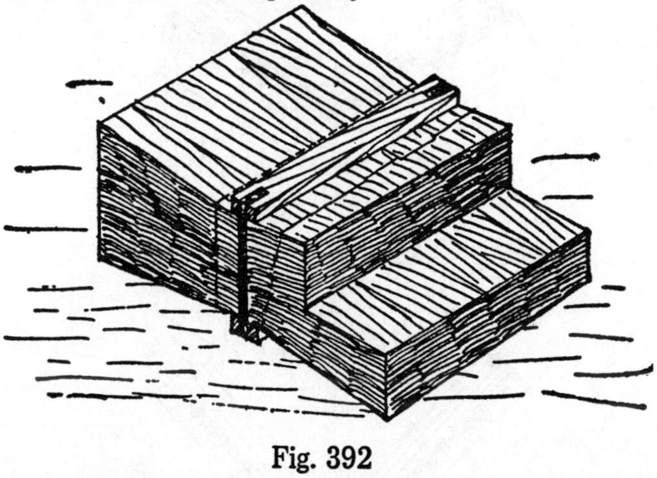

Fig. 392

Two faulty applications of shingles are exaggerated in Fig. 393. These faults can be discovered more nearly in the work of beginners rather than

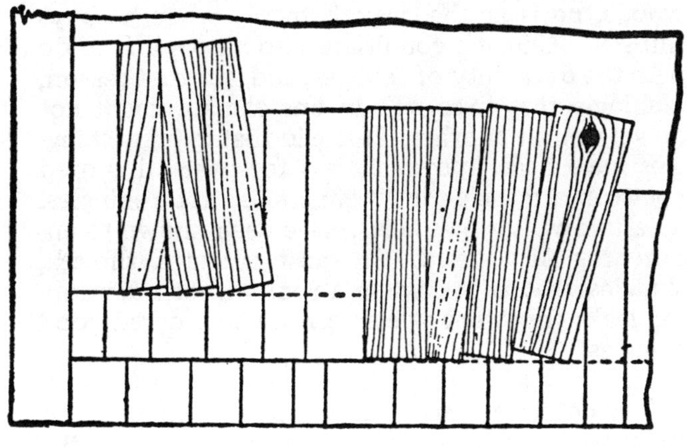

Fig. 393

by experienced carpenters. The trouble is caused by the workman keeping the perpendicular joints tight and ignoring the lining-up of the butts. A little trimming will correct this. Fig. 394 shows ten faults that often appear in shingling. All of these are due, first, to defective shingles and, second, to carelessness on the part of the workman who put them in place. Shingles with the defects indicated in this figure, if they are otherwise sound, can be trimmed and used or discarded.

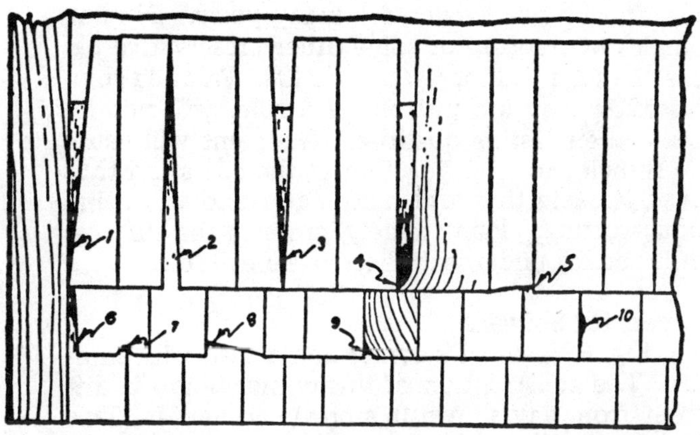

Fig. 394

Life of Shingles

The life of wood shingles depends on a number of things. The most important of these is the quality. Shingles made of a durable wood, with straight grain and free from knots or other defects, of course, are the best. Cypress, cedar and redwood are considered by good authorities as the leading woods for shingles because they endure well and give a pleasing appearance. White pine, yellow pine and spruce are also used for shingles, but they are not as substantial as the woods previously mentioned. Sap, knots, crossgrain, shakes, rot and poor milling all have to be considered in order to determine the rating of shingles.

Unless one is a good judge of the various kinds of wood and their suitability for shingles, and knows something of the lasting qualities of the different woods, he is at the dealer's mercy when he buys shingles. Climatic conditions also have much to do with the durability of shingles, and, for that reason, a shingle that lives well in one climate might not do so well in another. It is said that cypress shingles have been known to last for over a hundred years on a mansion in Virginia. Redwood shingles, a certain writer maintains, have been known to endure for fifty years. In most climates, though, shingles that will stay on a roof without developing leaks for twenty years can be considered good shingles.

Whether or not a roof has enough pitch to drain well is another factor to be considered in the life of shingles. Shingles used on the south and west sides of a roof always deteriorate faster than those on the north and east sides. The contrast is most noticeable between the north side and the south side of a roof. There is much less difference between the east side and the west side. This deterioration, of course, is caused by the rays of the sun. The south side is affected the most, next the west side and then the east side, while the north side is affected the least.

Ventilation is one of the best preservatives for shingles, and if such a roof is shaded, as by a tree, it will give service for many years. Dipping shingles in creosote or some other preservative before they are laid increases their life. Painting shingles after they are put on, as a rule, will not increase their lasting qualities. The paint will cause the shingles to stick together at the butts and thus cause pockets that will hold water and will bring about rotting. Paint merely protects the surface, while rotting underneath becomes inevitable.

Corners for Shingles

Fig. 395 shows a corner board used with shingles. The construction of this corner board is different from any shown in a previous chapter. This way of constructing a corner is no more appropriate for shingling than it is for stucco or siding – any other construction we have shown would be just as suitable. Fig. 396 shows a mitered corner. This gives satisfactory results if it is well done. The joints should fit perfectly. Open or partly open joints are objectionable.

A laced corner is shown by Fig. 397. The indicators point out the joints, which are placed alternately, one on one side and the next on the other. Another laced corner construction is shown by Fig. 398. Here the joints are all on one side, as pointed out. This method makes it possible to throw all the joints to the side of the corner that is least conspicuous.

A corner construction in reverse is shown by Fig. 399. Here the courses of shingles from one side are coped onto the courses of the other side. Ano-

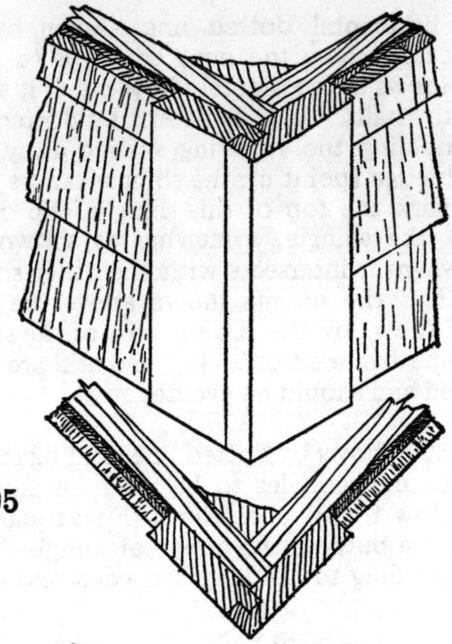

Fig. 395

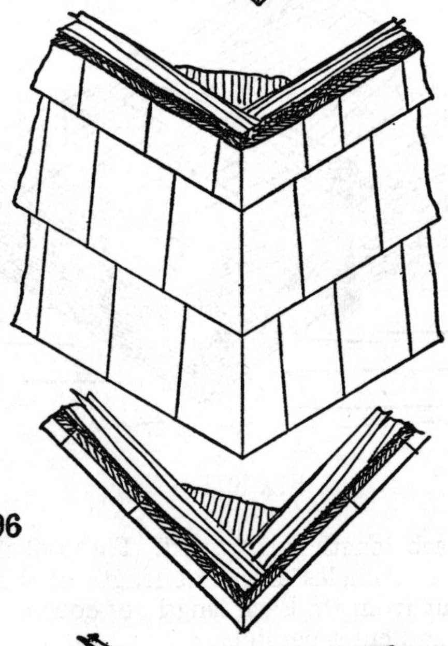

Fig. 396

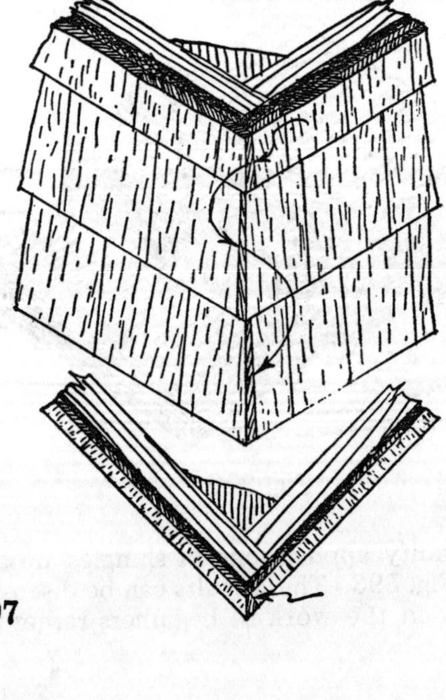

Fig. 397

Fig. 398

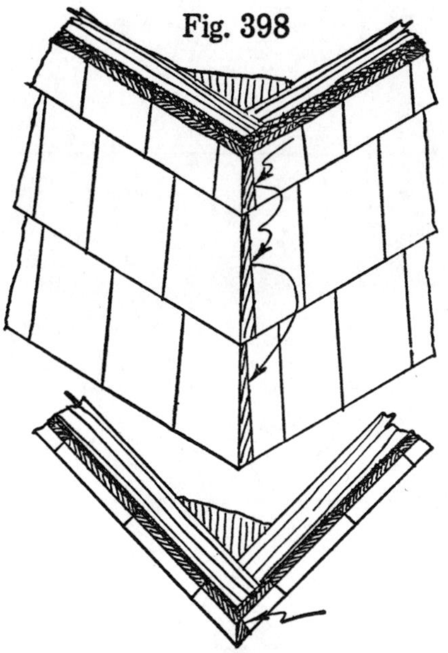

ther method of building a corner that is in reverse, or what is called an angle, is shown by Fig. 400, where an angle strip is used to receive the courses of shingles. This construction is more nearly suitable when corner boards are used for the regular corners.

There are two general classifications of wood shingles, the common shingles and the dimension shingles. The dimension shingles are made to a pattern. The width, as a rule, is 4 inches. The width of the common shingles runs from 2½ to 16 inches, every 4 inches of shingles is considered a shingle; a 16 inch wide shingle would be 4 shingles.

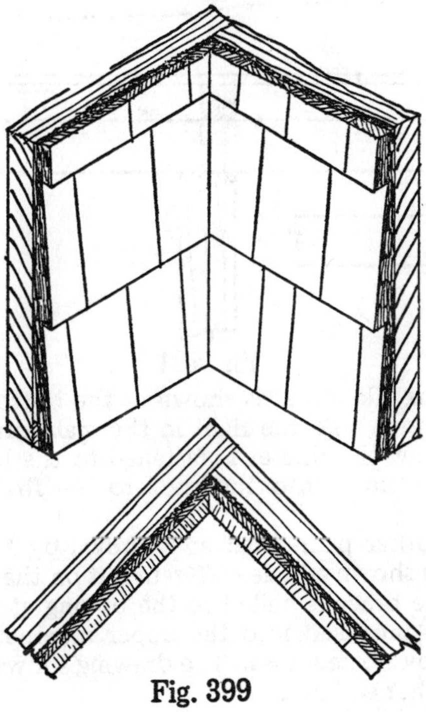

Fig. 399

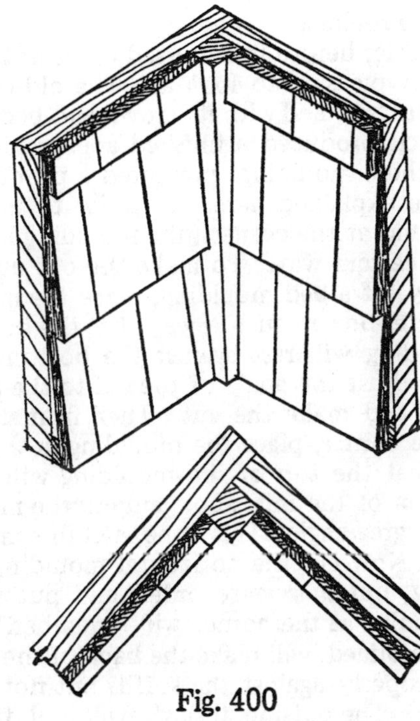

Fig. 400

28

Brackets and Gable Finish

Moulding Problems

A gable, before it is dressed up, is not very attractive or pleasing to look at. The old-time box cornice was trimmed with its crown and bed mouldings, which produced a finished appearance. Putting on these mouldings presented a problem that should be explained here. In order to make the joints sound at the corners, the moulding has to be cut in a certain way. To make the cut under the comb, say, of a bed moulding, place the moulding in the miter-box in such a way that the bottom of the moulding will rest against the bottom of the box and adjust the angle of the cut to the pitch of the roof, and make the cut. Then in making the cut at the corner, place the moulding in the miter-box so that the top of the moulding will rest on the bottom of the miter-box and cut the moulding on a 45-degree angle. This done, nail this moulding in place. Now cut the horizontal moulding in the same way, but in reverse order, and put it on so that the joint at the corner will "member." This, it will be noticed, will make the back of the moulding fit properly against the soffit, but not against the frieze. The outside, though, will look finished. The crown moulding is cut and put in place in the same way.

The bungalow gable has verge rafters and brackets for ornaments. These, of course, should harmonize with the rest of the building, in order to retain proper balance. This will be taken up again later in this chapter.

Brackets

Fig. 401 shows in the upper drawing one of the simplest brackets that can be used for supporting gable cornices. Besides its simplicity, it is about as serviceable as any, and, if used with a simple style of architecture, it looks good. The outside end of the bracket as shown in the main drawing is cut square, but if a little more fancy cut is desired,

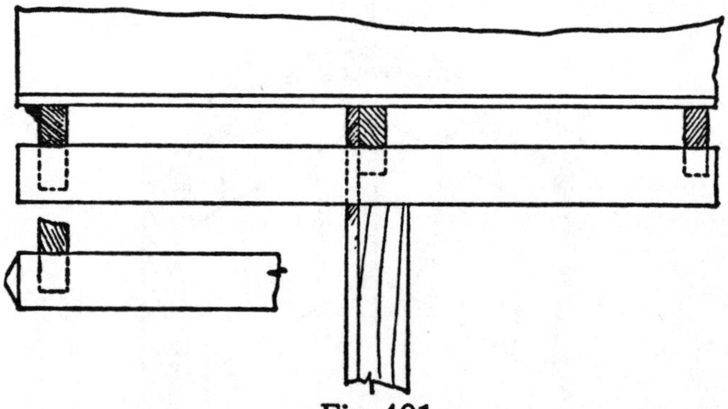

Fig. 401

something like what is shown at the bottom, left, can be used. To the right in the main drawing is shown how the timber is fastened to the building, extending back into the roof to the first rafter.

Another plain bracket is shown by Fig. 402. In this is shown a little different cut on the outside end. The brace is nailed to the boxing at the bottom and mortised into the upper member, as indicated by dotted lines. The drawing shows in part still another end cut.

Another step in the bracket evolution and another form of end cut are shown by Fig. 403. Here the brace is fastened at the top to the horizontal member and at the bottom to a perpendicular member.

A novelty in the form of brackets is shown by Fig. 404. This quarter-circle bracket can either be round throughout or it can be square, with the ends rounded one way, as shown. A square bracket of this form can be made easily by nailing together thinner pieces of material sawed to the proper curvature. Fig. 405 shows a bracket which is square and has the outside end cut square. The bottom end rests on a square lug. This bracket is excellent from both the standpoint of support and appearance.

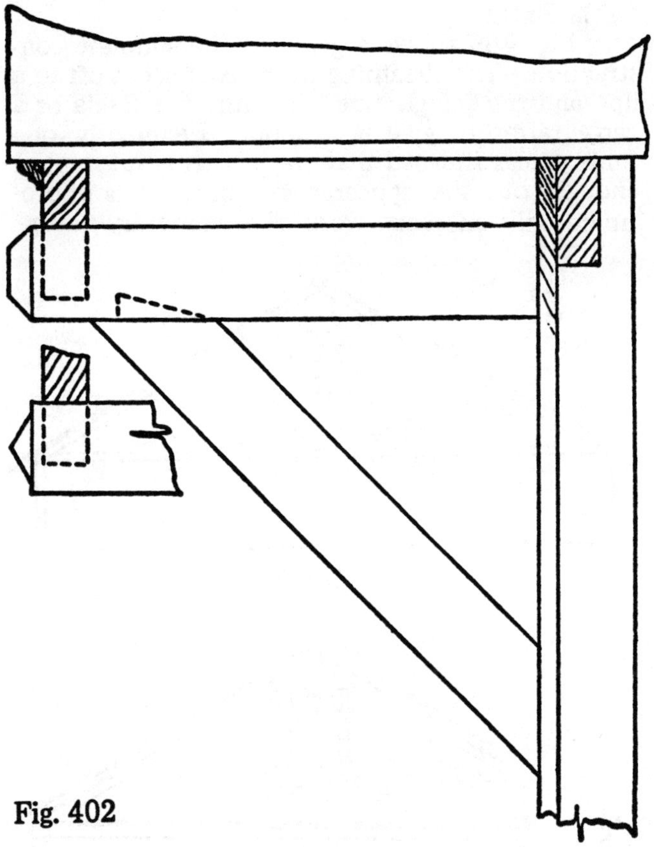

Fig. 402

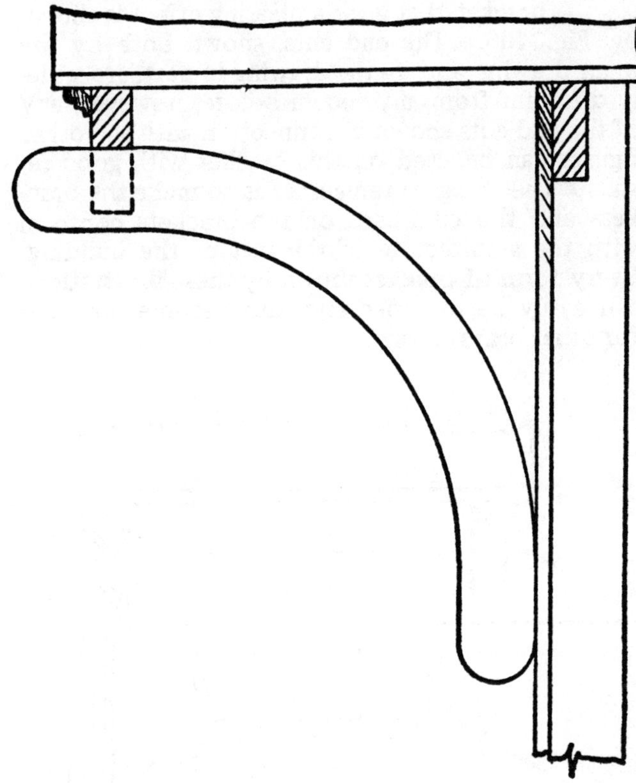

Fig. 403

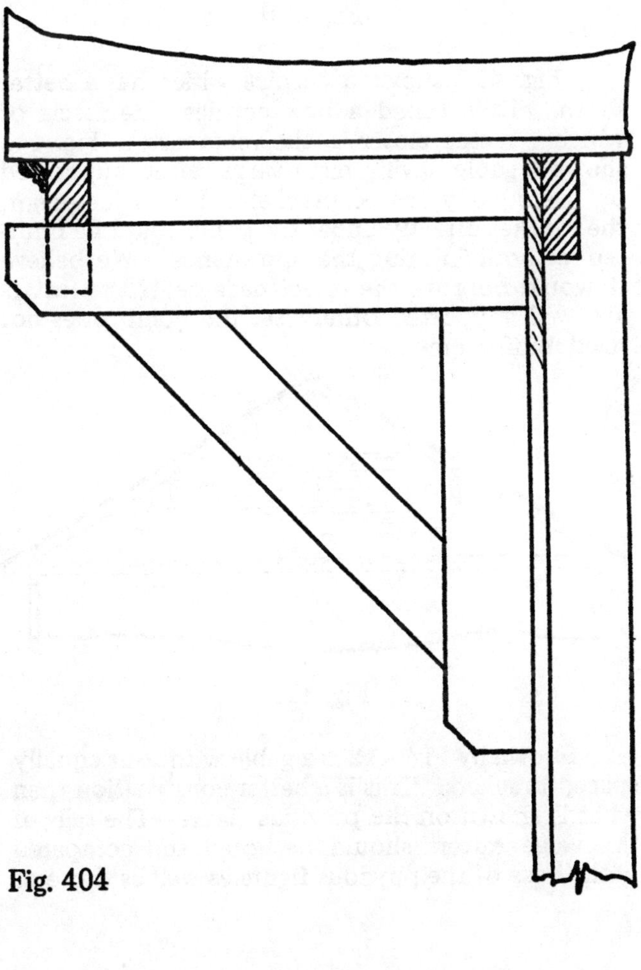

Fig. 404

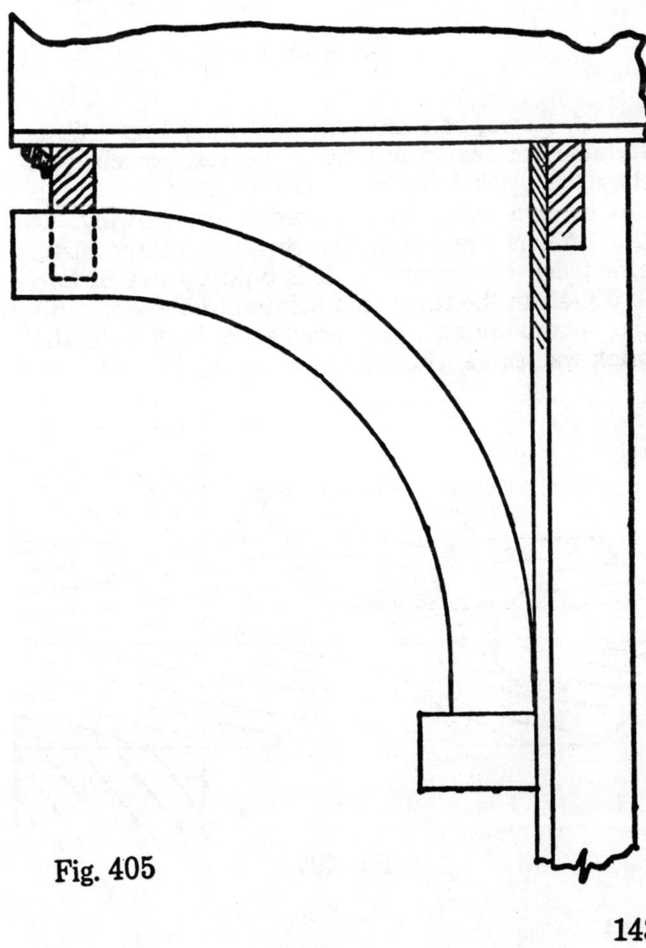

Fig. 405

A bracket that gives a pleasing effect is shown by Fig. 406. The end cuts, shown both by the main drawing and by the drawing in part, are widely different from any shown before, however, any of the end cuts shown in connection with the other figures can be used on this bracket with good results. The thing to remember is to make the brackets and the cuts used on the brackets conform with the architecture of the rest of the building. Every form of bracket shown by these illustrations can easily be modified and thus become the basis for other bracket forms.

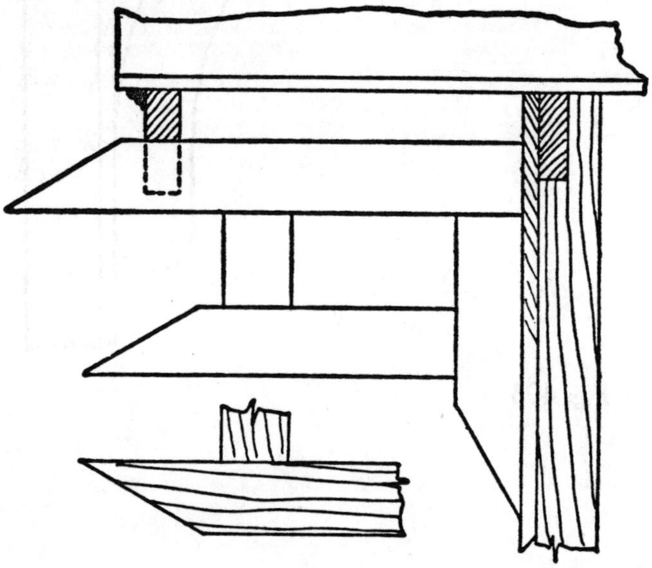

Fig. 406

A flower-box bracket, shown by Fig. 407, is suitable for use on apartment houses, especially if they are brick buildings. The bracket is built up out of two 4x6's, as indicated in the drawing. If the timbers are left in the rough, a rather rustic effect can be obtained. It is built up out of two 4x6's left in the rough and painted to blend in with the surroundings. The bracket is built into the brick wall and anchored.

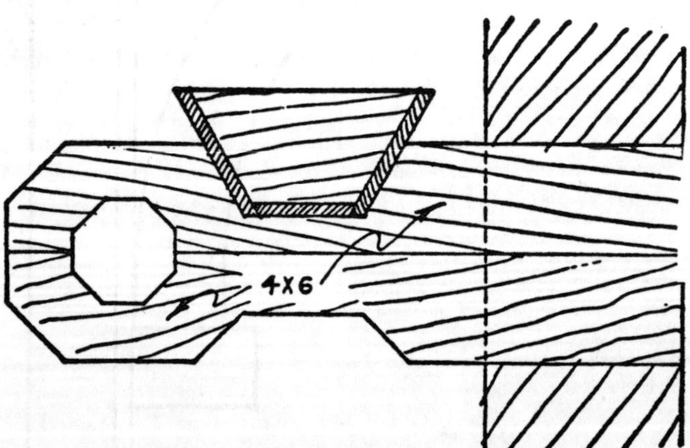

Fig. 407

Gable Finish

Fig. 409 shows a gable of the simplest construction. The sheathing of the roof is cut off to a line and the shingles are laid without a fascia or a verge rafter to give it support. Frequently one finds gables finished that way. This is substantial enough, but the appearance is bad. It is appropriate only on cheap work of a temporary nature.

Fig. 409

Fig. 410

Fig. 410 shows a cornice which has a better finish. It is called a box cornice, the fascia of which is a step closer to the verge rafter. Fig. 411 shows a gable having plain verge rafters supported by five equally spaced brackets. In our judgment, the bracket directly under the comb could be omitted without injuring the appearance. We believe it would improve the effect because, if the rafters are well supported otherwise, the comb does not need reinforcement.

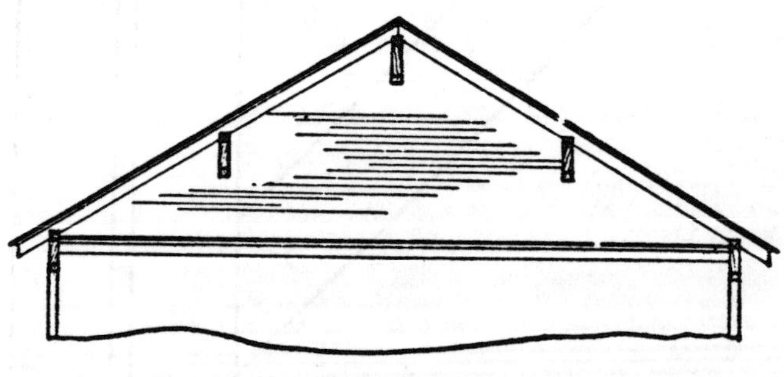

Fig. 411

Shown by Fig. 412 is a gable with four equally spaced brackets. This is a better construction than what is shown on the previous figure. The tails of the verge rafters should be noted and compared with those of the previous figure as well as those of

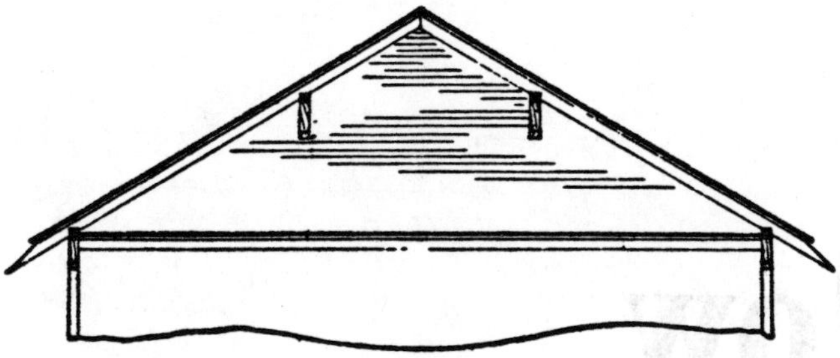

Fig. 412

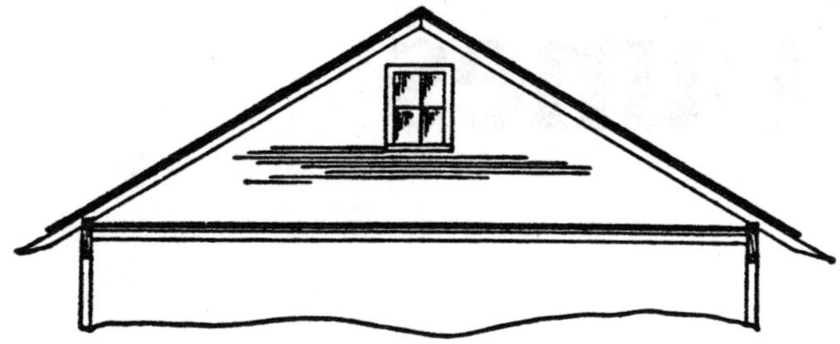

Fig. 413

Fig. 413 where the tail cut has been reversed. This figure shows only two brackets which is all right if the rafters are not too long.

Two details of verge rafter tails are shown by Fig. 414. Notice that the cuts are reversed, either of which gives a pleasing effect. It should be remembered that the bevels of these cuts can be either increased or decreased, whichever the circumstances require. More than that, the cuts can be improved, as for instance, what we show by Fig. 415. The hook, added here to the tail, puts just a little more action or life into the design.

The verge rafter or, as it is also called, bargeboard, often had variously designed curlycues -- a sort of architectural drapery, which was intended to beautify but in reality destroyed beauty. The verge board of today holds much of its decorative value in the simplicity of its design.

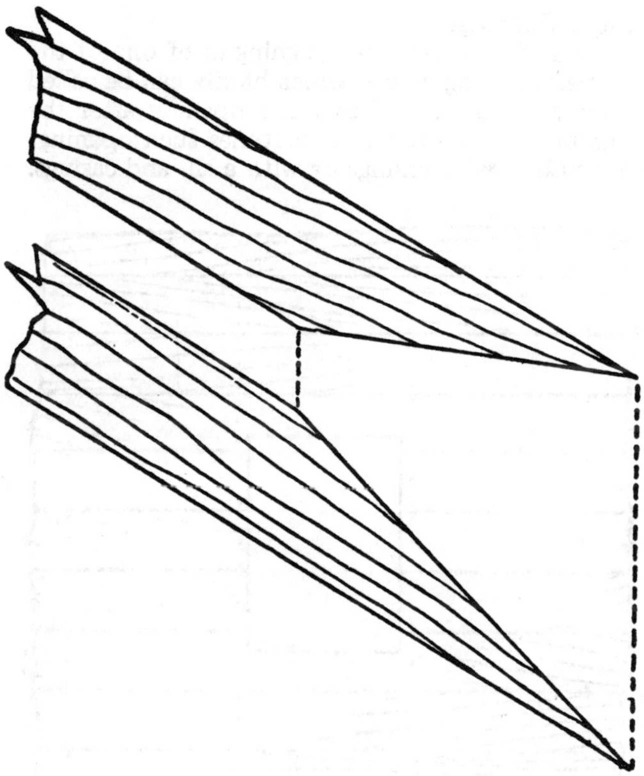

Fig. 414

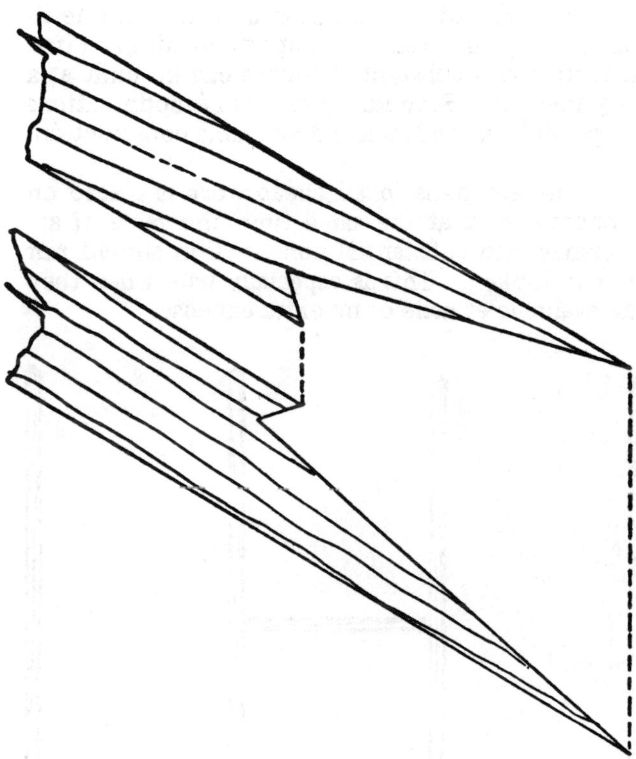

Fig. 415

29 Cheap Window and Door Frames

A study of the window and door frame illustrations will reveal two important things: First, attractive and substantial frames can be built at a very low cost. Second, a great many modifications are possible in window and door frame construction.

The emphasis in all cheap work is placed on economy, but, at the same time, the value of appearance and substantial construction should not be overlooked. This is especially true when they are available at little or no extra expense.

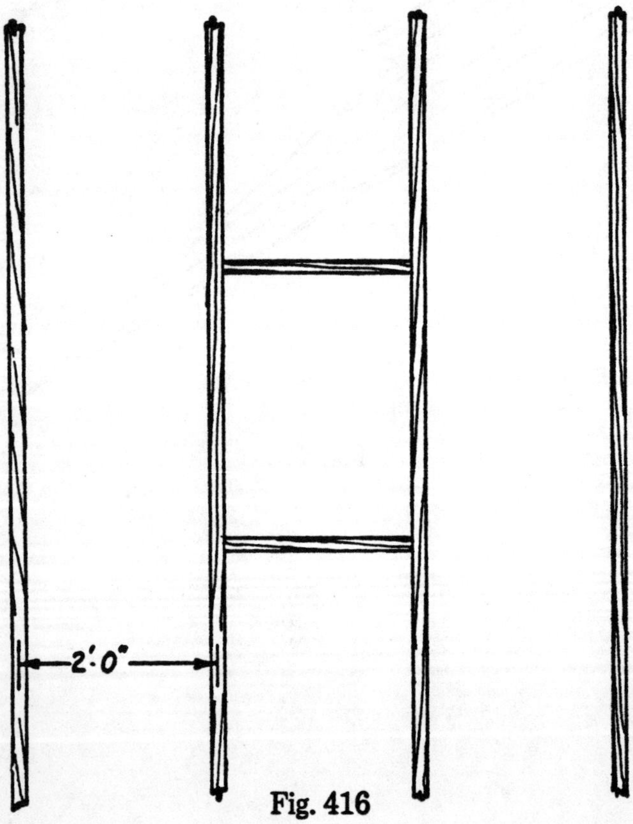

Fig. 416

Simple Openings

Fig. 416 shows the roughing-in of one of the cheapest openings in use, which hardly can be called a frame. Fig. 417 shows the opening after the boarding-up was done. Sometimes such openings are finished with casings or with a sill and casings.

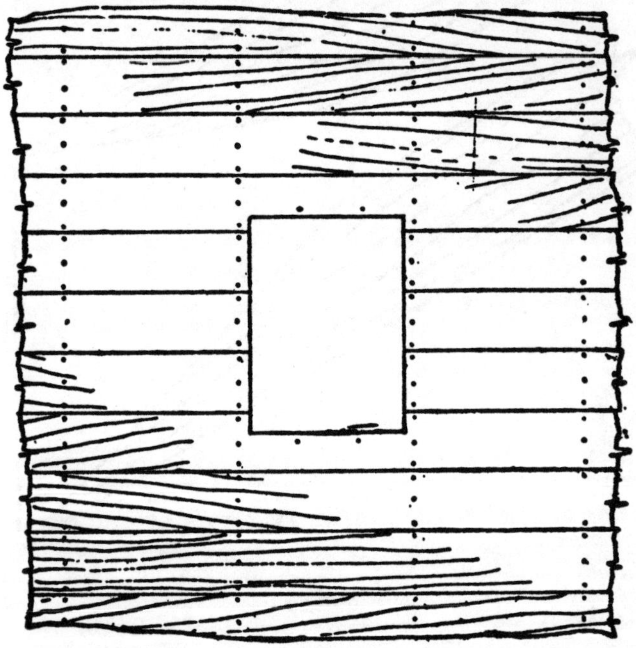

Fig. 417

Cheap Window Frames

Fig. 418 appears as if it were a roughed-in opening for a window, but in reality it is the beginning of a cheap window frame for a 30x24 double-hung window. The width of this frame is shown as being 34 inches, giving us 30 inches for glass and 2 inches for each of the two stiles. The height is shown as being 54 inches, providing for twice 24, or 48 inches for glass, 3 inches for bottom rail, 1 inch for parting rail and 2 inches for top rail, or in all 54 inches. This construction is suitable for cheap buildings such as cabins, bunkhouses or temporary dwellings. The roughing-in as it is shown in Fig. 418 constitutes the jambs, the yoke or head jamb, as we shall call it, and the sill. The sill is made of 2x6 material and the jambs of 2x4's. One

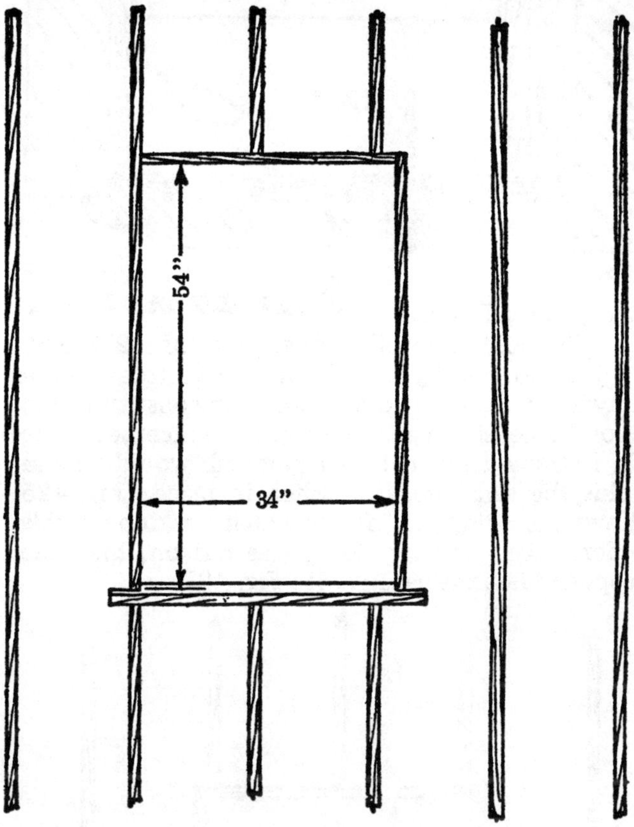

Fig. 418

end of such a sill is shown in Fig. 419, where the dotted line shows the width of a 2x6 sill, while the continuous line shows a sill made of wider material. To the upper left, in this figure, we indicate by dotted lines the relationship of the sill to the 7/8-inch inside finish, while at the bottom, somewhat shaded, is shown the upper edge of novelty siding. The 2x4 roughing-in jamb is shown in its right relationship with the sill. The completed frame is shown by Fig. 420.

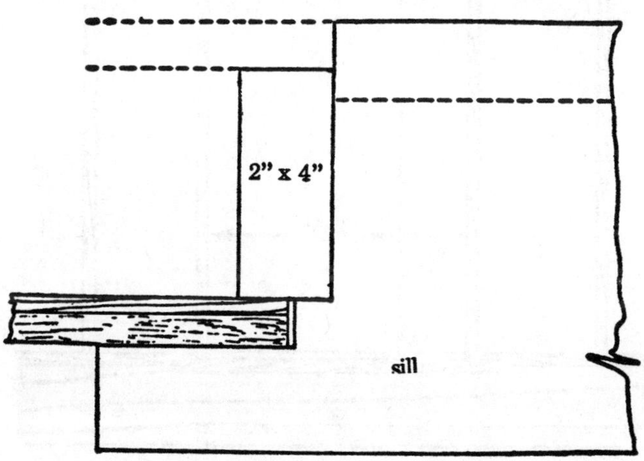

Fig. 419

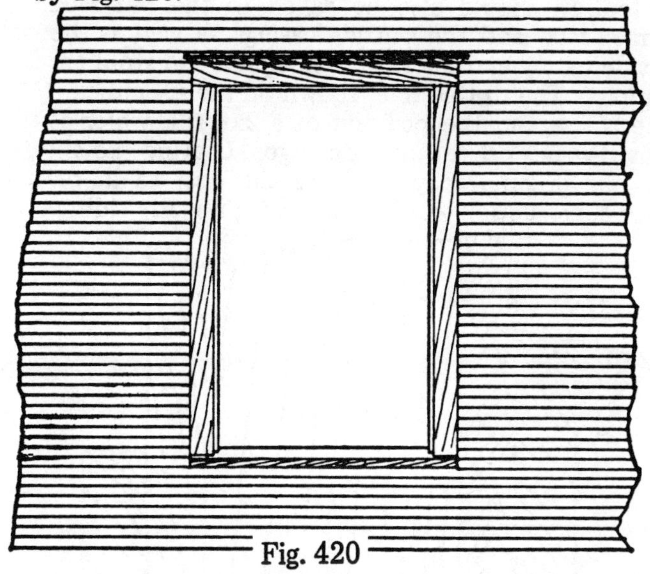

Fig. 420

In Fig. 421 is shown, in part, two elevation details of the corner constructions of the frame.

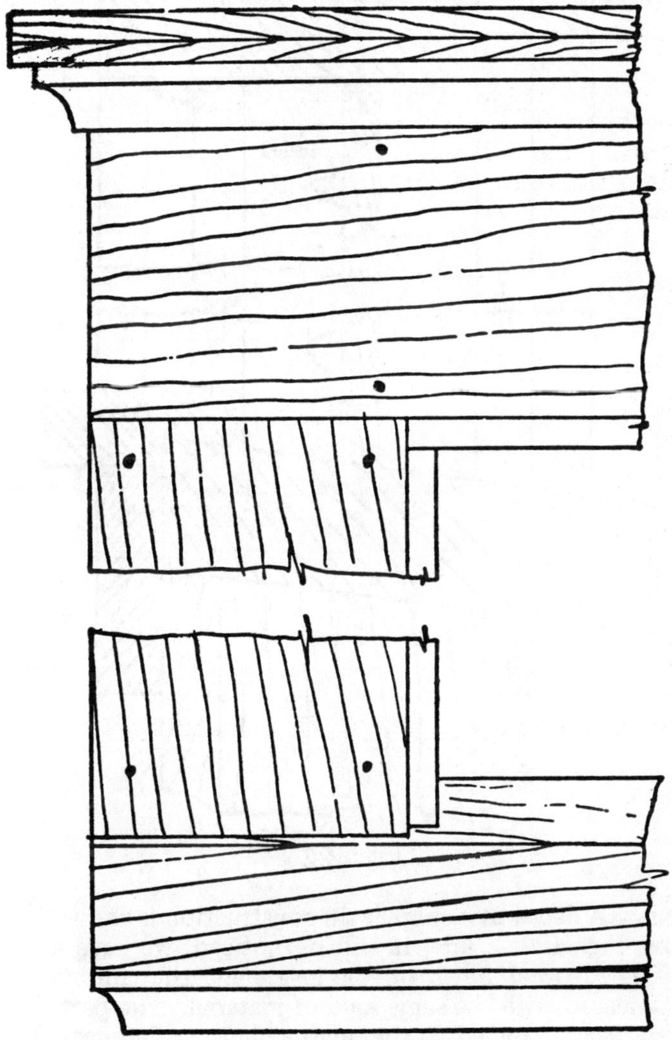

Fig. 421

Fig. 422 shows cross sections of the head and of the sill as well as the inside finish. It should be noted that the parting bead is nailed to the 2x4 jamb. Attention should also be called to the blind or screen-stop and the outside casings as well as the inside casing and the stop. The apron and stool are simple. The dotted lines shown on the sill indicate the shape and the position of a 2x6 sill, which is not beveled on the outside edge. Leaving the sill in this original shape reduces the cost of labor. This construction works in.well with novelty siding.

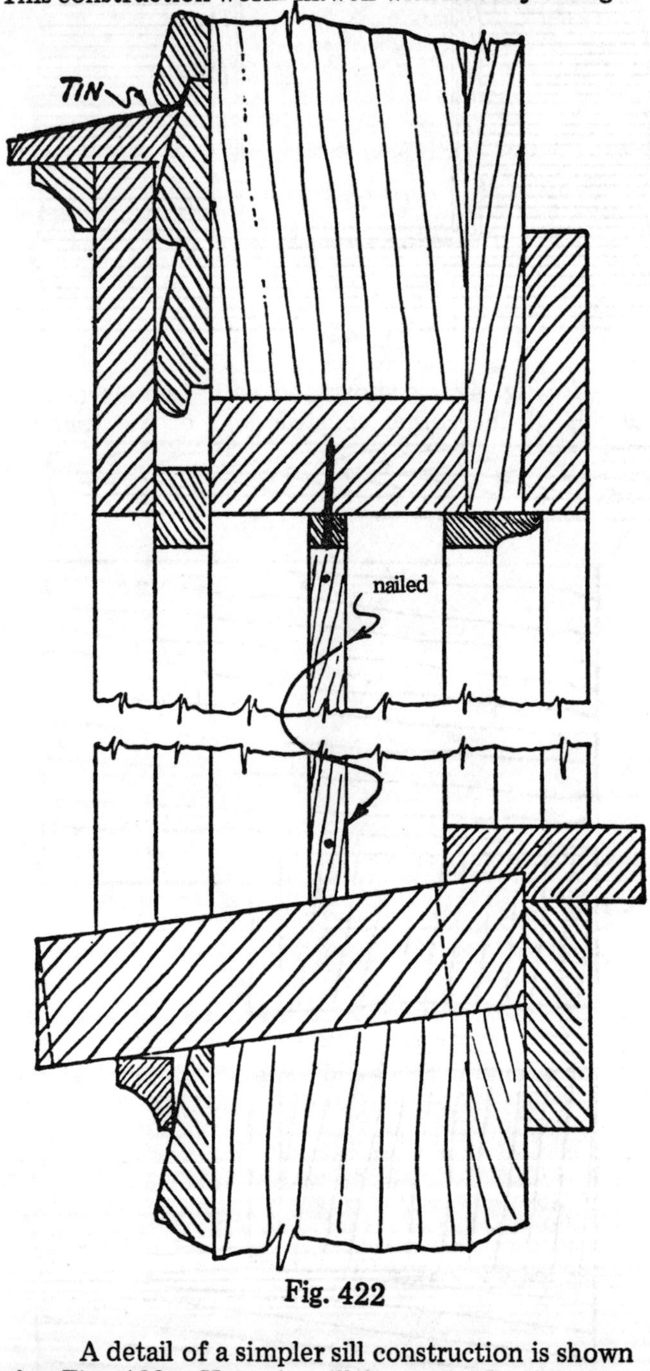

Fig. 422

A detail of a simpler sill construction is shown by Fig. 423. Here, it will be noticed, we have a pressed-wood finish on the inside and the jamb is veneered with the same kind of material. The parting bead is nailed to the jamb as in the other case, but the frame does not have a screen-stop.

Fig. 423

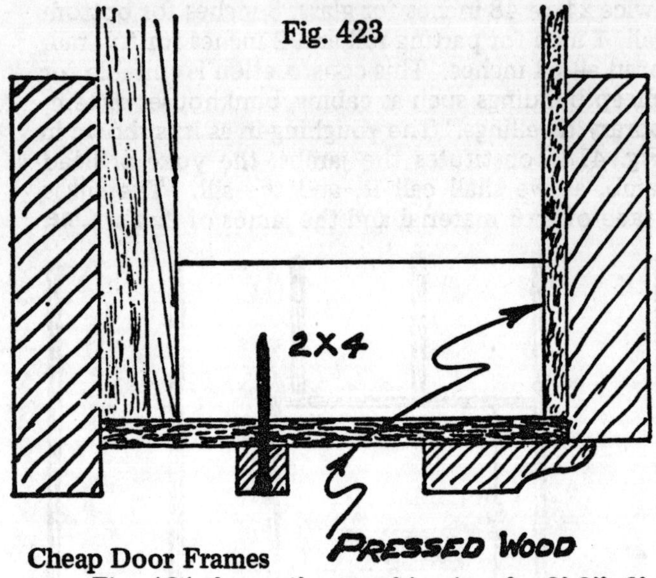

Cheap Door Frames

Fig. 424 shows the roughing-in of a 2'-8"x6'-8" door, which, if you will notice the figures on the drawing, means that the roughing-in constitutes the door frame, so far as the jambs are concerned. The sill is framed much as a regular sill would be and holds the same position when in place. Fig. 425 shows the completed frame which was done in this order: The novelty siding was put on, then the stop, and last the casing. See Fig. 427.

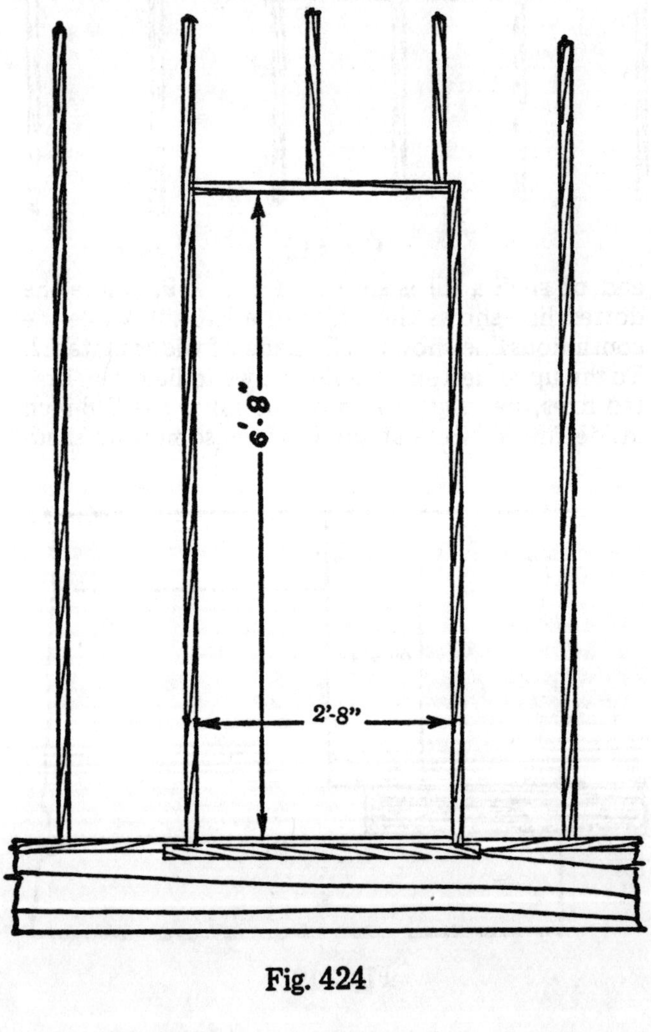

Fig. 424

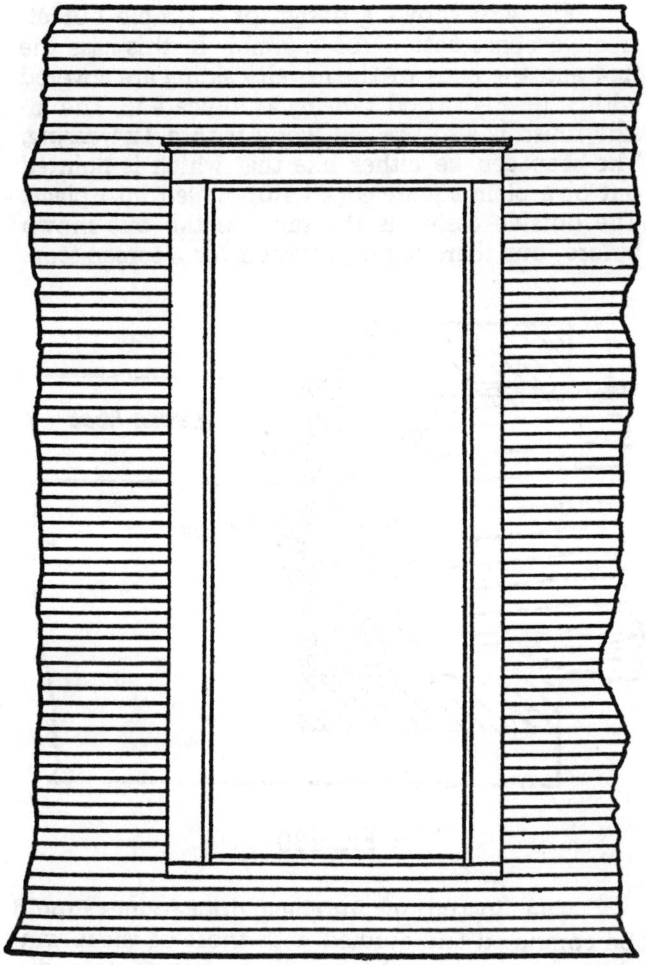

Fig. 425

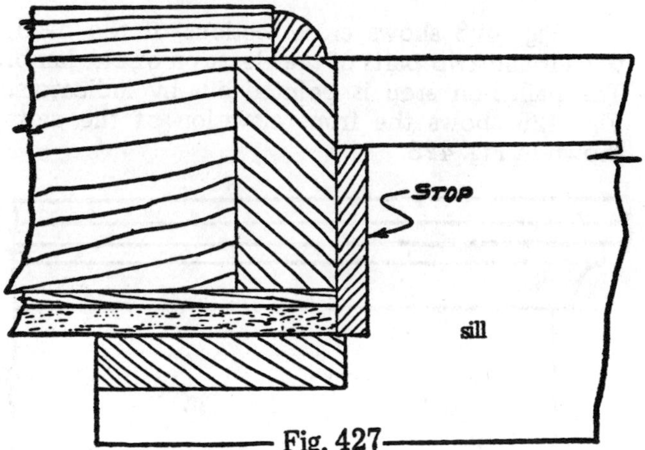

Fig. 427

Fig. 426 is a plan view of one end of a sill showing its relationship to the 2x4 studding, which at the same time answers for the jamb. (Compare this with Fig. 424.) The same part of the sill is shown, Fig. 427, as it would be when completed. Here the stop is pointed out with an indicator. The 2x4 jamb, the inside finish, the novelty siding and the outside casing are shown shaded.

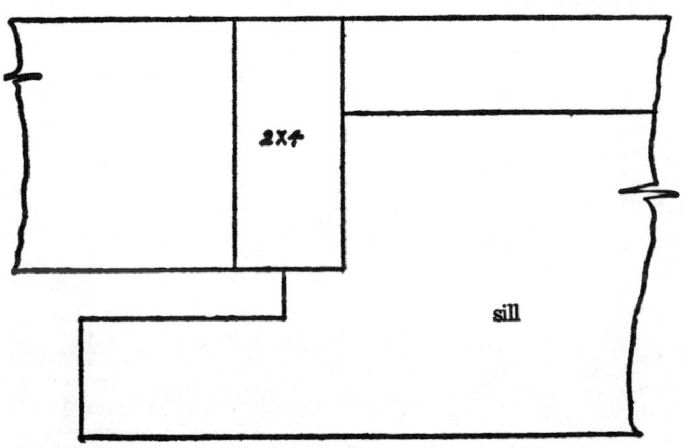

Fig. 426

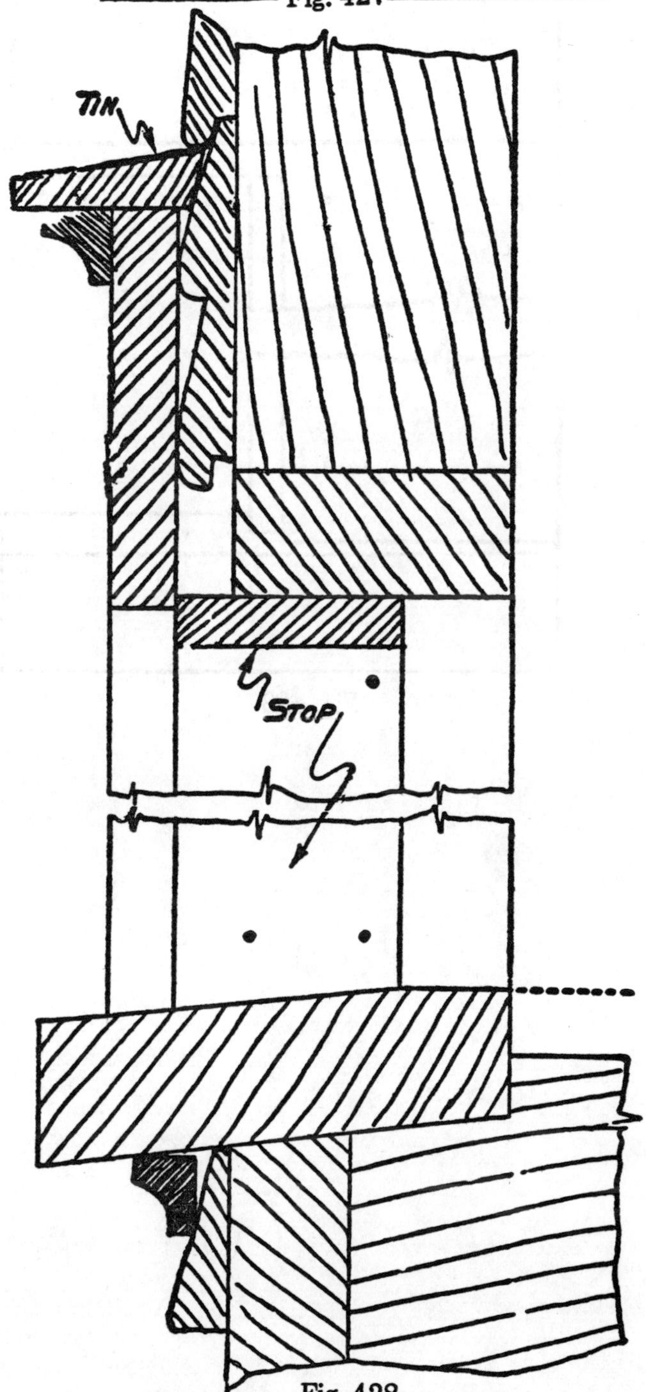

Fig. 428

Fig. 428 shows cross sections of the head, the sill and two parts of the elevation of one jamb. The nailed-on stop is pointed out by indicators. Fig. 429 shows the front elevation of the parts shown in Fig. 428.

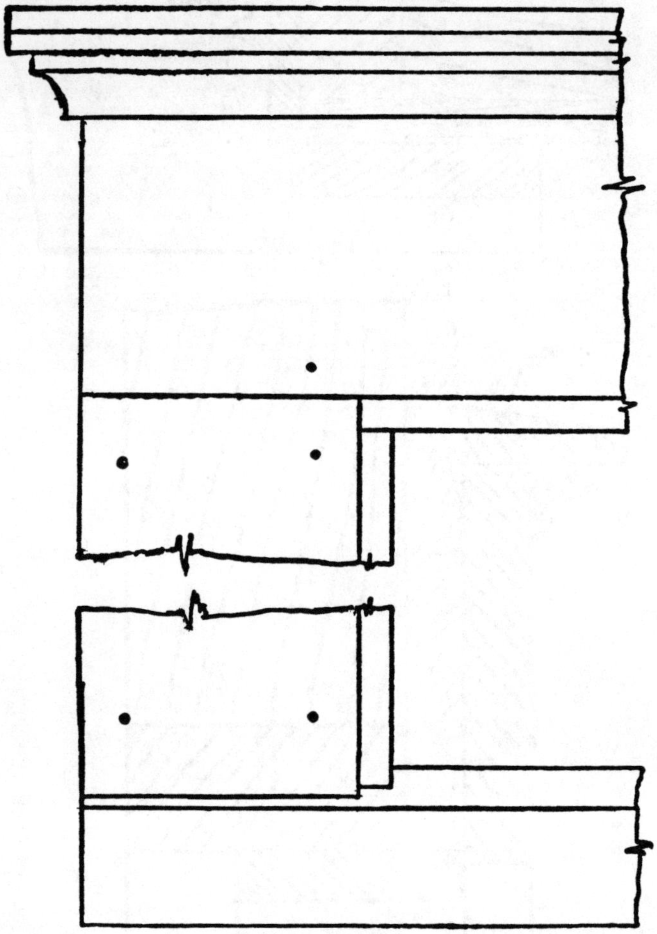

Fig. 429

Fig. 430 shows a design of a simple but attractive construction for a jamb. In this case the 2x4 and the ends of the novelty siding are covered with a pressed-wood veneer, as indicated. The inside finish is also pressed wood with a 1x4 casing. The stop can be either like that which is pointed out or a plain square-edged stop nailed into place. The outside casing is the same as the one shown before, but there is no provision for a screen-stop.

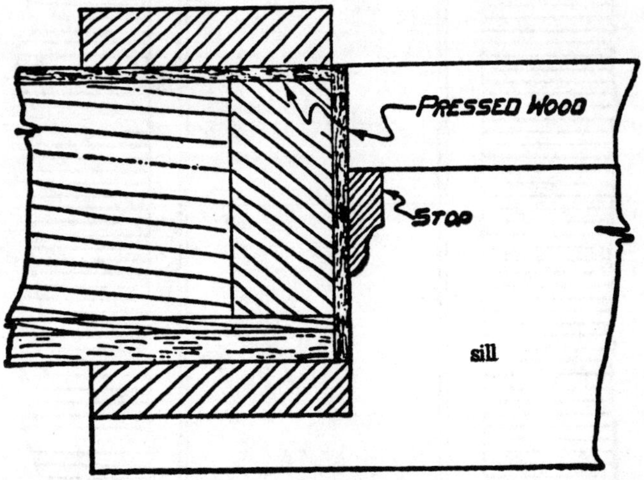

Fig. 430

The illustrations for cheap door frames should be compared and studied with those of cheap window frames. In each of these frames two distinct designs are shown, and in their practical application, the design for the window should harmonize with the one for the door. For instance, if the veneered jamb is used for the window, it should also be used for the door, and vice versa.

30

Door and Window Frames

Doors

In its relationship to a building, the door is one of the most important, if not the most important, feature of the whole structure. Its size should be governed by the purpose for which it is intended. Where the traffic is heavy, as in large public buildings, the width of the entrance door should be sufficient to prevent traffic jams. Such doors should not be less than 6 feet wide. A rule adopted for classic structures proportions the width to the height of doors for main entrances, as 1 to 2. That is, the door should be twice as high as it is wide. In its application to public buildings this rule should not be taken as a hard-and-fast one. For instance, there are many small public buildings where a 3-foot-wide door is sufficient, but who would want such a door only 6 feet high. On the other hand, there are public buildings that need such wide entrances that the width necessarily must exceed the height, or at least, be on a par with it. The same authority gives this rule: For single doors of residences or office rooms, the proportion should be 1 to 2½; or, the height should be 2½ times the width. Applying this to stock sizes, as 2'-8" by 6'-8" and 3'-0" by 7'-0" doors, we again see that such rules are not strictly reliable -- perhaps such rules should be taken as basic. The best rule for governing the height of doors, especially for residences and offices, is to make all the doors the same in height, regardless of the width. But here again, one must not stick too firmly to the rule, for there are circumstances that often come up which render all rules useless. The best rule that we have found, whether they are written rules or pocket rules, is to use them for what they were intended, measuring sticks, rather than guides.

Rough Door Openings

Fig. 431, to begin with, shows a rough door opening for a 3'-0" by 7'-0" door, giving in figures correct dimensions. These figures are based on jambs that are 1¼ inches thick. Where thinner material is used, the opening should be reduced in size accordingly.

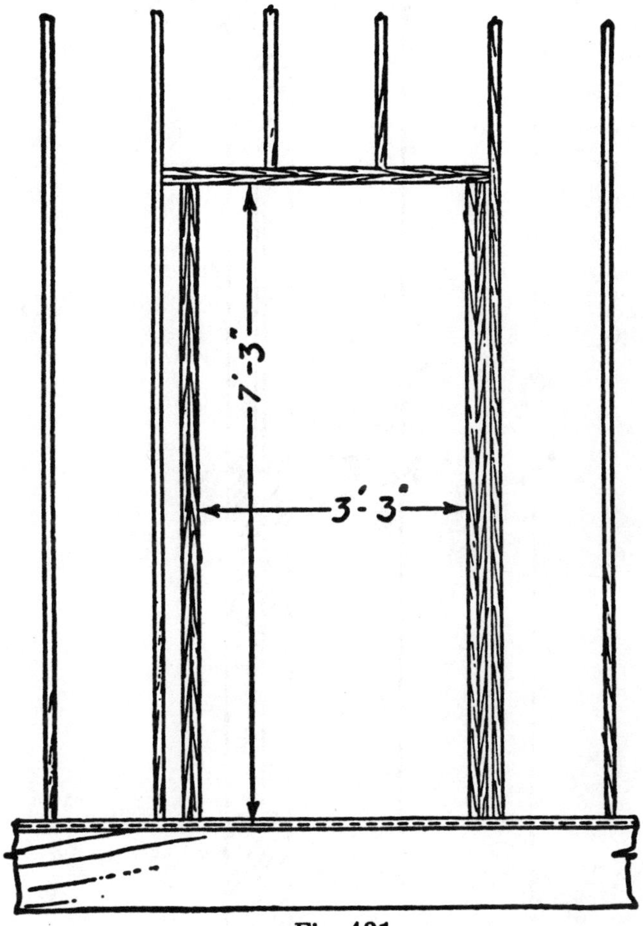

Fig. 431

Door Frame

Figs. 432 and 433 are not drawn to scale. In this way we are able to bring out the points we want to make more clearly. The first of these illustrations, Fig. 432, shows why 3 inches are added to the width of the door in order to get the necessary width of the rough opening. Here we have ¼ inch on either side of the door for plumbing clearance, 1¼ inches on either side for jambs, and 3

Fig. 432

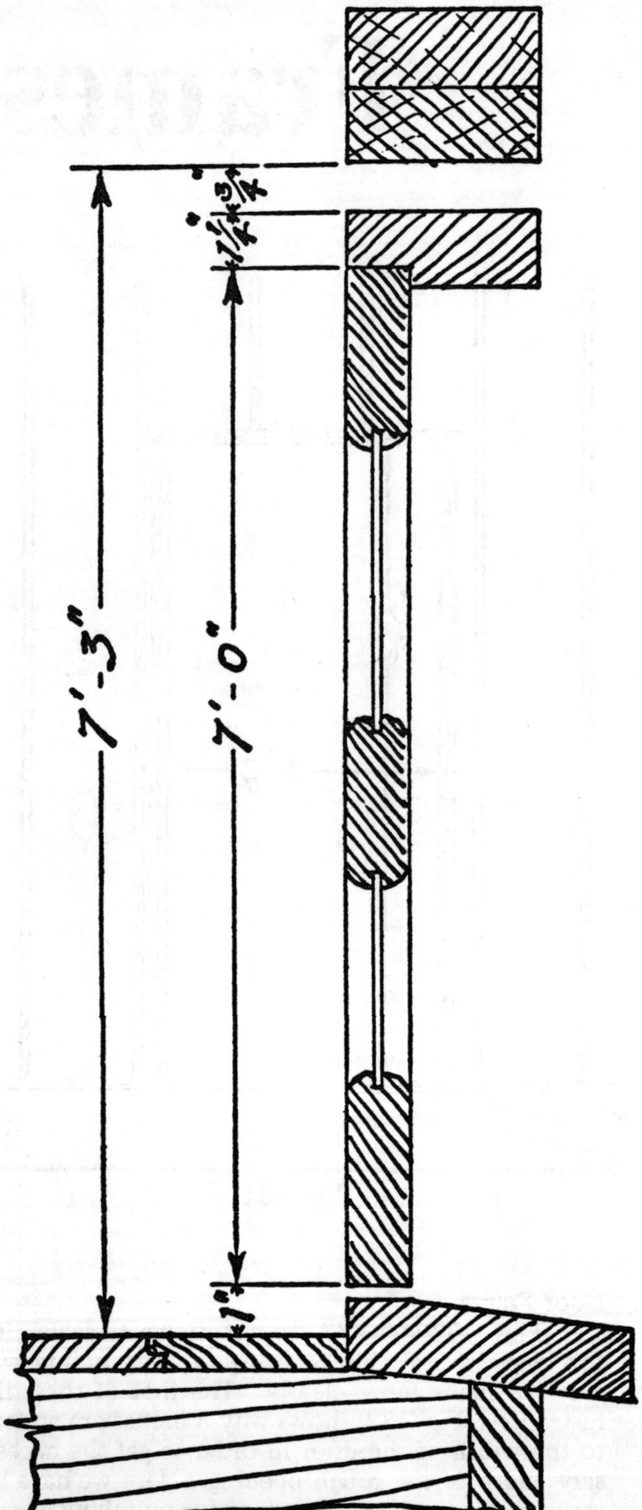

Fig. 433

feet for door -- a total of 3 feet, 3 inches. Fig. 433 shows why 3 inches are added to the height of a door to get the correct height of the rough opening. Starting at the top and reading down we have ¾ inch for clearance, 1¼ inches for head jamb, as it is commonly called (the right name is yoke), and 7 feet for door -- to which must be added 1 inch for floor and threshold clearance -- a total of 7 feet 3 inches, as shown by figures. These unscaled drawings should be studied in connection with the scaled drawing of the completed door frame shown by Fig. 434.

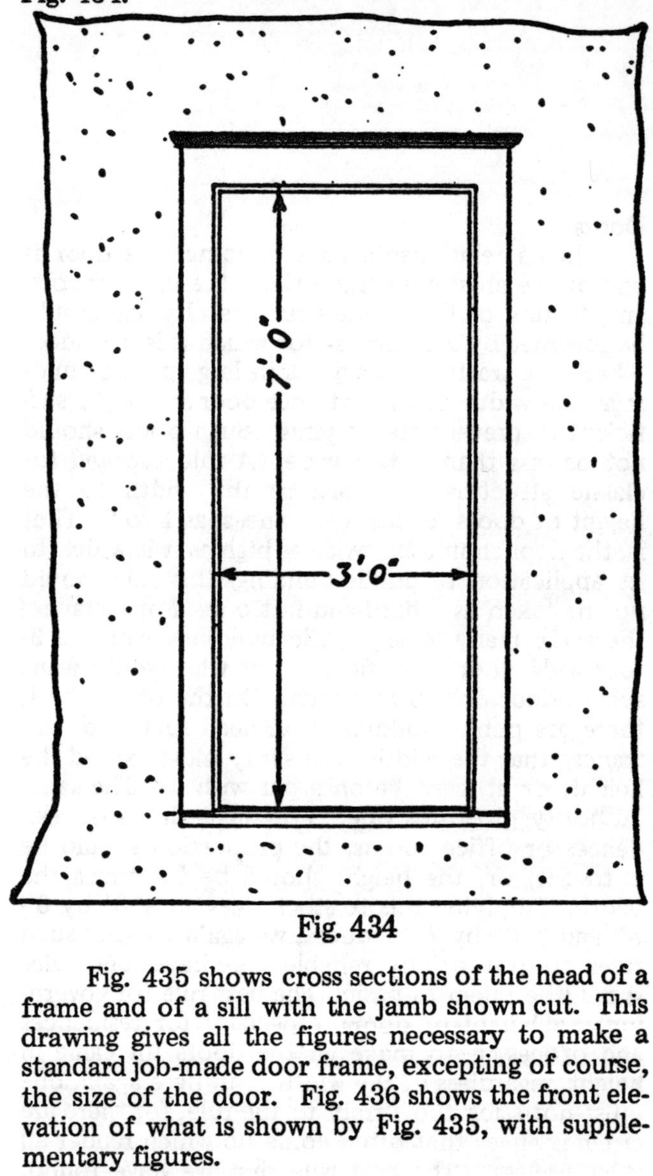

Fig. 434

Fig. 435 shows cross sections of the head of a frame and of a sill with the jamb shown cut. This drawing gives all the figures necessary to make a standard job-made door frame, excepting of course, the size of the door. Fig. 436 shows the front elevation of what is shown by Fig. 435, with supplementary figures.

Fig. 437 shows a head and a sill joint ready to be made, and by the projections, other details are shown. The inside edge of the top and the bottom parts of a jamb are shown at *A*; *B* shows the face of these parts. *C* gives the inside edge of a part of the head, while *D* shows the face of it, and *E* shows the inside edge of the sill. *F* is a detail of the jamb, showing the rabbet. These parts put together are shown by Fig. 438 where *A*, *E* and *C* represent the same parts they do in Fig. 437.

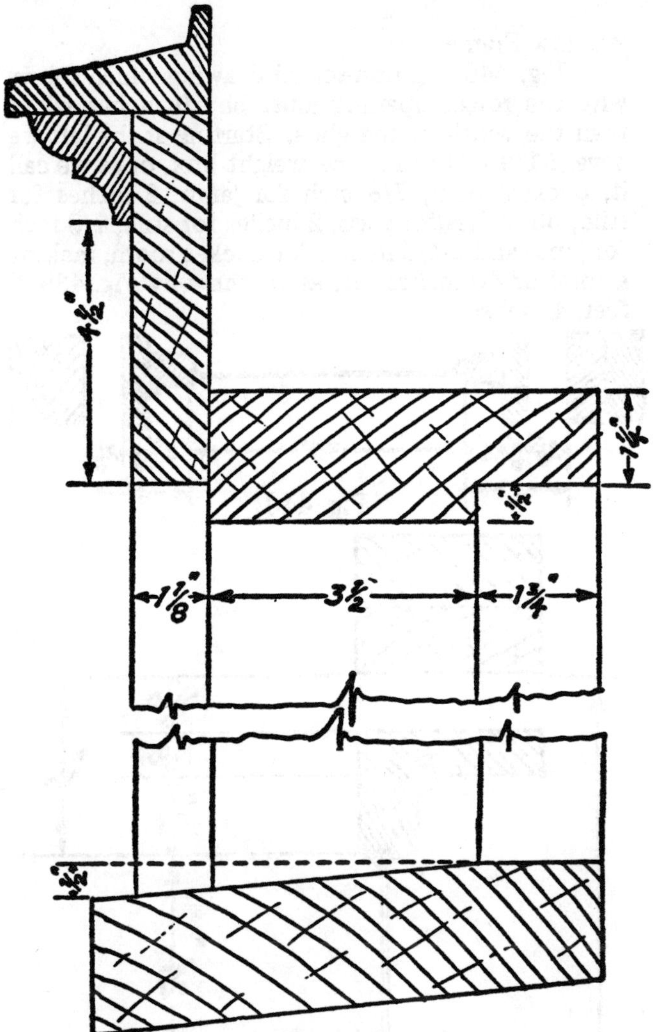

Fig. 435

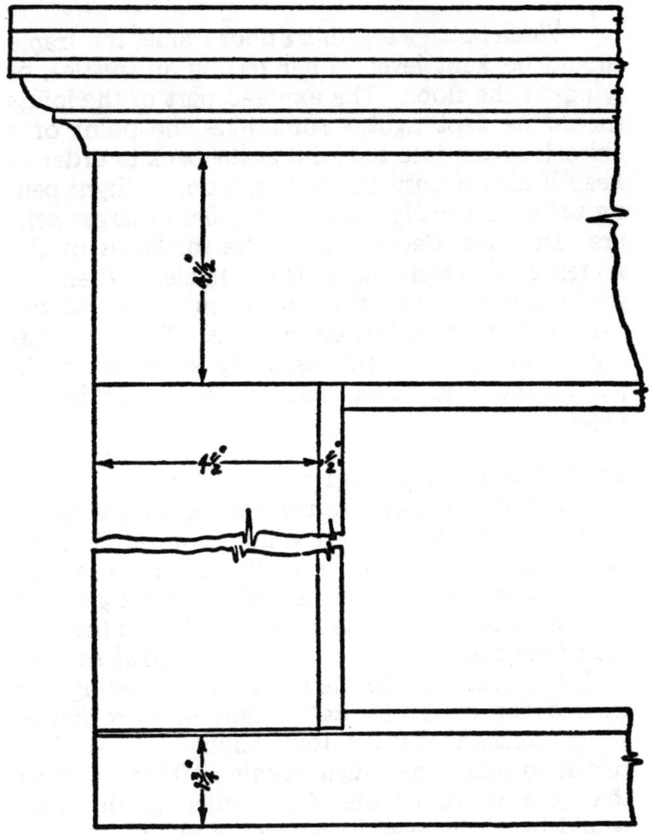
Fig. 436

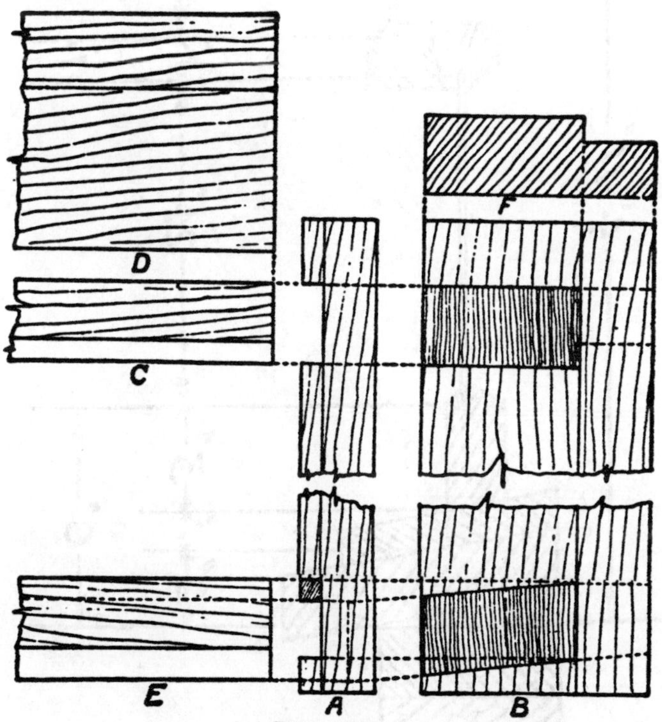

Fig. 437

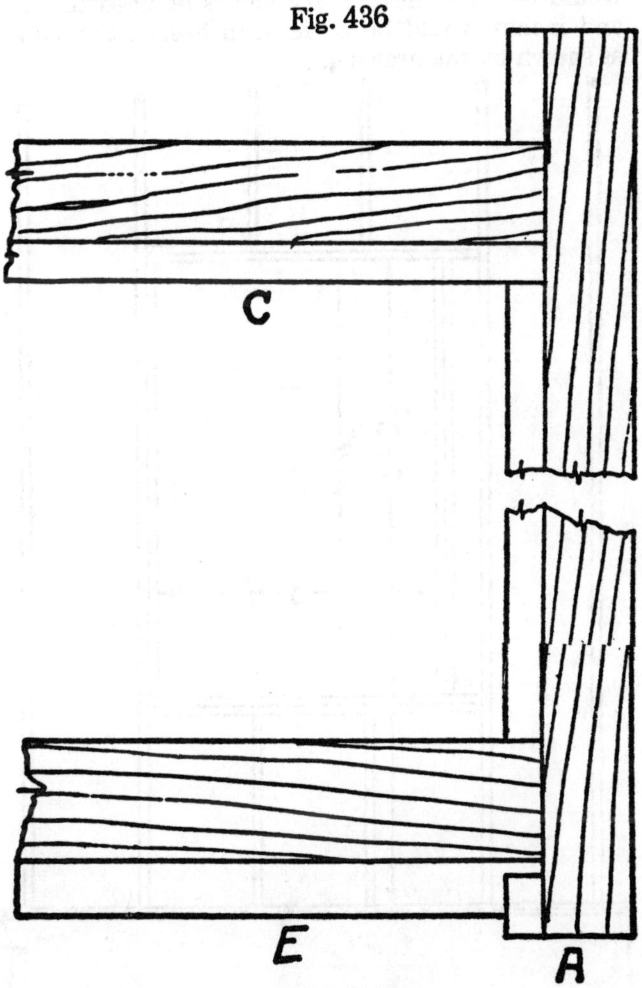

Fig. 438

When putting together a door frame, the frame should be kept level, either resting on trestles, or lying on the floor. The exposed part of the joints should be kept tight – sometimes the point of a chisel is driven into the joint at the back in order to keep it closed until the nailing is done. Eight penny nails are usually used but smaller or larger nails are also used, depending on the thickness of the material of which the frame is made. When the jambs are together, the side casings are put on, nailing them with 10d casing nails. The head casing is put on last. (Occasionally a carpenter will put on the head casing first and then the side casings).

Rough Window Openings

Fig. 439 shows a rough opening of a window ready for the boxing. It will be noticed that the height of the opening from the top of the rough floor is exactly the same as that which we gave for the rough opening. The reason for this is that the top of the doors and the top of the windows should, so far as practical, be the same. The opening was framed for a 30" by 24" 2-light window frame, which means that it is a double-hung window. The rule is to make the rough opening 10 inches wider than the width of the glass, which in this case would be 40 inches. Forty inches reduced to feet and inches would be 3 feet 4 inches, or the same as shown by the drawing.

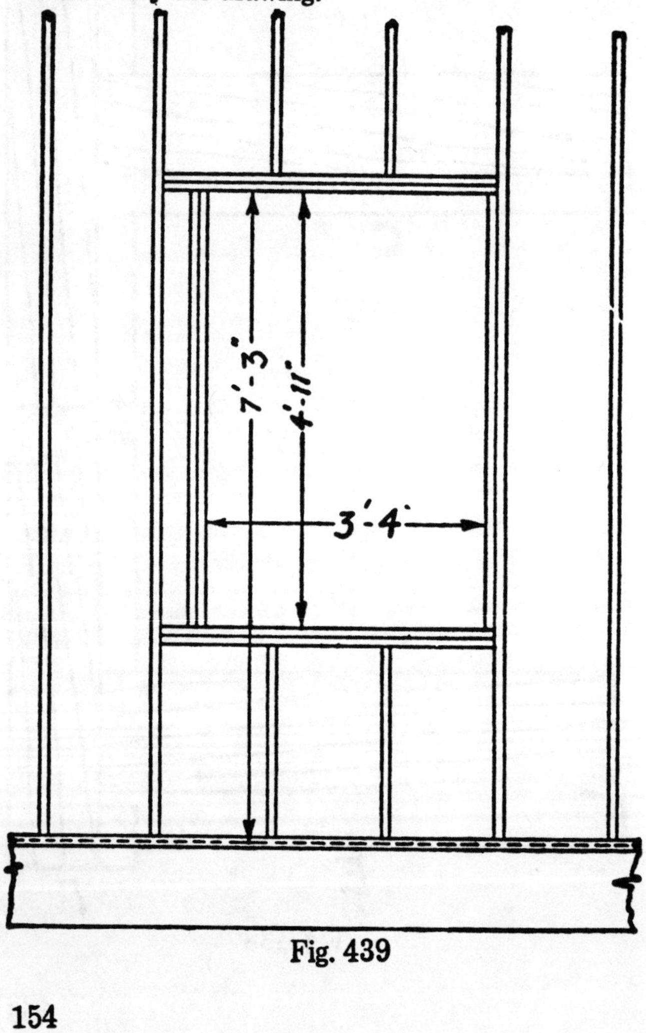

Fig. 439

Window Frame

Fig. 440 is a non-scaled drawing which shows why the rough opening must be 10 inches wider than the width of the glass. Starting at the left we have 2-1/8 inches for the weight box, or as we call it, pocket room, 7/8 inch for jamb, 2 inches for stile, 30 inches for glass, 2 inches for stile, 7/8 inch for jamb and 2-1/8 inches for pocket room, making a total of 40 inches, or, as we show by Fig. 439, 3 feet, 4 inches.

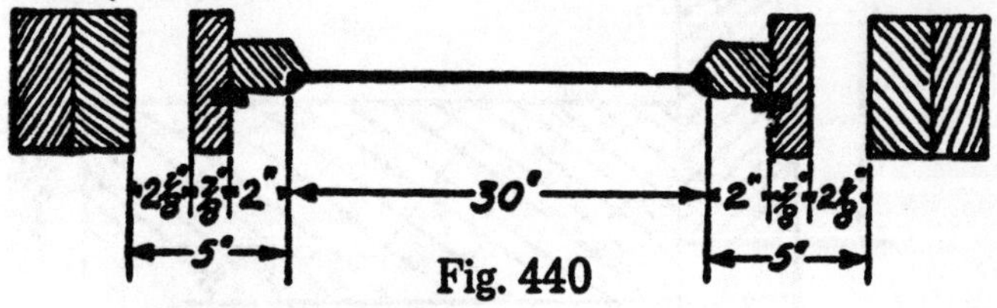

Fig. 440

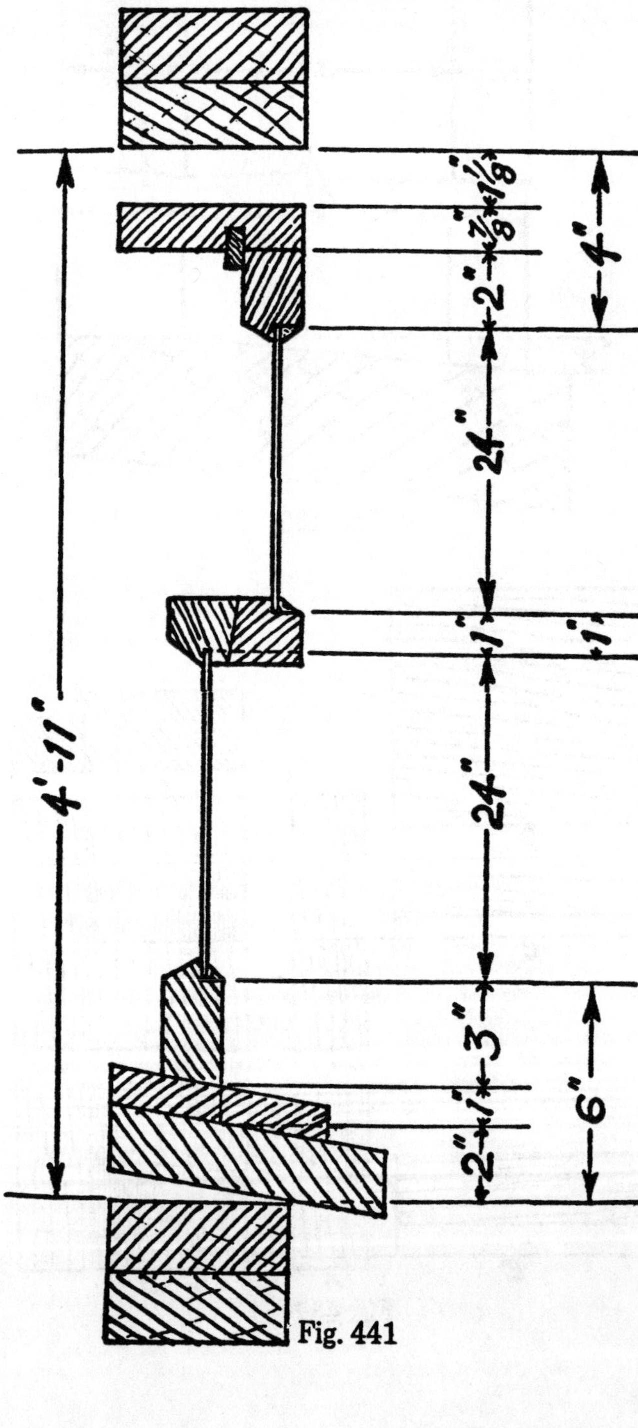

Fig. 441

The height of the rough opening for a frame with a sill and a subsill must be 11 inches more than the height of the glass for both the upper and the lower sash. This is illustrated by Fig. 441, where, reading from the bottom up, we have 2 inches for sill, 1 inch for subsill, 3 inches for bottom rail, 24 inches for glass, 2 inches for top rail, 7/8 inch for head and 1-1/8 inches for clearance; this makes a total of 59 inches, or as we show on the drawing, 4 feet, 11 inches. For frames that do not have subsills the rule is to add 10 inches to the height of the glass.

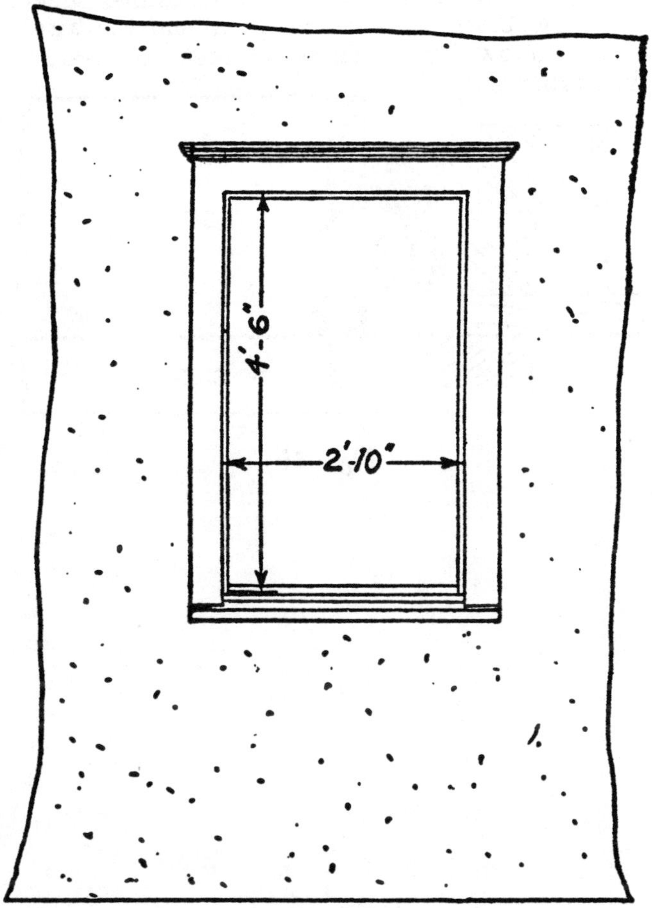

Fig. 442

In Fig. 442 is shown a front view of a window frame set into an opening such as shown by Fig. 439. The measurements of this frame, where the sash are, is shown on the drawing.

Details of the head and sill, with the jamb shown cut, are shown by Fig. 443. Reading the figures given on the drawing from left to right, we have 1-1/8 inches for casing, 7/8 inch for screen stop, 1-7/16 inches for pulley stile, ½ inch for parting bead and 2-7/16 inches for the inside pulley stile and stop. (The stop is not shown.) To the right we show the allowance for lath and plastering or wallboard as ¾ inch. A front view of what is shown by this figure is shown by Fig. 444. The width of the casing, the screen stop and the thickness of the sill are shown in figures --- When wallboard or some other manufactured material is used, the thickness of such material must be used in determining the width of the jamb.

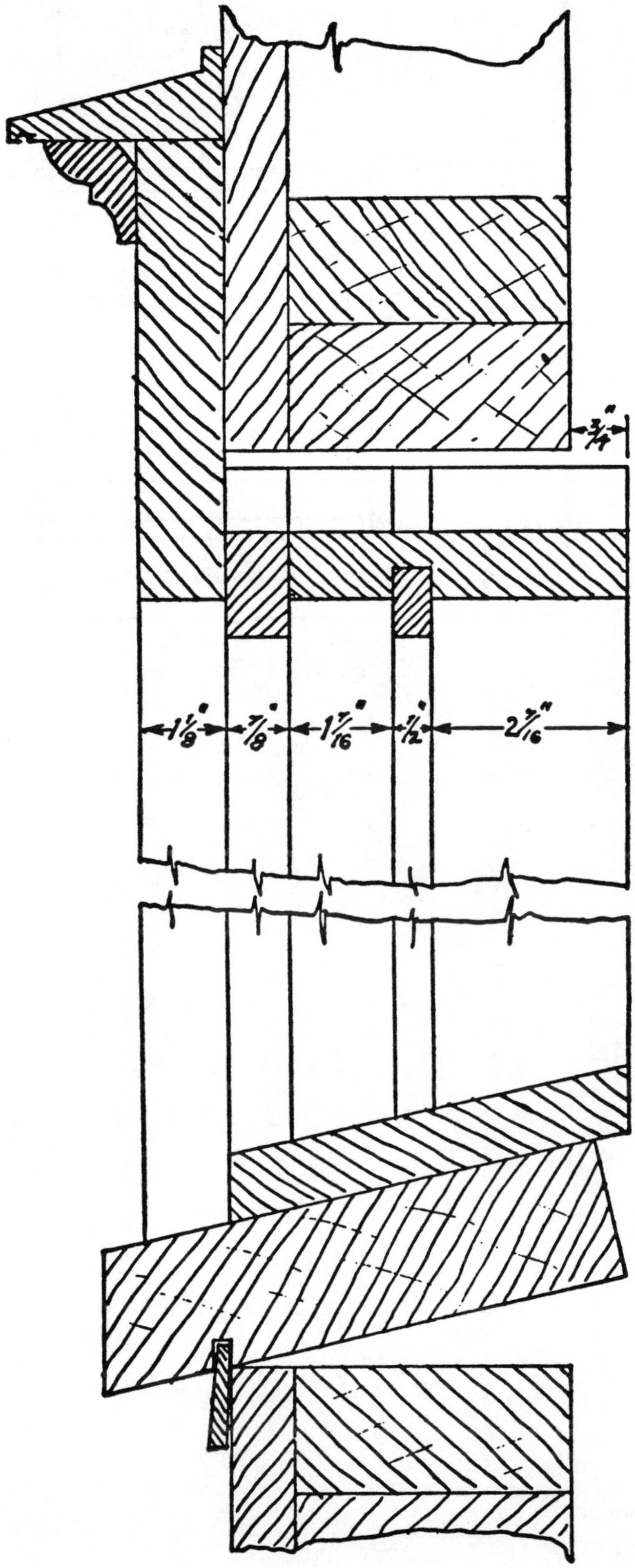

Fig. 443

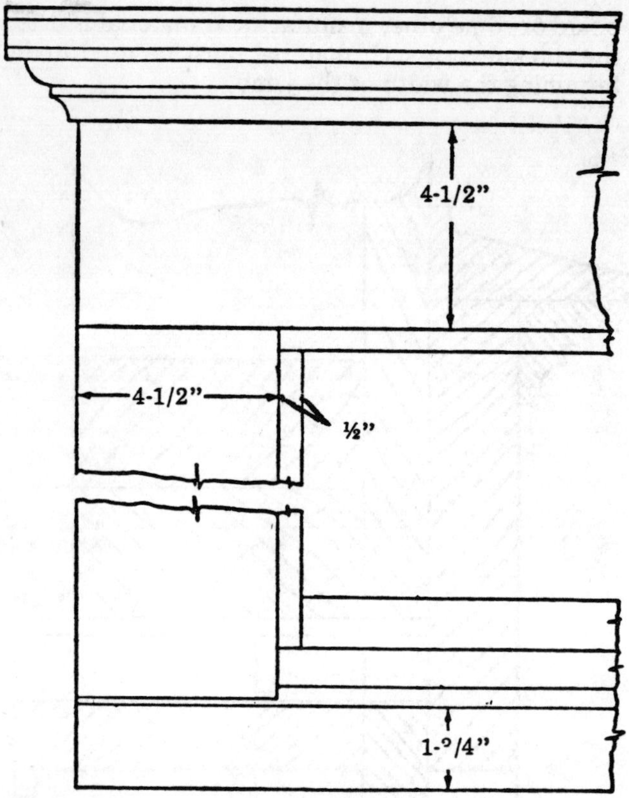

Fig. 444

A detail of the jamb and pocket room is shown by Fig. 445. The shaded discs indicate the window weights. The pocket room is given as 2-1/8 inches. The casing, screen stop, jamb and parting bead are shown shaded. The doubled trimmers and the boxing are also shown shaded to the left. The sill and subsill are named on the drawing.

We are dealing with what is called a skeleton window frame, which is the kind of frame that is usually made on the job, with more or less modifications. Besides the skeleton frame, there are box frame (used in brick, stone and other similar buildings), the casement window frame and the basement window frame. These we intend to treat in their order.

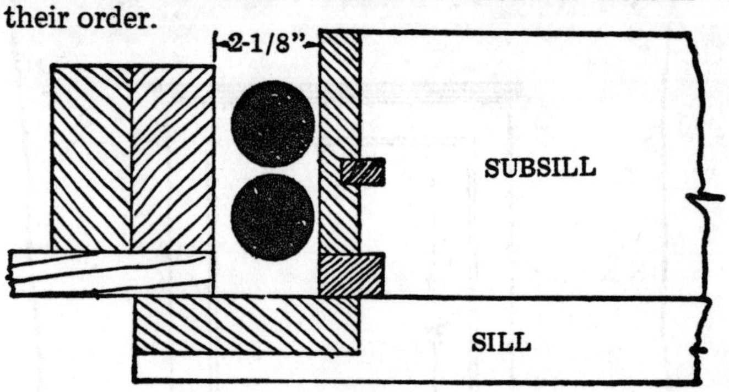

Fig. 445

31

How To Make Window Frames

Making Window Frames

Let's take a window frame of the size used in the last chapter: 30" by 24" 2-light frame. We will start with the main sill, as shown in Fig. 446. At *A* is timber 46 inches long, this is a little longer than the sill will be when completed; the extra length is needed for cutting. To the left is a cross section of the timber. The dotted lines indicate how much of the piece must be taken off in order to obtain the proper bevel (also shown by the cross section to the left). The dotted lines marked *B* on the drawing show how the timber must be marked for cutting the ends after the beveling is done. All the figures necessary to do this marking are shown. The sill completed is shown at *C* in plan. The section to the left shows the slope to which the sill is set. The corners marked *X* and *X* are important. When these are cut away, as they often are, the frame will leak at those points. The dotted line at either end shows where the lugs are cut off. The horizontal dotted line indicates what the sill would be like if a narrower piece of material were used.

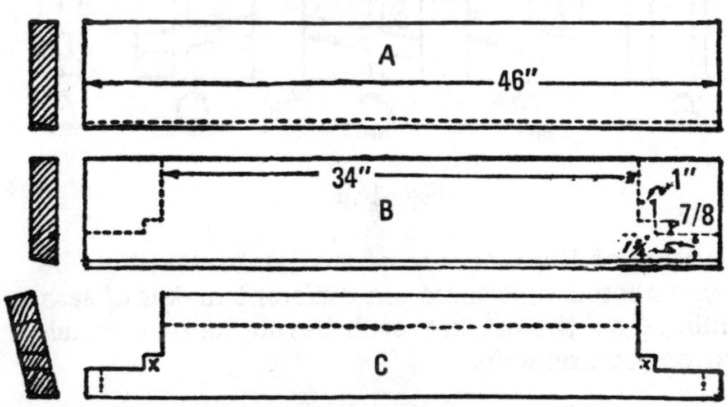

Fig. 446

The subsill in the same order of development as the main sill is shown by Fig. 447. In this case a 1x6, 36½ inches long is shown at *A*. The dotted lines indicate how it must be worked over in order to get the proper bevels and width. At *B* the dotted lines show how the ends are cut, leaving a small lug at either end. This sill is cut ¾ of an inch longer than the main sill, as the figures will show, to allow for the housing of the jambs. (Some carpenters add an extra 1/8-inch in order to reduce the dressing of the sash when installed.) *C* shows the subsill completed, setting at the slope shown by the cross section to the left. The lugs of the subsill answer the same purpose that the corners marked *X* on the main sill do. These, besides preventing leaks, aid in putting the frame together.

While we are dealing with just one frame here, in practice there are usually several frames to be made of the same size. In such cases, all the sills and subsill material is cut to size and then assembled.

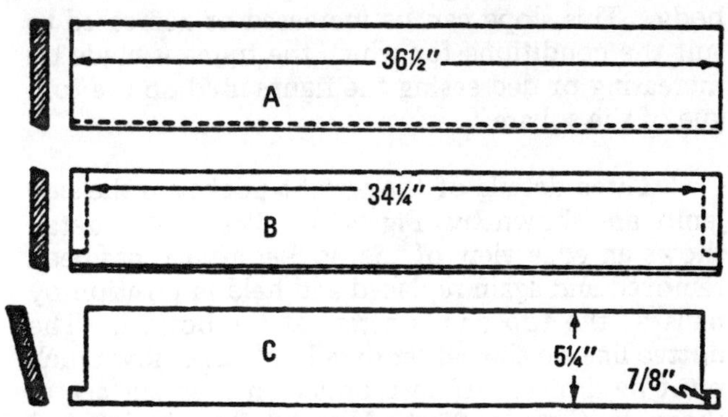

Fig. 447

A, Fig. 448, shows the piece for the head jamb cut to the proper length and marked for the width. The length, it will be noticed, is the same as that of the subsill, not taking into consideration the lugs. Dotted lines at *B* show how the head jamb must be plowed for the parting bead. The figures show 1-7/16 of an inch for play, assuming the sash is 1-3/8 inches thick. The 1/16-inch play is sometimes overlooked by the beginner. *C* shows the head jamb completed.

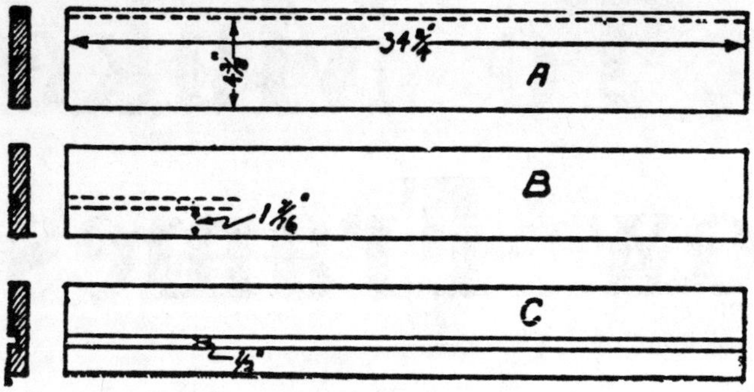

Fig. 448

A, Fig. 449, shows the side jamb developed to where we left the head jamb in Fig. 448, *C*. The length is 58 inches. This allows for the lugs above the head jamb and below the subsill, as we show at *B*, where the housing has been completed. The exact length between the head jamb and the subsill is 54 inches measuring on the inside of the parting bead. This figure applies to a window jamb for a 30" by 24" 2-light window.

At *C*, Fig. 449, the jamb is shown marked for the pulleys and for the weight pocket. The figures are given here but can be changed to suite the owner or the workman. At *D* we show the mate to the jamb shown at *C*, and the pulleys are shown in place. At *E* is shown an edge view of a pair of side jambs with the pocket cuts marked and the gaining for the head jamb and the subsill completed. Fig. 450 illustrates how the bevel for the sill is obtained by taking 5 on the tongue of the square and 24 on the body. This slope can be increased or decreased to suit the conditions for which the frame is made by increasing or decreasing the figure used on the tongue of the square.

Three details of the weight pocket in the side jamb are shown by Fig. 451. The center detail shows an edge view of the pocket after it has been removed and again replaced and held in position by nails at the top and a screw at the bottom. The dotted line on the center detail indicates how much must be dressed off to produce an even surface as shown by the edge view of the detail to the left and the face view of the detail to the right.

Illustrated by Fig. 452 are, *A*, the jamb mortised for the pulleys; *B*, the face view of the pulleys in place; *C*, an edge view of the jamb with the pulleys installed; and *D*, the back of the jamb showing the pulleys and how they are fastened. The dotted lines show the location of the plowing and housing of the face side of the jamb. It should be remembered that there are different makes of sash pulleys on the market, each having its own installation requirements.

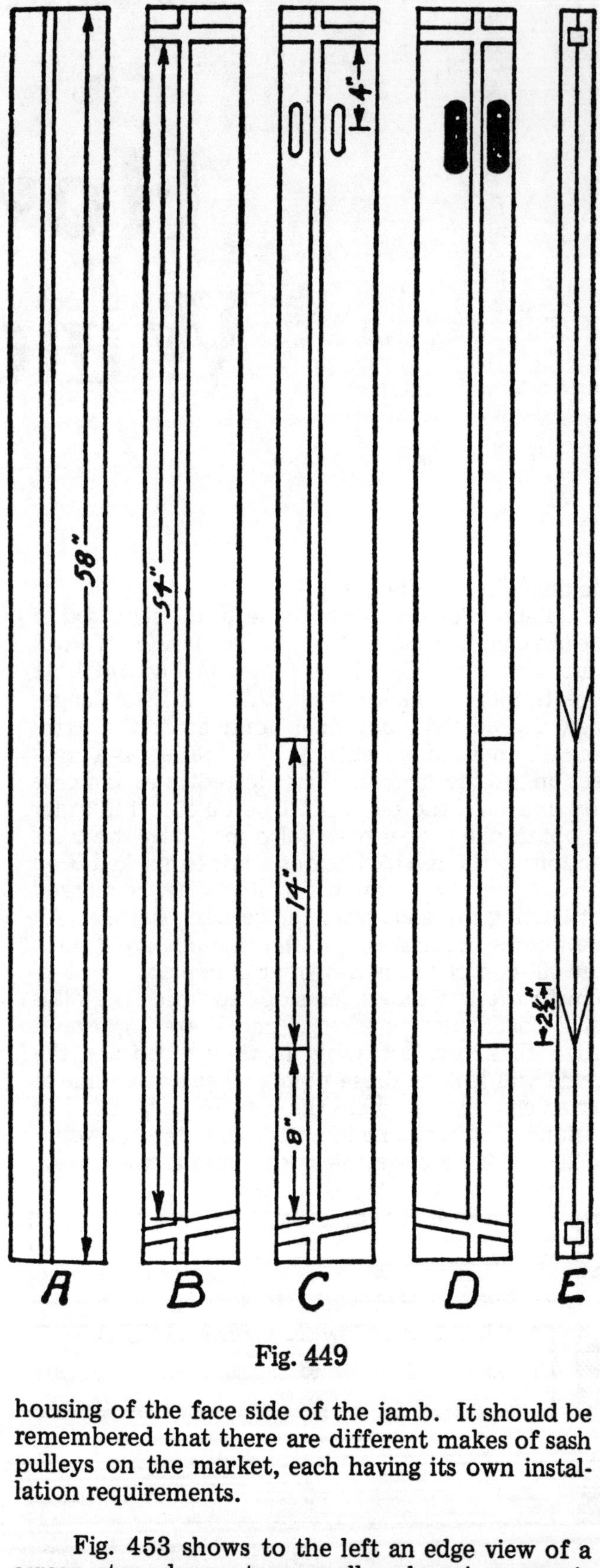

Fig. 449

Fig. 453 shows to the left an edge view of a screen stop drawn to a small scale. A square is shown applied to the edge view giving the figures to be used in getting the bevel. This bevel is the same as the one we gave in Fig. 450; 5 on the tongue of the square and 24 on the body. The face

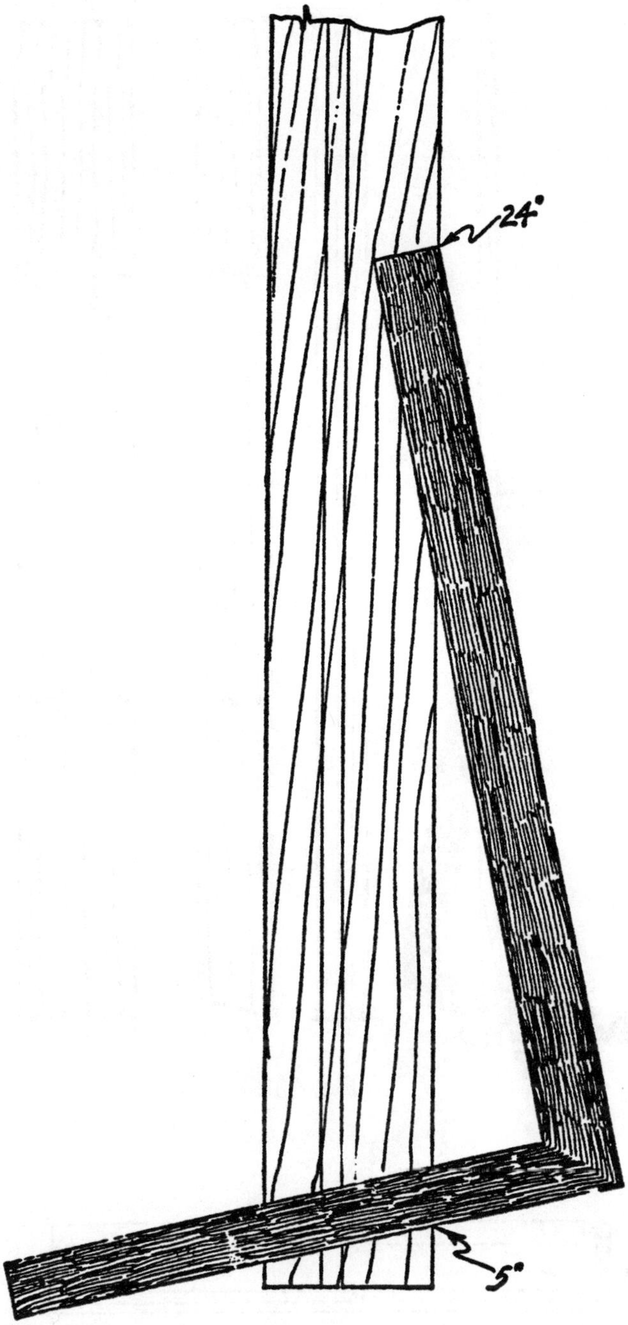

Fig. 450

view (second from the left) gives the length of the screen stop as 55¾ inches. This allows for the necessary projection beyond the head jamb. The drawings to the right show details of the top and bottom ends of the screen stop. To the extreme right is a face view giving the width in figures. The second from the right is an edge view showing the bevel and giving the thickness in figures.

It might be mentioned that a 1x6 board can be ripped so that it will make a jamb and a screen stop as shown in Fig. 454. A dotted line indicates how it must be ripped in order to make a head jamb and a screen stop. The length of the head and the

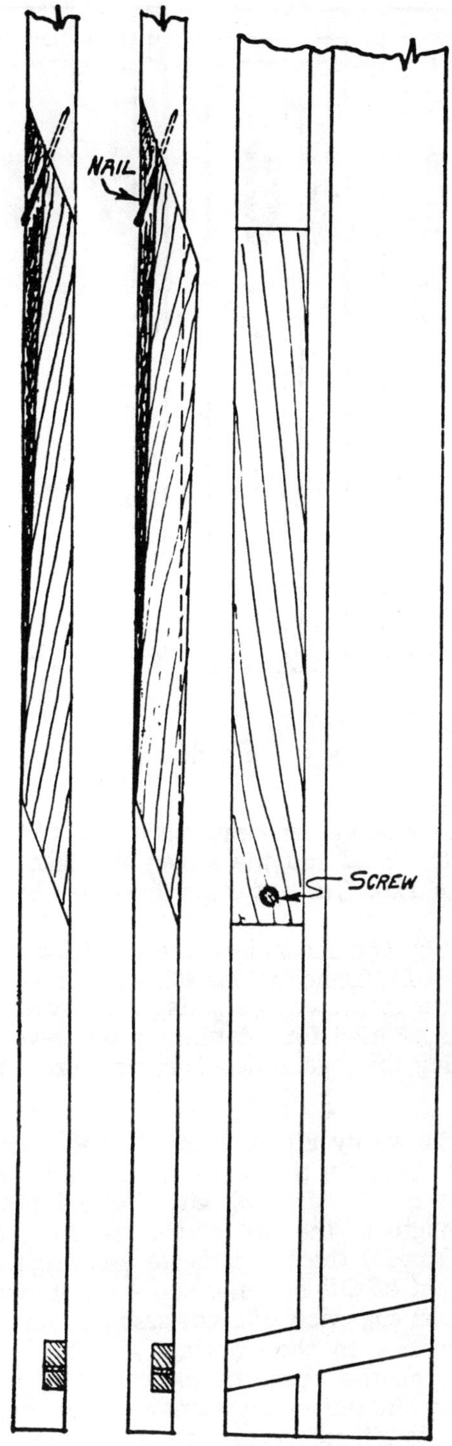

Fig. 451

screen stop are given in figures. The screen stop is 1¾ inches shorter than the head jamb – three-fourths of an inch is taken up by the housing of the jambs and 1 inch by the side screen stops. At *b* is a parting bead for the head, which, as will be noticed, is as long as the head jamb. A detail of the parting bead and a section of it is shown at *A*. At *B* and *C*, respectively, are shown sections and details of the head and of the screen stop. By dotted lines is indicated the plowing for the parting bead.

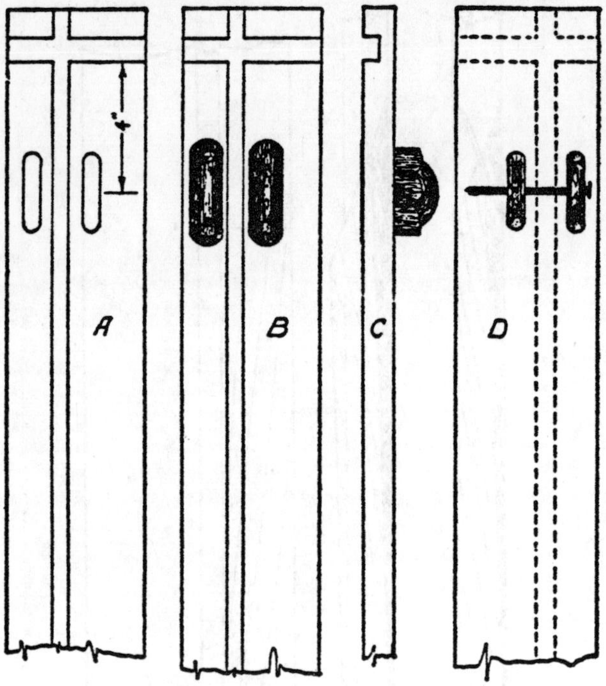

Fig. 452

A little study of the drawings in Fig. 454 will show how the bead and the screen stop can be ripped from a 1x6. This is further illustrated by Fig. 455.

Fig. 456 shows how the bevel for the various pieces of a frame are transferred to a bevel square, which is then used in marking the bevels. The steel square is used for establishing the bevel, but for marking, the bevel square is much more convenient.

Shown by Fig. 457 are, *1* an edge view of the parting bead for the jambs, and *2*, a side view. The figures on the drawing give the width, thickness and length of the side parting bead. The shaded parts toward the top indicate cross sections. To the right of this figure, *3* and *4* are shown respectively, an edge view of a side casing and a face view. The figures on the drawings give the thickness, width and the length of the casings. The shaded parts toward the top represent cross sections. There are just four more pieces left to complete the frame. These are shown in Fig. 458 -- first individually, and then put together.

Number *1*, Fig. 458, gives a top view of the watertable or cap. The length is given in figures and also the width. The shaded part represents a cross section. A short distance from either end are pointed out saw kerfs that should be cut into the back of the cap so that, when the siding or shingling is done, the cap can be cut out and the outside finish can be slipped behind it. A face view of the cap is shown at *2.* At *3* is a face view of the moulding and also a section. Number *4* shows the head

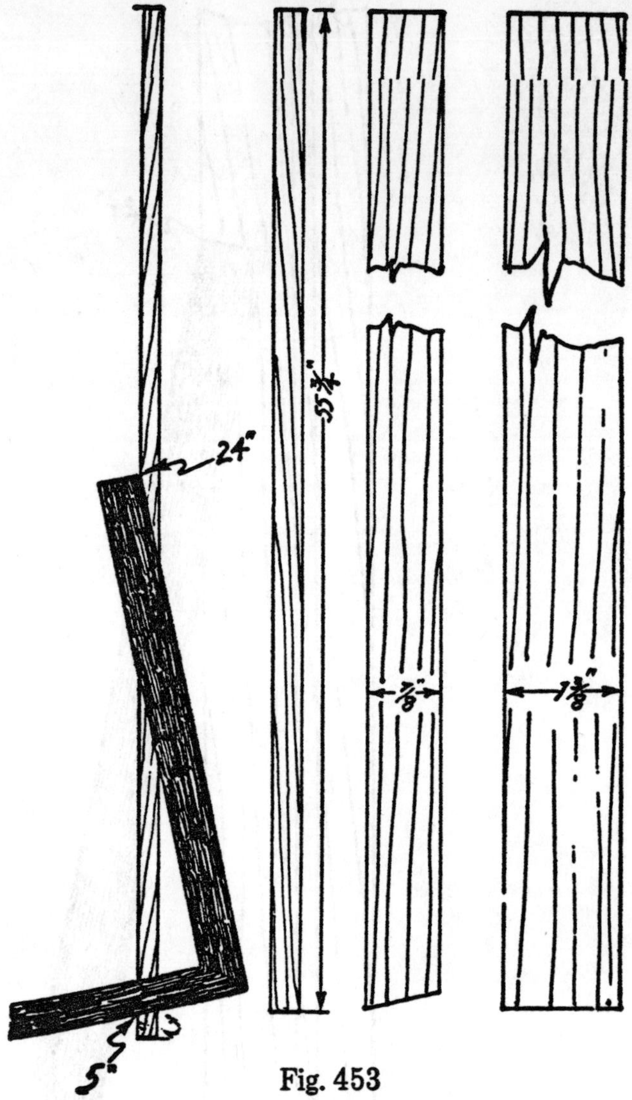

Fig. 453

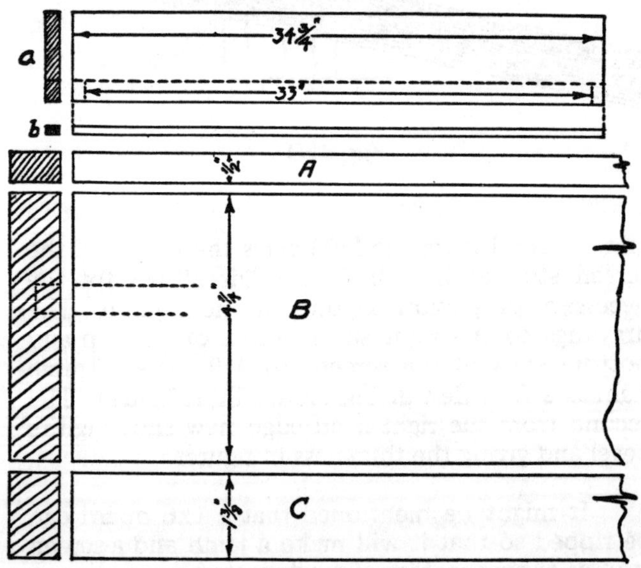

Fig. 454

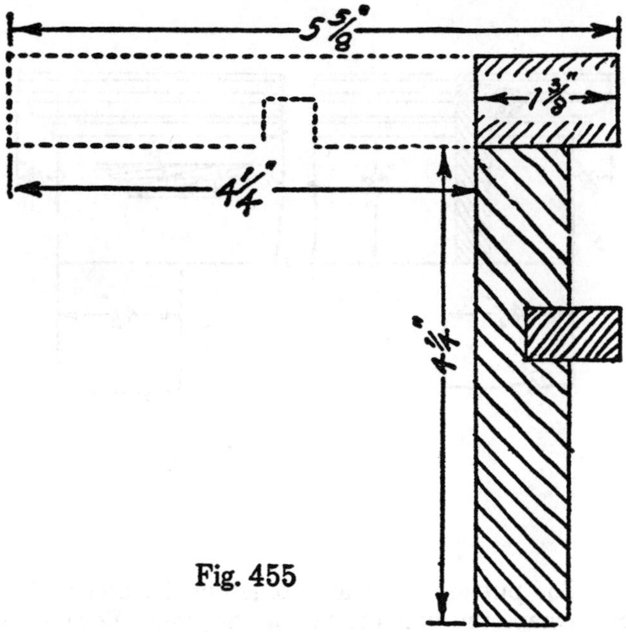

Fig. 455

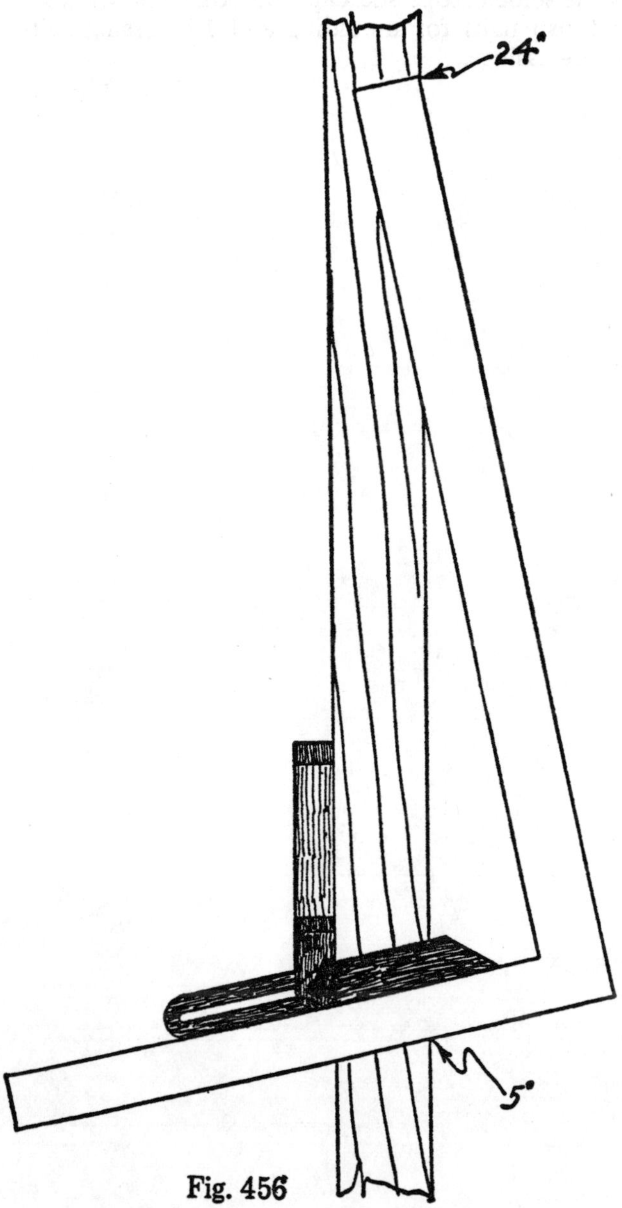

Fig. 456

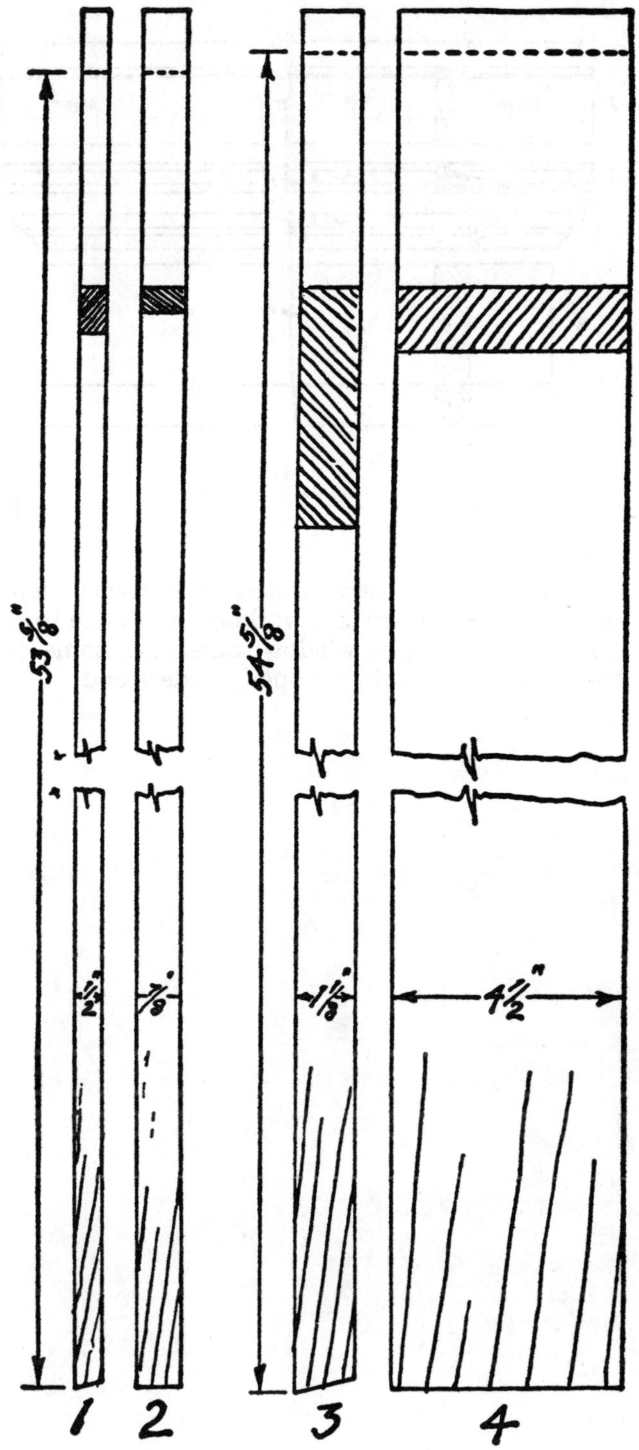

Fig. 457

casing, giving the width and length in figures. A cross section is also shown, shaded. Number *5* shows these pieces put together. To the right, we show how a short piece of moulding cut on either end to a 45-degree angle should be attached to the head casing so that when the siding is done the workman can use it for the return mouldings. The figures shown on the drawings given in Fig. 458 are sufficient to complete the head casing of a window frame or of a door frame.

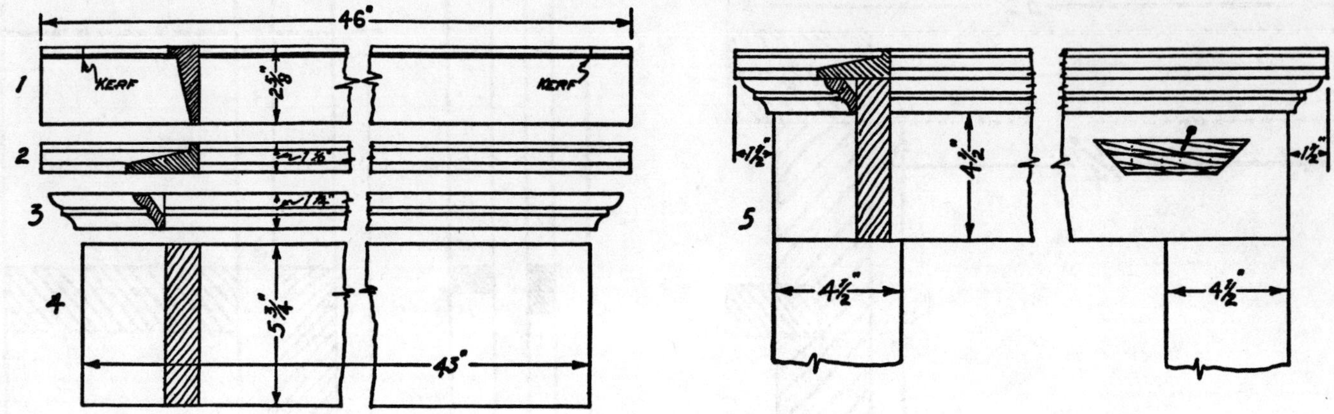

Fig. 458

It should be remembered in studying these drawings that we are still working on a frame for a 30" by 24" 2-lights window; otherwise, some of the figures could not be properly understood.

For putting the frame together, 8d box nails are perhaps the best for nailing the jamb. For nailing the screen stops and caps, 6d box nails will do -- 6d finish nails for moulding and 10d casing nails for the casings.

32 Details of Window Frames

Details of Window Frames

Fig. 459 shows a design for a frame to be used in a building without the walls boxed. This design is suitable for cheap buildings and for buildings located in mild climates. No provisions are made for screens, although such frames can easily be provided with screens. The same kind of design is shown in Fig. 460, where the frame is used in a building that is boxed. The shaded parts indicate the parts of the frame that are different from the frame in the previous figure. The moulding receives the outside wall finishing. This design can be used with shingles, siding or stucco. The siding laps

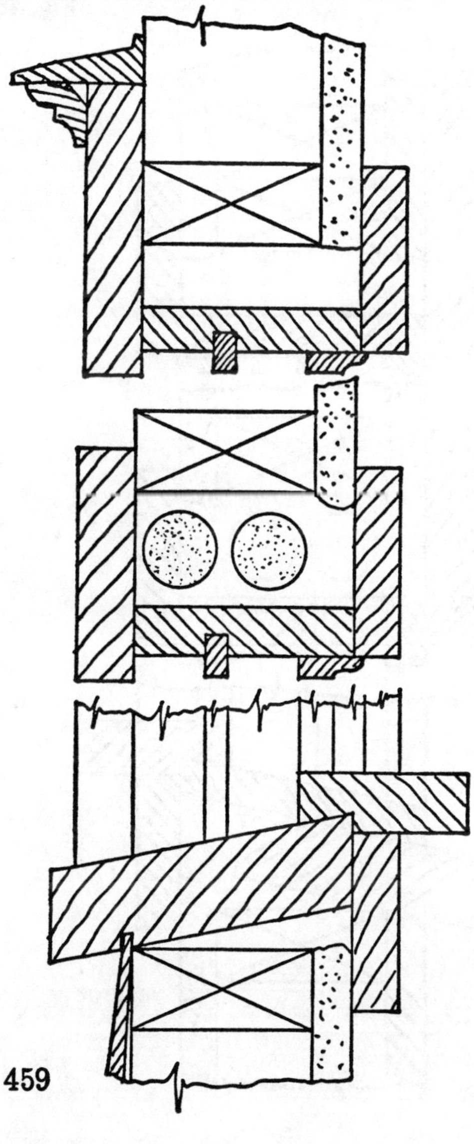

Fig. 459

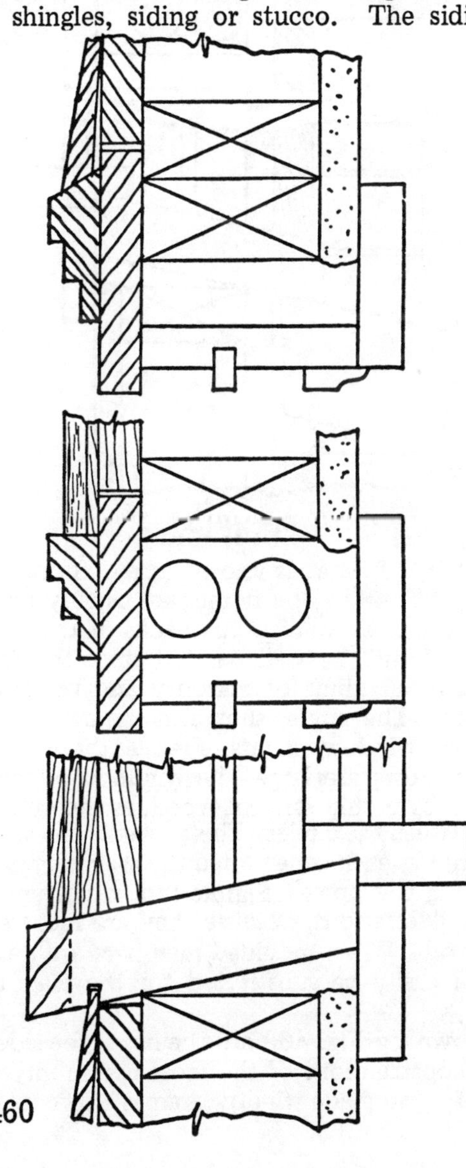

Fig. 460

over the joint between the boxing and the blind casing, as shown, and makes a wind-proof joint, especially if building paper is used.

Another modified form of this design is shown by Fig. 461 where the frame is used in a building with brick-veneered walls. The change is indicated by shading. In all of these illustrations we show a detail of the sill construction at the bottom, a detail of the side at the center and a detail of the head at the top.

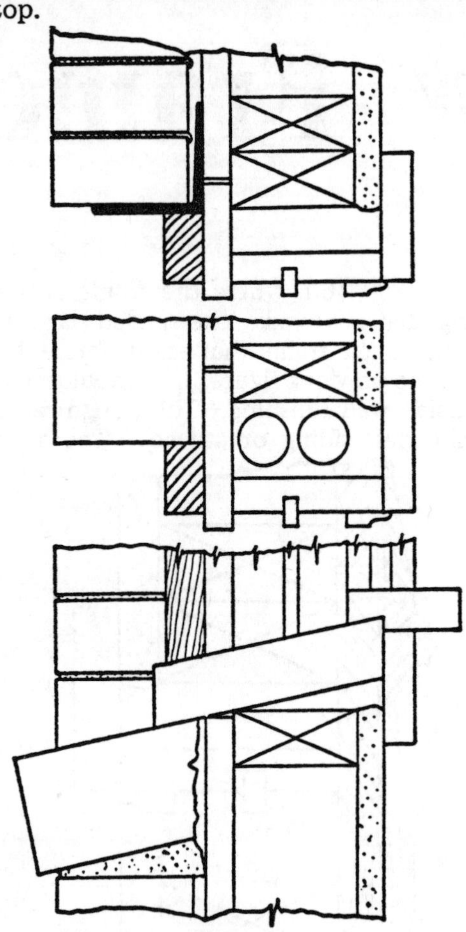

Fig. 461

Details of what is known as a box frame are shown by Fig. 462. Box frames are usually housed, and the joints as well as the backs of the various members should be well painted since the frames are used in all kinds of masonry and in concrete structures. The circles shown in the center detail indicate the window weights. Frequently, especially if the windows are large, the window weights are separated by a thin strip of wood, as shown in this detail between the circles. This prevents the weights from striking each other when the sash is operated, eliminating the noise. Simple mouldings are used in these illustrations, because they are more easily understood. When moulded members are desired, they can easily be substituted for the plain ones.

Shown by Fig. 463 are the head, the side and the sill constructions of the most commonly used designs for job-made window frames. Two mullion

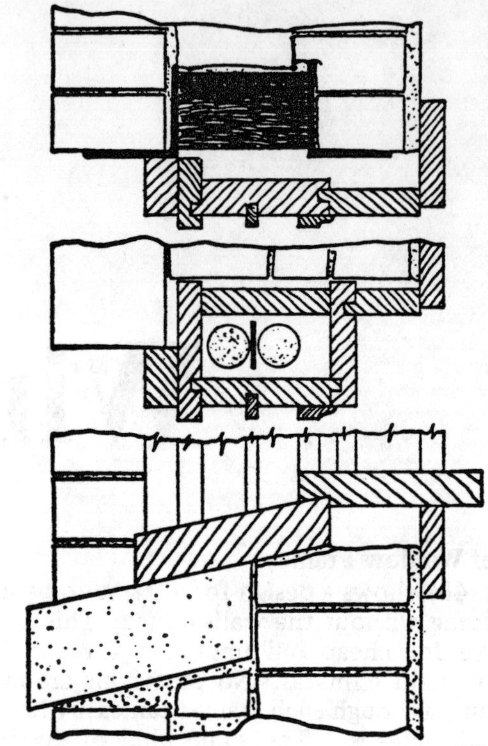

Fig. 462

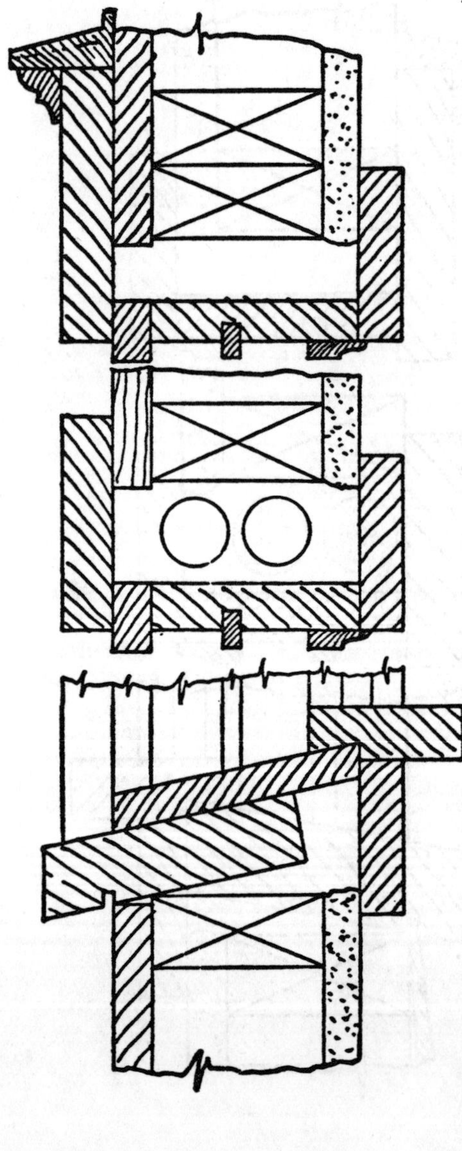

Fig. 463

designs are shown by Fig. 464. The design at *A* has the outside casing nailed onto the screen stops; the design at *B* has the outside casing nailed onto a blind casing. The blind casing is shaded to emphasize the difference. Mullions should always conform with the sides of the frame.

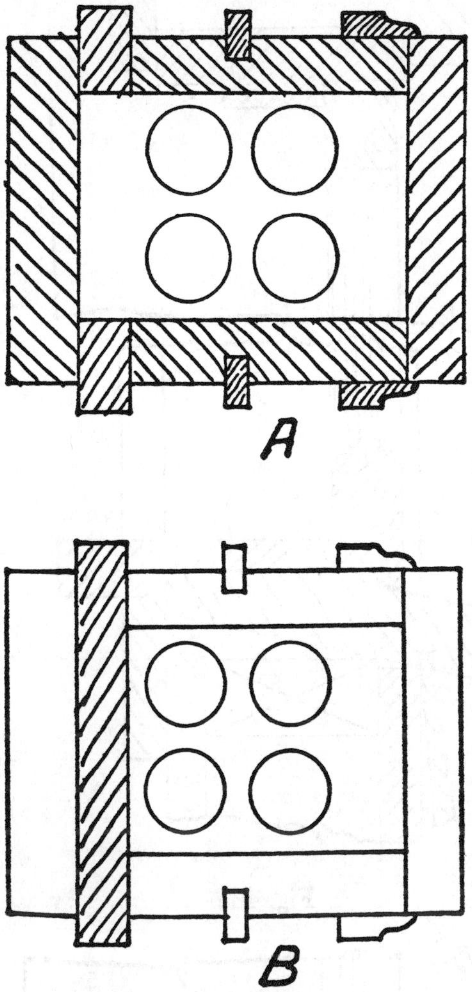

Fig. 464

Fig. 465 shows the sill construction and the head of a jibhead window frame. The frame is made for a single sash which is hung in such a way that it slips up into a pocket provided for it. The bottom of the sash comes flush with the head stop when it is up. Wide stops eliminate the use of parting beads. When parting beads are used, the stops conform with those used on the double-hung windows.

Material for Frames

The material for all kinds of window frames should be the best available -- and what is true of window frames is also true of door frames and all other kinds of frames exposed to the weather. White pine, as a rule, is preferred for frame making, but Douglas fir, cedar, larch, cypress, redwood and sugar pine are also used. Which of these woods to use depends largely on the locality in which the frames are made. The cost of transportation is the chief factor in determining the price of lumber and therefore establishes, to a great extent, the kind of wood used for frame-making in different localities.

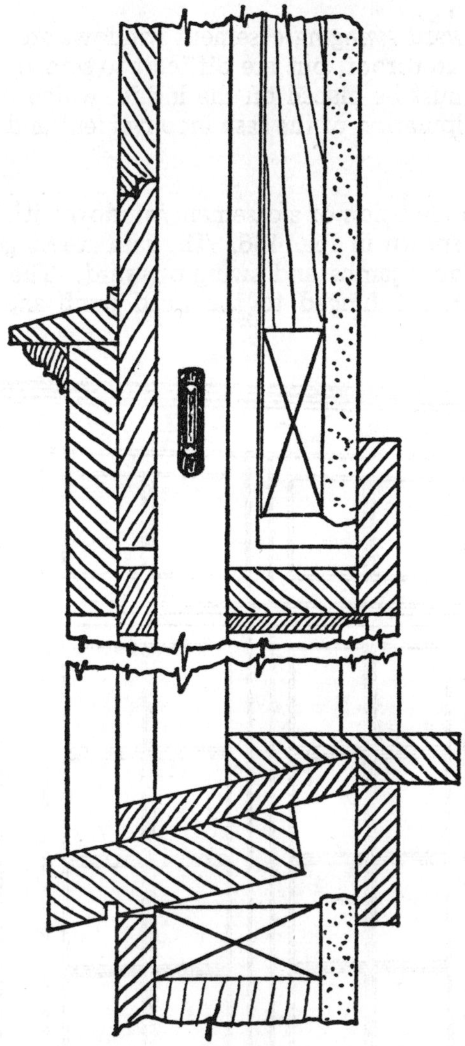

Fig. 465

Double-Hung Windows

While double-hung windows are perhaps the most practical of all the windows that are in use. Double-hung windows are easily screened but less than half of the window space can be used for admitting air. The sash must have enough play to permit easy operation. This necessary play in cold weather will admit cold air into the room and increase the cost of heating. Weather stripping, especially approved metal stripping, remedies this fault to a large extent, but not altogether. Windows with pivoted sash can be opened so as to use practically all of the space for ventilation.

Casement Window Frames

Casement windows have their own advantages and disadvantages. While all of the window space can be used for admitting air, it is almost impossible to make the inward-swinging casement windows waterproof. However, they are easily screened.

The outward swinging casement windows are more nearly waterproof, but are difficult to screen. The screen must be placed on the inside, which makes the manipulation of the sash inconvenient and awkward.

An elevation of a casement window with transoms is shown in Fig. 466. The sash are hinged to the window jambs and swing outward. The transom sash are hinged to the head jamb and also

Fig. 466

swing outward. Fig. 467 shows, at the bottom, the sill construction of the window shown in the previous figure, and at the top the side construction of the same frame. The transom-bar is shown by Fig. 468 at the bottom, and the head of the frame is shown at the top. Fig. 469 shows the mullion of the same frame. Provision for the screen is pointed out. Self-storing aluminum storm windows and doors are available, ready to set into place. They are expensive, but give good service.

A sill construction of an inward-swinging casement window is shown by Fig. 470 at the bottom. At the top is the side of the same layout. Fig. 471 shows the transom-bar at the bottom and the head at the top. The transom sash is hinged at the top and swings inward. A mullion of an inward-swinging casement window is shown by Fig. 472. Provision for the screen is pointed out on the drawing.

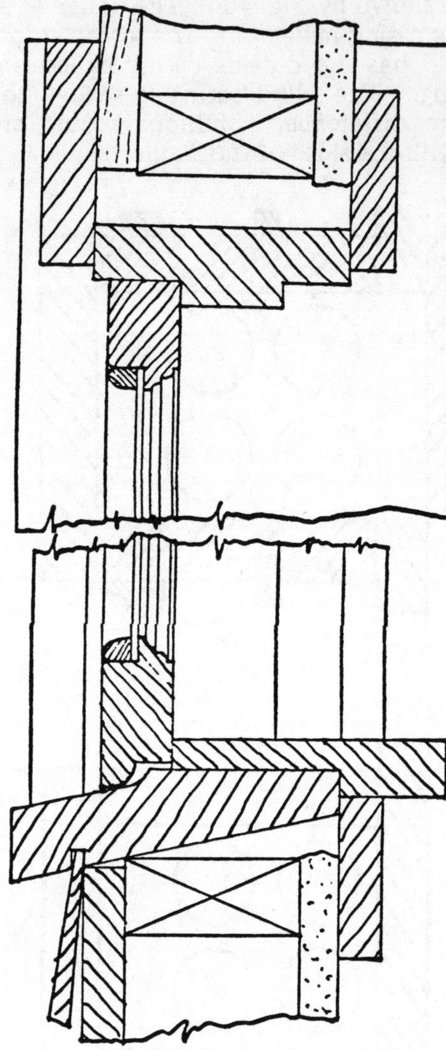

Fig. 467

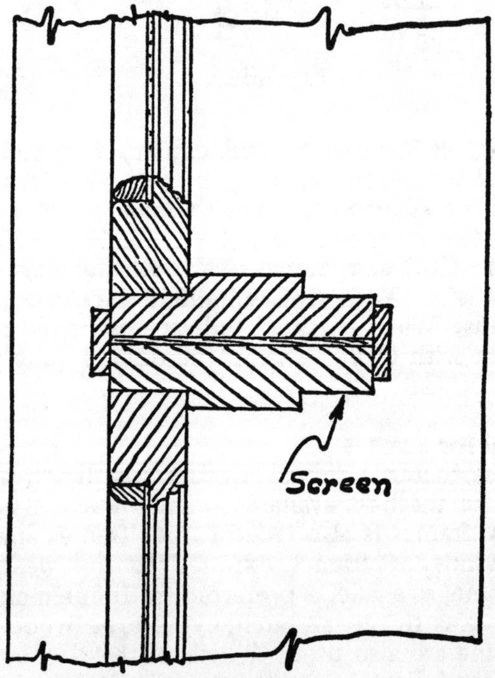

Fig. 468

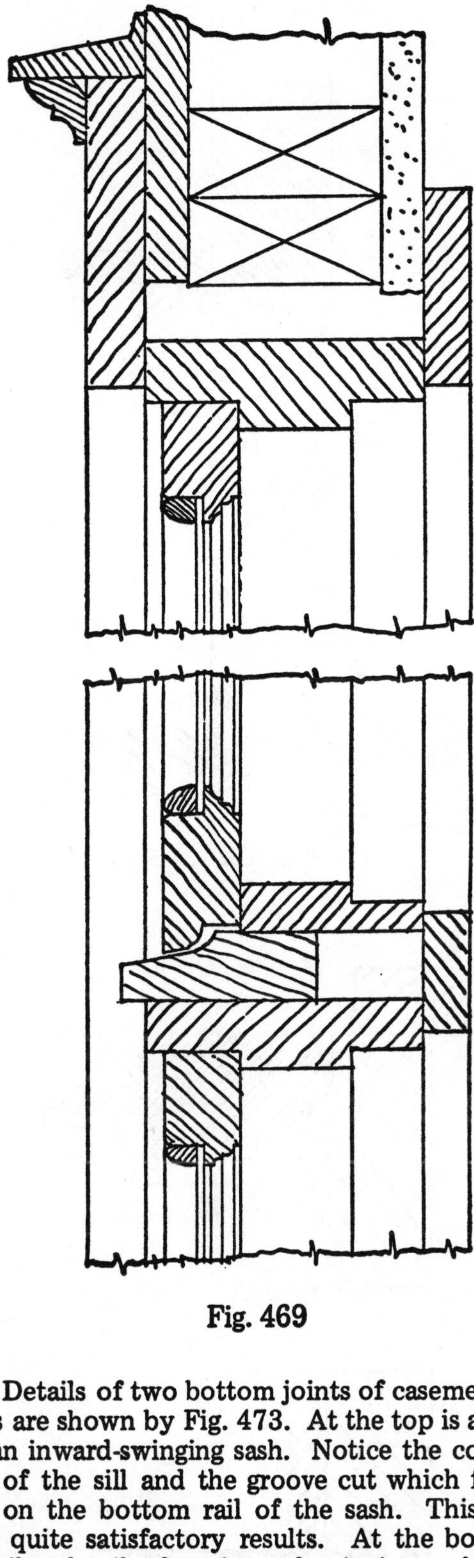

Fig. 469

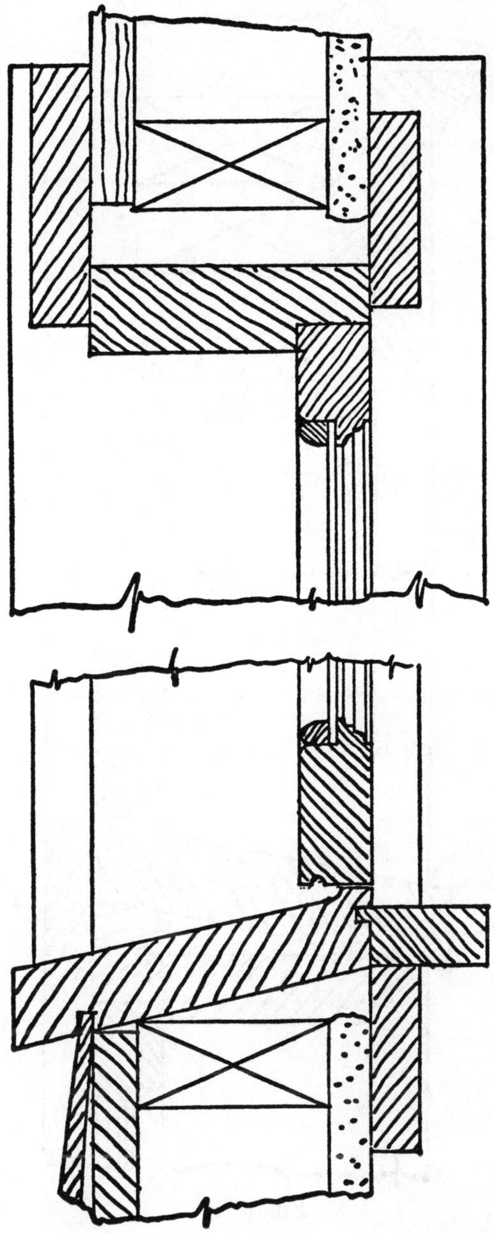

Fig. 470

Details of two bottom joints of casement windows are shown by Fig. 473. At the top is a design for an inward-swinging sash. Notice the construction of the sill and the groove cut which forms a drip on the bottom rail of the sash. This design gives quite satisfactory results. At the bottom is a similar detail of an inward-swinging sash. This construction is much better than the one shown in Figs. 470 and 471 for it has a double drip. The drip-moulding is housed into the bottom rail, as shown by the drawing.

Details of two improved side constructions of casement window frames are shown in Fig. 474. At *A* is a detail of an outward-swinging sash, while at *B* is a form of an inward-swinging sash.

Five details of center sash joints of casement windows without mullion are shown in Fig. 475. Number *1* is a commonly used joint that gives fairly satisfactory results. Number *2* is a double astragal joint which is more expensive than the one just shown. Number *3* is a housed joint which requires both sash to be opened or closed together. Number *4* shows a rabbeted joint somewhat similar to the joint shown by number *1*, except that astragals are fastened to the sash, as shown. Number *5* is another form of a housed joint which requires both sash to be opened and closed at the same time.

Light Rules

There are several rules for lighting rooms which should be observed: First, for every 100 cubic feet of room-space not less than one square foot of win-

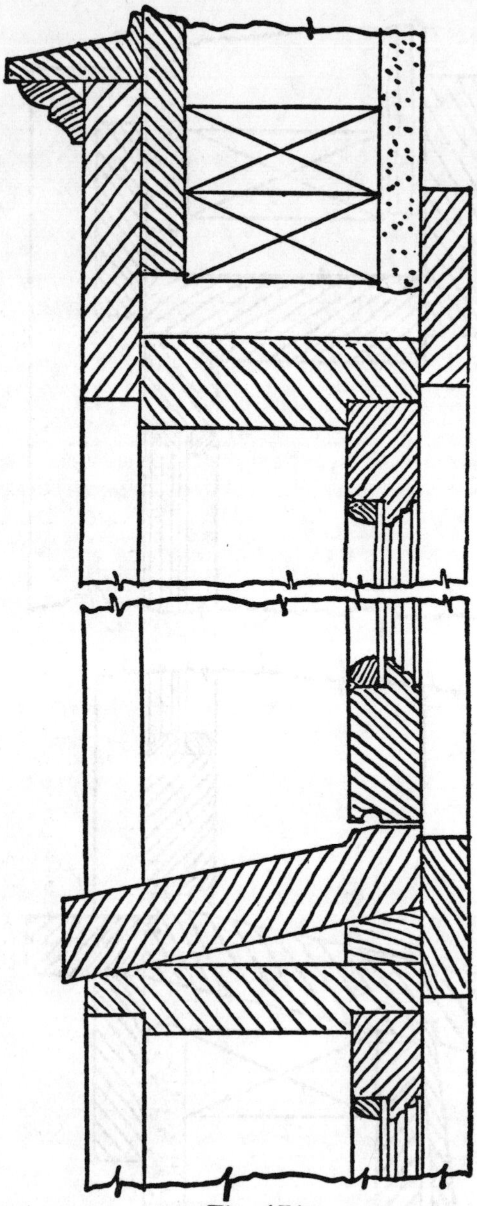

Fig. 471

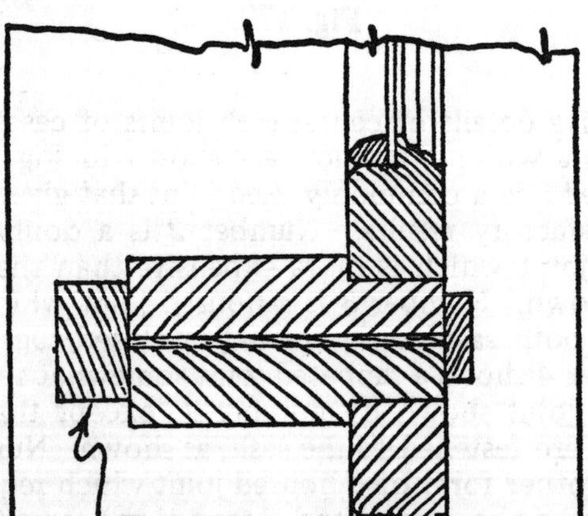

Fig. 472

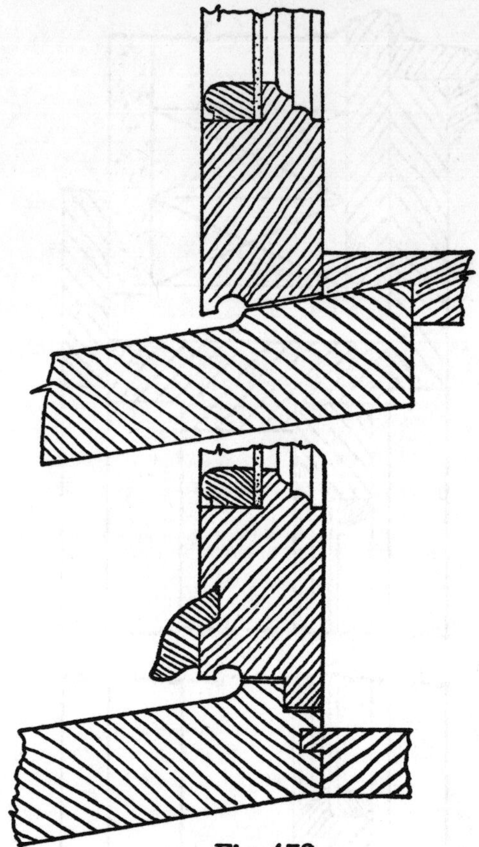

Fig. 473

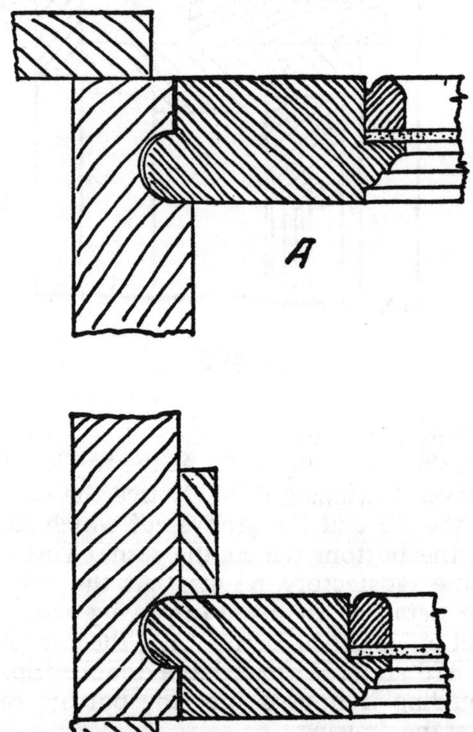

Fig. 474

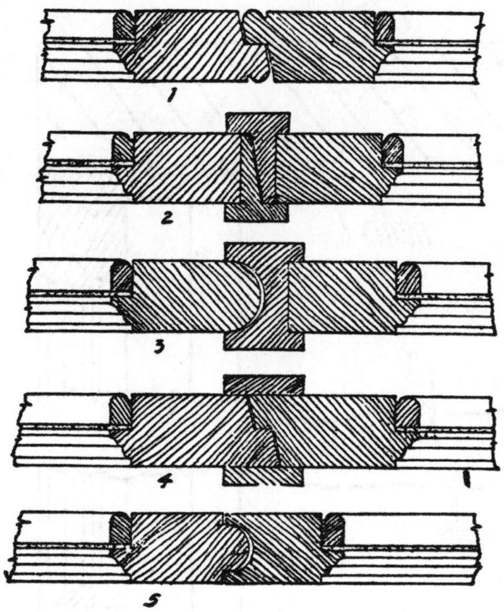

Fig. 475

dow glass should be provided. Second, the window glass surface should be equal to one-eighth of the wall surface. Third, divide the floor space, in square feet, by 10, and the quotient will equal the number of square feet of window glass required. Some discretion must be used when these rules are applied, but errors should be made in favor of more light for later adjustments can be made with blinds or shades.

Basement Windows

Basement windows should be located wherever possible directly under the window openings of the superstructure. The width of the basement window does not have to equal the width of the windows above, but the center of one window should be in line with the center of the other. Basement windows are frequently narrower than the windows above, but never wider. In cases of extremely large windows in the building superstructure, a well-balanced set of basement windows can be used instead of a single window.

Basement windows are sometimes double-hung. In such cases the frames are made much on the order of any other frame that is built into masonry. It should be remembered, though, that basement walls draw more moisture than do the walls of the structure above, and for that reason, double precaution should be taken as to the quality of wood used for the frames and the kind of preservative used for the wood. Sometimes it is necessary to build retaining walls for basement windows to keep the earth from washing against them.

Basement windows should never be used for coal chutes excepting in cases of emergency. Coal chutes should be provided when the basement wall is erected.

In basements of store buildings, especially in small towns, the windows are often provided with prism glass for the purpose of throwing the light back into the room as far as possible. Prism glass also has the advantage of obscuring the view; it will admit light but will not permit observation from either side.

Cellar Windows

Fig. 476 shows a front elevation of a cellar window frame with the sash in place. This figure shows side and head casings, lightly shaded. Fig. 478 shows a frame with similar casings. Fig. 477 shows a top and a face view of a cross section of the head. To the left is the face view of one of the jambs, and below is a cross section of the sill. At the bottom is a face view of the sill. The rabbeted edges of the various members of this cellar frame are shown in projected position ready to go together -- note and study the dotted lines. At *A* is an enlarged detail of lugs which, when used, are embedded in the masonry. These lugs are left on either end of the head and of the sill.

Fig. 476

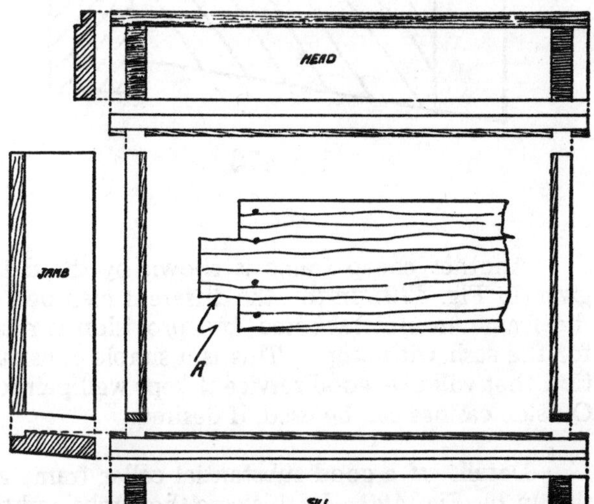

Fig. 477

Soft wood that will endure exposure to the weather, such as white pine or cypress, is the most suitable for cellar frames. It should be well painted both in the joints and on the back before the frames are put together and installed. Fig. 478 shows a construction of a rather cheap frame. However, this frame, kept well painted, will give good service. The sides and top of the outside edges are cased as shown by the elevation in Fig. 476. The rabbet for the sash is formed by the inside casings. The casings are made of material equal in thickness to the thickness of the sash.

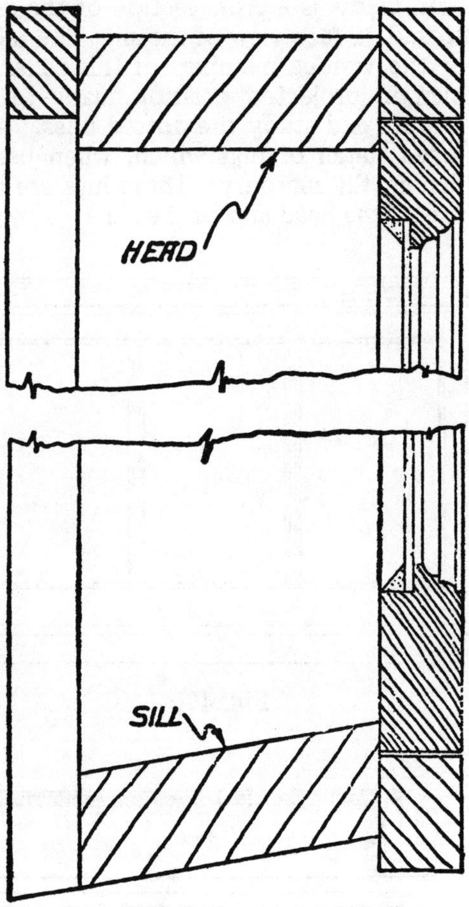

Fig. 478

Another cheap frame is shown by the detail given in Fig. 479. In this the different members of the frame are not rabbeted, but provision is made for the sash with stops. This is a simple construction that will give good service if kept well painted. Outside casings can be used, if desired.

Details of a good substantial cellar frame are shown by Fig. 480. In this case the jambs and the sill are rabbeted. Notice the rabbet on the sill and on the bottom rail of the sash. The groove cut into the bottom rail of the sash provides a drip. To the left is a rabbet which provides for screening. Details

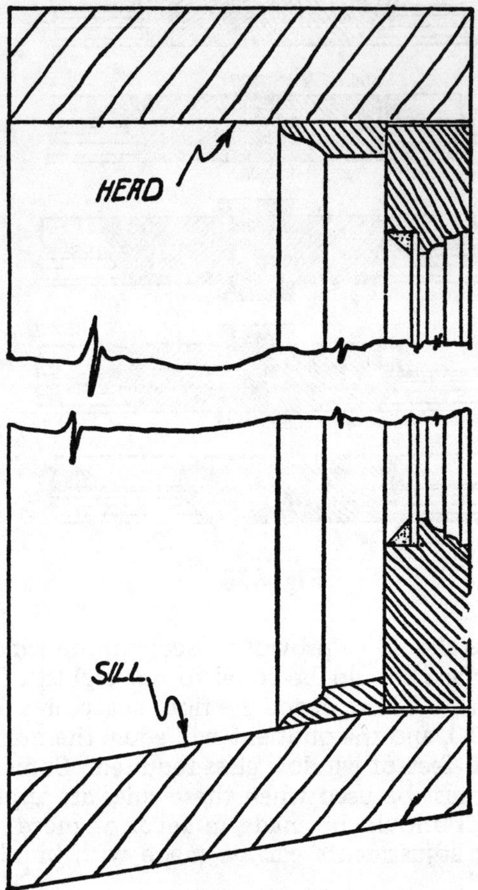

Fig. 479

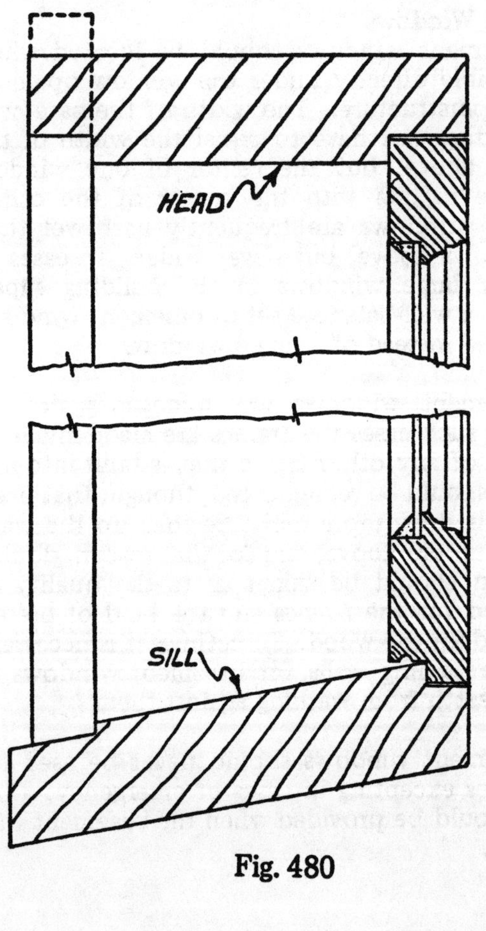

Fig. 480

of a similar, but much better cellar frame are shown by Fig. 481. This frame has a brick moulding at the outside, is cased on the inside and is rabbeted both for the sash and for screening. The drip and the shoulder against which the bottom rail of the sash strikes should be studied and compared with the previous figure. A 2x4 on the inside under the sill provides nailing for the apron. This is one of the best cellar frame constructions that can be used.

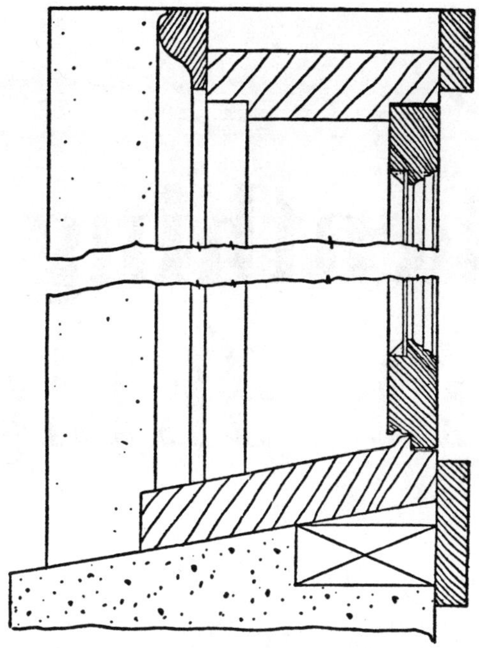

Fig. 481

33 Sheathing and Shingling

Spreading Sheathing

There are two reasons for spreading the joints of roof sheathing. One is that there is a saving in material. The other, which is the most important, is that spreading the sheathing provides ventilation for the shingles. Without this ventilation many wood-shingle roofs would rot, especially in wet climates and during periods of heavy rainfall. Solid sheathing for roofs, however, is very common because of the use of plywood and the extensive use of asphalt shingles. Copper, tin and roll roofing also require solid sheathing. In these cases there is no danger or rotting and therefore no need for ventilation.

Cutting Sheathing

Fig. 482 shows a section of a plain roof with the sheathing nailed on. The ends have not yet been cut. Assuming these ends are to project 18 inches beyond the first rafter, we would proceed to

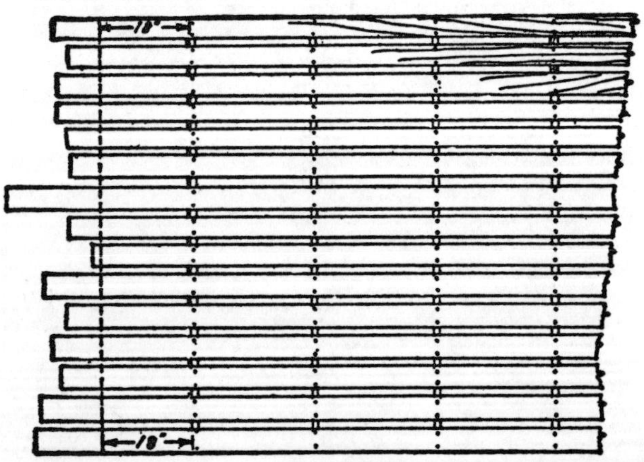

Fig. 482

measure 18 inches from the outside of the rafter both at the top and at the bottom and mark the sheathing as indicated on the drawing. Then strike a chalk line from point to point as shown by the dotted line. Next, the ends should be cut off. This will leave the section of roof as shown by Fig. 483. If a plain fascia is to be nailed to the ends of the sheathing, then the boards should be cut off square, but if a moulding is to be nailed to the ends, then the cutting should be done so that the boards will fit the moulding.

A right-handed man and a left-handed man, we might mention, make a good team for cutting off the ends of sheathing, for then both can start at the top and work down. If the work is done by one man, he has to work up on one side and down on the other, unless he can use both right and left hands to good advantage.

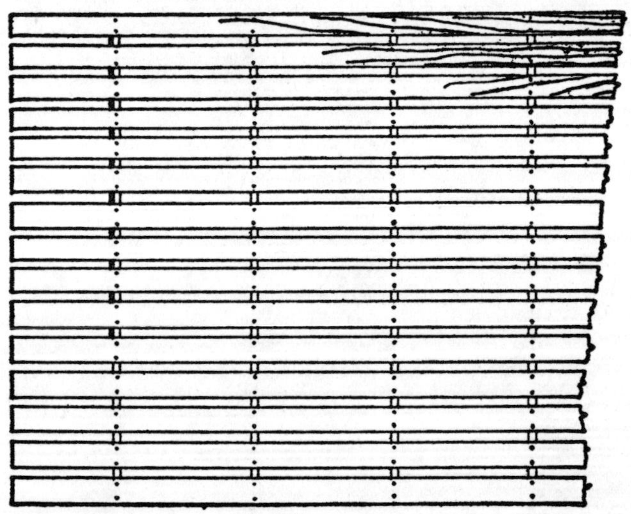

Fig. 483

Sheathing Problems

Fig. 484 shows a corner section of the roof with the edges ready for shingling. The sheathing boards, in this case, are 1x6's, and the spread between the boards is 2½ inches. It is not necessary to measure to be sure the spread is right. Every carpenter soon learns to give sheathing the proper spread by means of his judgment. A rule that is often used, especially by beginners, is to make the spread as wide as the length of an 8d nail. After a few boards are nailed in this way, the rest of the spreading can be done by approximation.

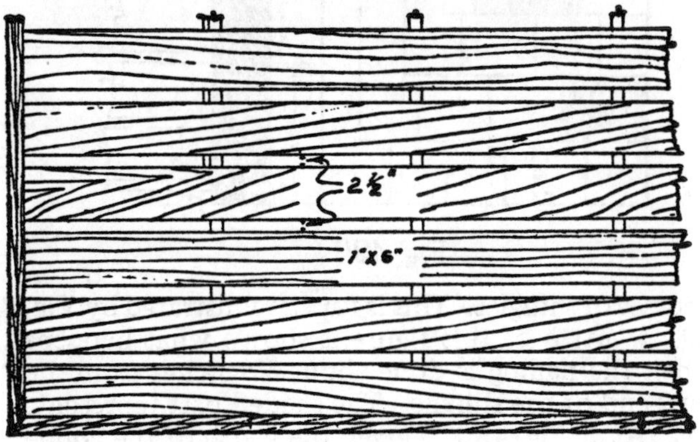

Fig. 484

The sheathing of a dormer seen from above is shown by Fig. 485. The problem here is to simplify the cutting. By referring to the drawing we find that one of the boards numbered *1* is the longest of one side, and the other is the shortest of the other side. These two boards, if placed end to end, would be 16 feet long. Consequently, they could be cut from a 16-foot board. The same is true of the two boards numbered *2*, and the two numbered *3* and so on to number *7*. The workman should establish the difference in the lengths of one board and the next one to it, and then start with the longest and cut each board that much shorter than the one before. Having cut the number of boards required for one side of the roof to the proper bevel and lengths, the pieces that are cut off will make the sheathing for the other side. The dotted line shown to the left shows how the ends are marked for cutting.

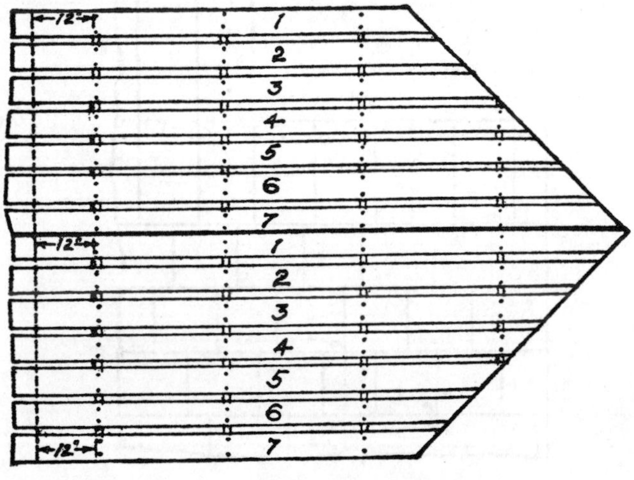

Fig. 485

A drawing of the sheathing boards for one side of a hip roof, cut and ready to nail on, is shown by Fig. 486. The board numbered *1*, at the bottom, has enough cut off to make the top piece, also numbered *1*. The board numbered *2* at the bottom will have enough cut off to make number *2* at the top, and number *3* at the bottom will leave enough for number *3* at the top and so on to number *8*.

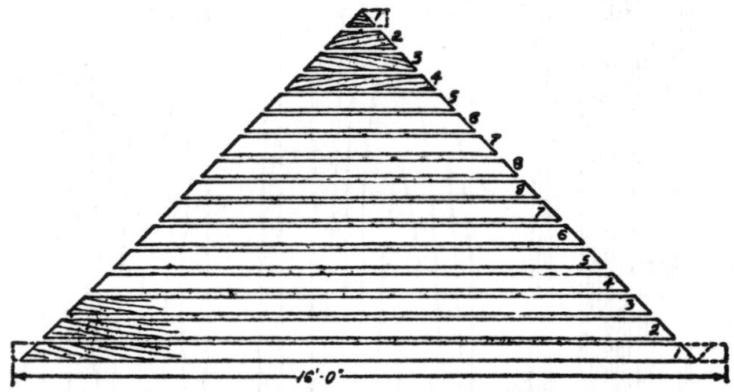

Fig. 486

At *B*, Fig. 487, is shown a set of cripple sheathing boards to be used between the valley and the hip of a hip and valley roof. These have been cut to a pattern which makes them uniform. At *A* is shown a set of cripple sheathing boards that lack uniformity.

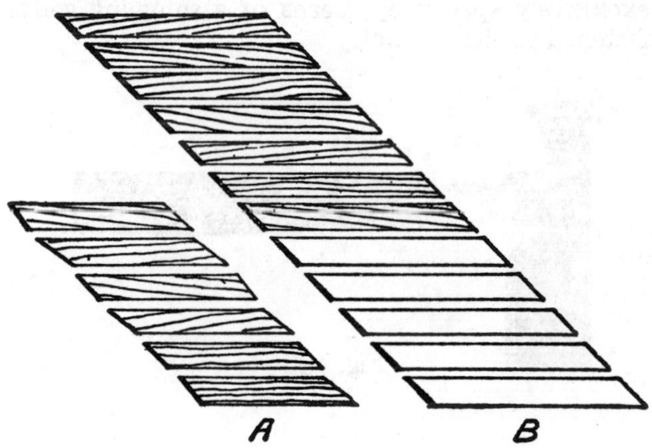

Fig. 487

At *A*, Fig. 488, is shown how the grain should run if a shingle is to give maximum service. This grain is called a vertical grain. At *B* is shown a shingle with flat grain. Such a shingle will crack easily after it is exposed to the weather. Flat-grain shingles not only crack easily, but they shrink a great deal more than vertical grain shingles. At *C* is a shingle with a large knot. Shingles like this should not be used except on the cheapest kind of roof and then only when it is known that the roof is only temporary.

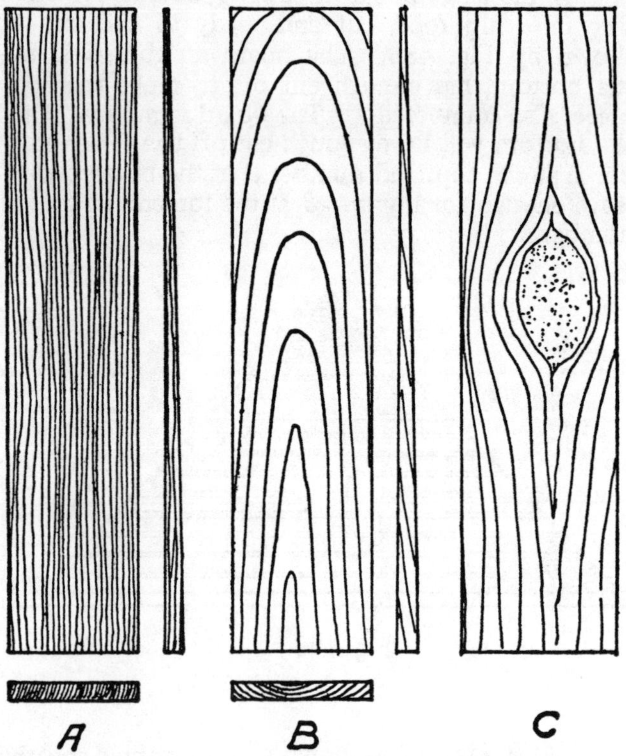

Fig. 488

Shingling Problems

Fig. 489 shows a shingling hatchet with a shingling gauge attached. Wood shingles are almost exclusively spaced by means of a shingling gauge fastened to the hatchet.

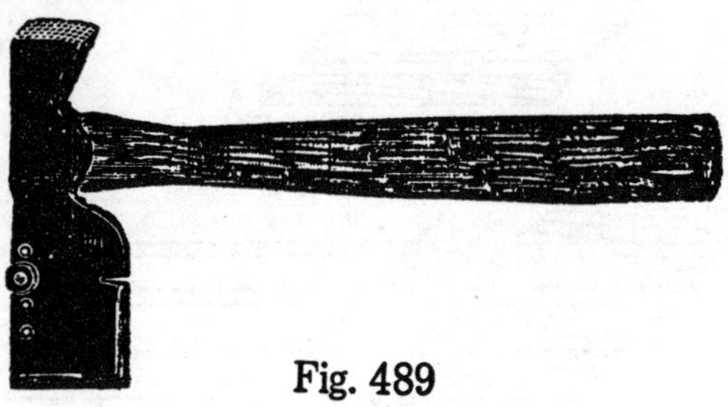

Fig. 489

A corner of a roof partly shingled is shown by Fig. 490. The shingle shown to the right, shaded, is cross-grained, as indicated by the shading, and should be discarded unless it can be used at the bottom of the double course, as illustrated. The dotted lines to the left show how much the shingle to the left laps over the adjacent shingle. The heavy dot at *g* shows how the bottom shingles of the double course should be nailed at each edge. At *a*, is shown how the upper corner of the edge shingles should be clipped off. At *b*, the width of the edge projection is given. At *c* and *c*, we give in figures the distance the nailing should be done from the butt-end of the shingle when the shingles are laid 4½ inches to the weather. This protects the nails from moisture, for the nails are covered with two

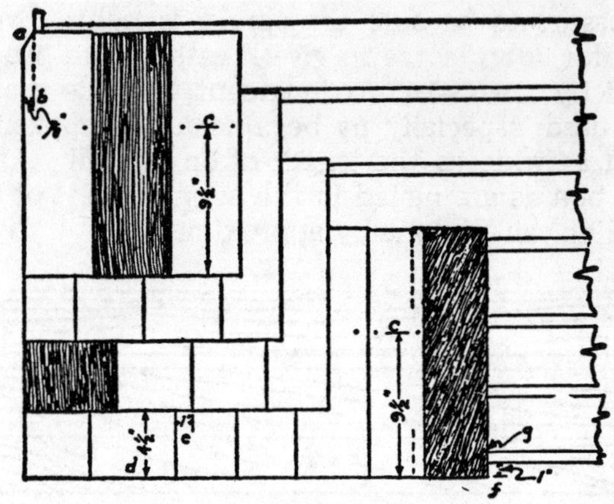

Fig. 490

courses of shingles. This is further illustrated by Fig. 491 *a*. At *b* of this figure is shown what happens when the nailing is done about 6½ inches from the butt-end of the shingle. The shaded part represents rot. At *d*, Fig. 490, is given the width of the courses and at *e* is shown how much the shingles should lap over each other. The projection at the eaves *(f)* is 1-inch, though this distance can vary somewhat.

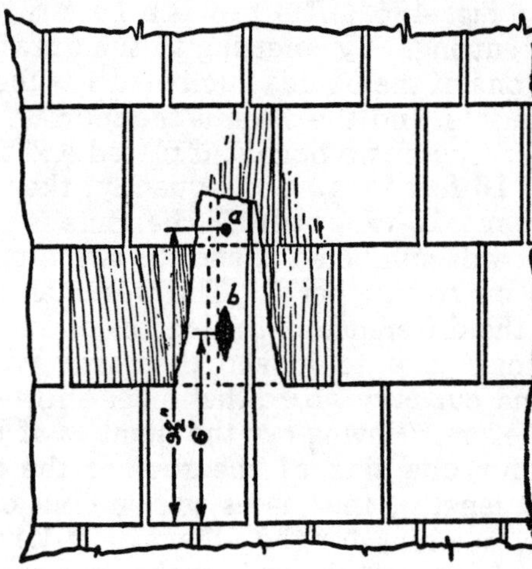

Fig. 491

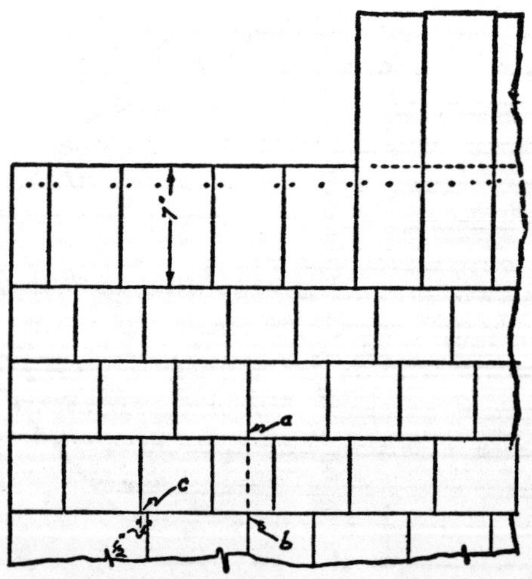

Fig. 492

A section of roof near the comb, with some of the shingle-tops sawed off, is shown by Fig. 492. The dotted line shows where the cutting is yet to be done. The last course should be 7 inches from the comb if an ordinary metal ridgeroll is used. This course should receive more nails than the others, as we show by the dots, which prevent splitting when the tops are sawed off. Joints coming directly over each other with a course between, as shown at *a* and *b*, should be avoided. One-half-inch lap, as shown at *c*, should never be permitted.

Another section of a roof near the comb, is shown by Fig. 493. Notice that the top shingles are nailed even more securely than those shown in the previous figure. This is necessary because the shingle-tops are broken off with the hatchet. The part marked *1* is sawed in line with the comb and then split off as indicated by the irregular dotted line. Then with the blunt end of the hatchet (not the cutting edge) hit the part marked *2* a quick hard blow just a trifle above the comb. Number *2* out of the way, strike number *3*, then number *4* and so on until the job is done. If the shingles are dry and well nailed to the sheathing the tops can be broken off without injuring the parts that are to stay on the roof. There is a trick to this that only practice will make perfect.

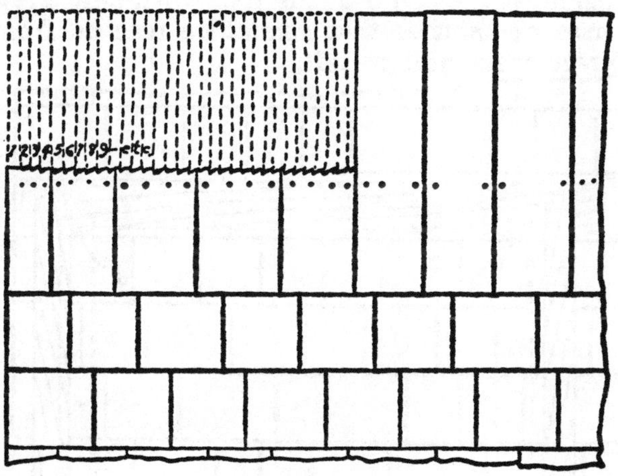

Fig. 493

Fig. 494 shows a metal ridgeroll at the top. Metal ridge is commonly used because it is inexpensive and makes a water-proof job. Just below this we show a comb-board in place. This kind of comb finish was much in use before metal ridgeroll became popular and in some instances is still used. At the bottom we show three designs of comb-boards. Number *1* is a good one and is waterproof. Number *2* looks good but will permit water to pass through the joint after exposure to the weather. Number *3* is used only on the cheapest roofs.

Two methods of putting on the double first course of shingles are shown by Fig. 495. Point *A*, in the upper drawing shows blocking necessary if

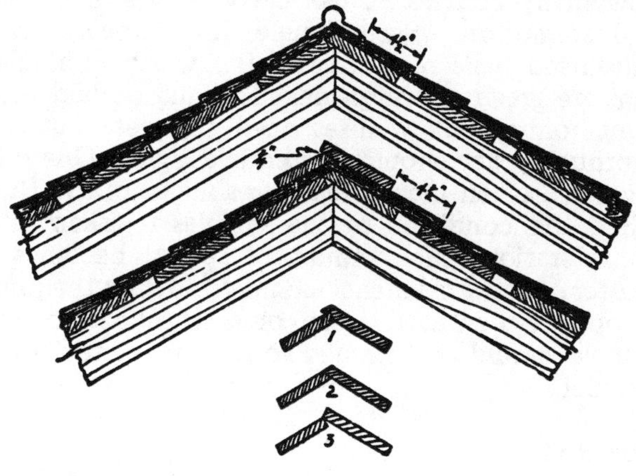

Fig. 494

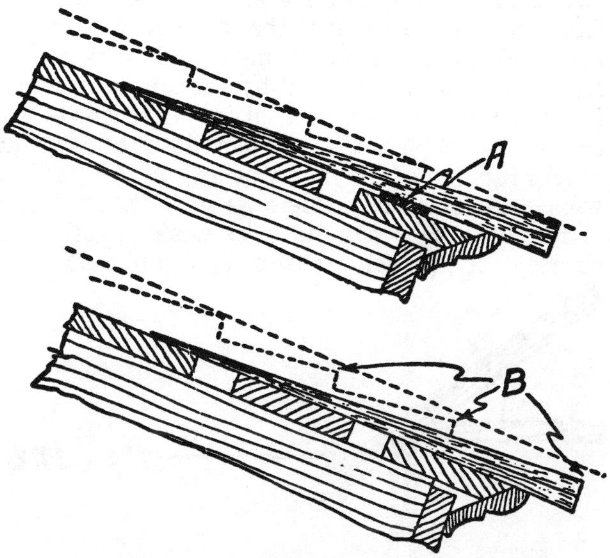

Fig. 495

the high points of the shingle courses are to be in alignment. This alignment is indicated by the dotted line at the top. At *B* in the bottom drawing the indicators point out the drop of the first three courses when the blocking is not used. The moulding shown in the upper drawing is higher then that of the bottom drawing. The upper method is technically correct, while the bottom one is more practical and more commonly used.

Miscellaneous

While one occasionally still sees the chalkline and rule used for lining and spacing courses for shingling, the major part of the spacing for shingling is done with a shingling gauge fastened to the hatchet. The line, though, must still be used in starting and frequently for straightening the courses or in adjusting the courses so that they will come out right at the comb.

Some shinglers use a shingling stool, but we believe the toe-holt is the safest and most practical for the average carpenter. Six, eight, ten or even twelve courses can be carried across the roof at once. Each shingler must determine for himself how many courses he can carry with a minimum of lost motion. Each shingle should receive two galvanized nails regardless of the width. Shingles that are green or water-soaked should be laid with tight joints, but well-seasoned shingles that are thoroughly dry should not be laid tight. One rule gives 1/8 of an inch spread for shingle joints. But, when the condition of the shingles is taken into consideration, the amount of spread becomes a matter of the workman's judgment. When the joint is open either at the top or at the bottom, the shingle should be trimmed so as to make the edges parallel.

Toe-Holts

Fig. 496 shows the height of the scaffolding for shingling the eaves of a roof. The distance between the staging and the edge of the roof varies from job to job because the distance must be determined by the man or men who work on the scaffold. This is indicated by the line called "Hip-Line" on the drawing. That is, the ledger boards should be so placed that when the runboards are on the hip line of the men working on the scaffold, will be approximately where we show it. (If a short man and a tall man have to work together, of course, a compromise must be agreed upon.)

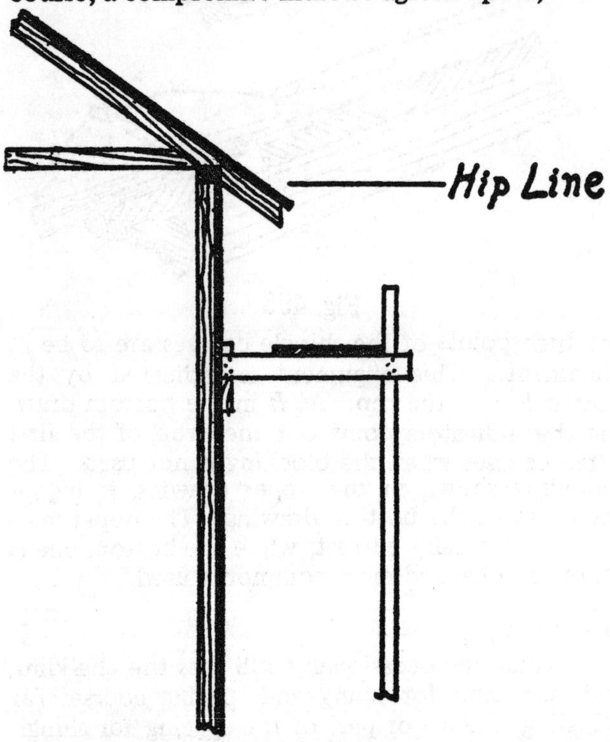

Fig. 496

A good method of fastening toe-holts is shown by Fig. 497. How the tie-shingles are nailed to the 2x4 is shown by the position marked *1* and *1*. After these shingles are on, the 2x4 is given a half turn which puts the shingles in the position shown at *2* and *2*. Points *a* and *a* show by dotted lines how the tie-shingles are cut off after the shingling is finished and the toe-holts are taken down.

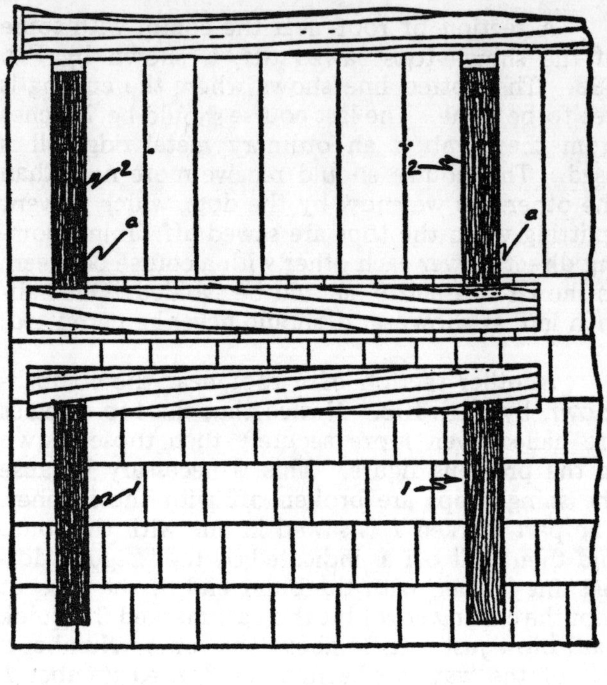

Fig. 497

Fig. 498 shows the same layout with extension cleats nailed to the regular toe-holt and also to the extended toe-holt marked *c*. This method of extending toe-holts is often used for the last lap of shingling at the comb of the roof when only a few courses of shingles that cannot be reached from the regular toe-holt are left.

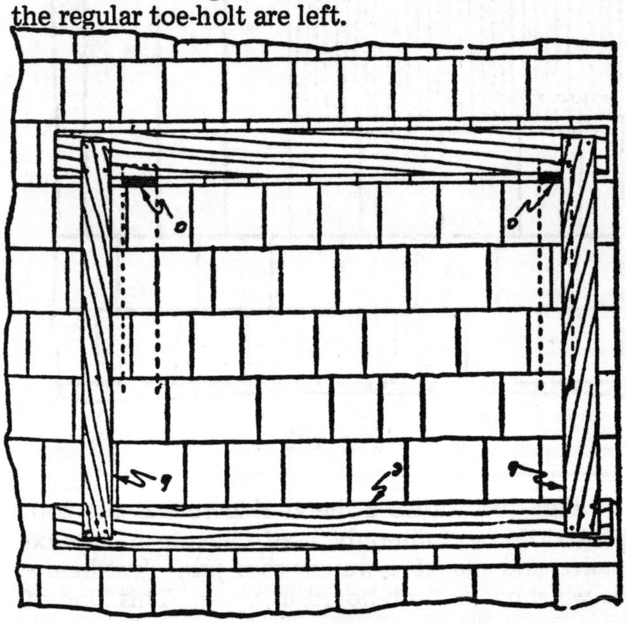

Fig. 498

Fig. 499 shows at *a* and *a* the butts of the tie-shingles after they have been cut off. Points *b* and *b* illustrate how much of the ends of the tie-shingles are covered by the third course of shingles. This lap is sufficient to prevent any damage because the shingle has been cut shorter than the original length. The dotted lines indicate the spacing of the shingle courses.

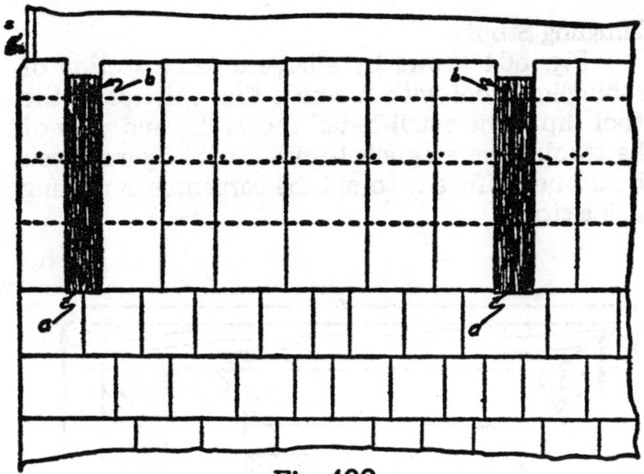

Fig. 499

A commonly used method of fastening toe-holts is illustrated by Fig. 500. In this case the position *1* and *1* represents the position of the tie-shingles after they have been nailed to the 2x4. Numbers *2* and *2* show the position of the shingles after the toe-holt has been fastened to the roof. Fig. 501 shows by dotted lines and dots the effects on the roof after the toe-holt has been removed.

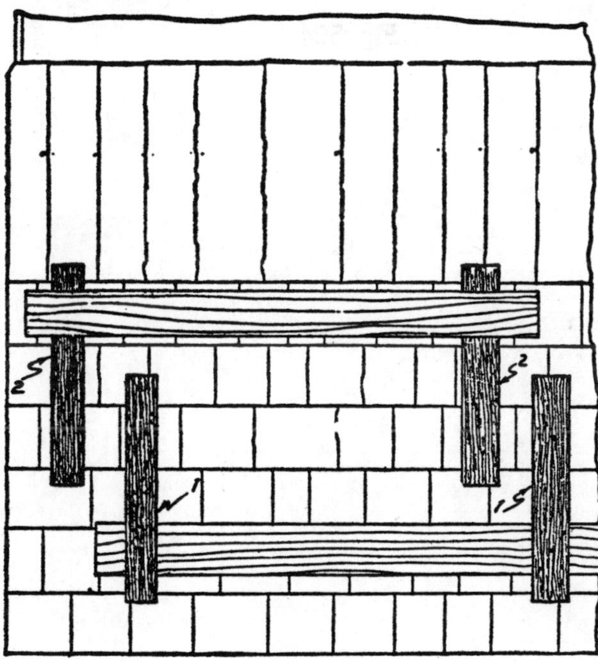

Fig. 500

The two shingles marked *X* and *X* have been driven up about ½-inch in order to close up the nail holes where the toe-holt had been fastened to the roof. This makes a fairly water-tight job.

A toe-holt fastened to the roof by means of wire ties is shown by Fig. 502. This is perhaps the most satisfactory way to fasten toe-holts since it does not injure the roof in any way. When the toe-holts are to be removed the wires are clipped off in line with the course of shingles that cover them. The twisting of the wire at *a* and *a* is very important -- also the fastening of the wire to the sheathing

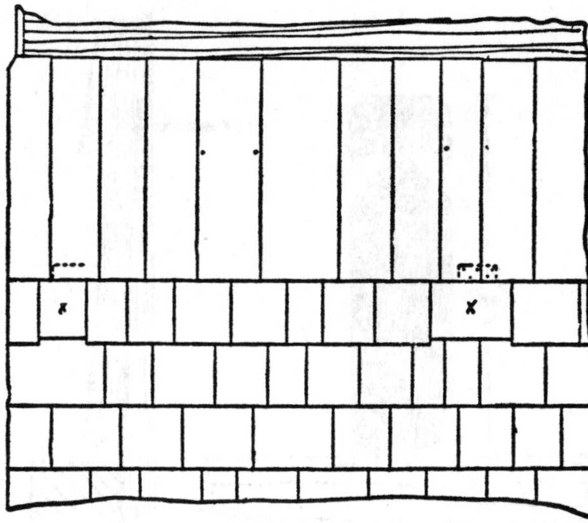

Fig. 501

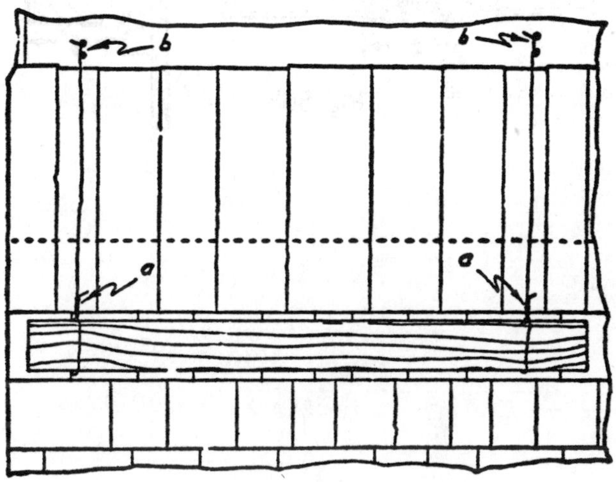

Fig. 502

shown at *b* and *b* where two nails are driven into two half-hitch loops for the purpose of fastening the wires securely.

Fig. 503 shows two details of a toe-holt hook sometimes used by shinglers. This hook is long enough so that the nails can be driven above the second course of shingles as we show in the drawing to the right. The drawing to the left shows the fastening of the hook to the sheathing and the toe-holt to the hook. The hook is shown fastened with only one nail. This is sufficient if the roof is not steep and the nail is large enough to carry the load. The sheathing into which such nails are driven should be perfectly sound. When two nails are used for fastening the hook to the sheathing, on removing it, the bottom of the hook must be worked back and forth like a pendulum. This movement will spread the nails so that the hook can be removed.

Shingling Stool

Fig. 504 shows by a large detail a section of a shingling stool with toe-nails that will not let the stool slip. The small detail shows the end view of the stool. These details together with figures given should be sufficient to aid the carpenter in making such a stool.

Fig. 503

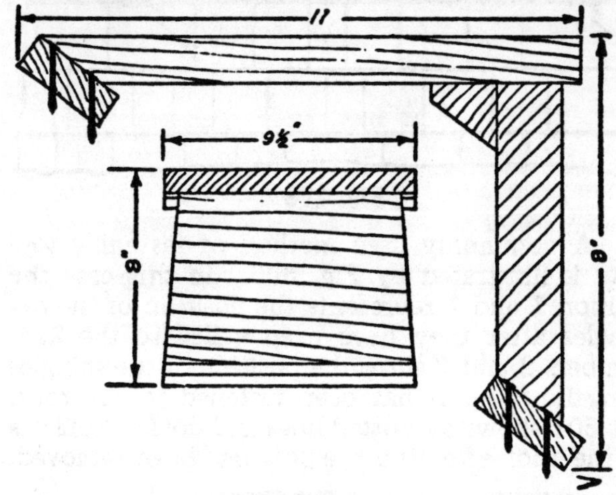

Fig. 504

34

Asphalt Shingles

Asphalt shingles have advantages over wood shingles. They eliminate much of the fire hazard and therefore may decrease the cost of insurance. They can be made to order, both in color and in shape, and can be laid on sheathing and over old shingles, with a saving on labor cost. The appearance of asphalt shingles is pleasing, and they do not require as much pitch in the roof as wood shingles do.

Method of Applying

Fig. 505 shows an oblong figure marked "roof", which represents one side of a plain roof. This roof has been divided into five sections, as indicated by the dotted lines in Fig. 506. Each of these five divisions represents a definite number of courses of shingles. This should be determined by the workman's reach. For example, if the workman can reach 8 courses from the scaffold, the first line from the bottom should be set for 8 courses. For the second division, if he can reach twelve courses from the toe-holt, the line should be set for that. In the same way the lines for the rest of the divisions can be established. This out of the way, the toe-holts should be placed somewhat as shown by the drawing.

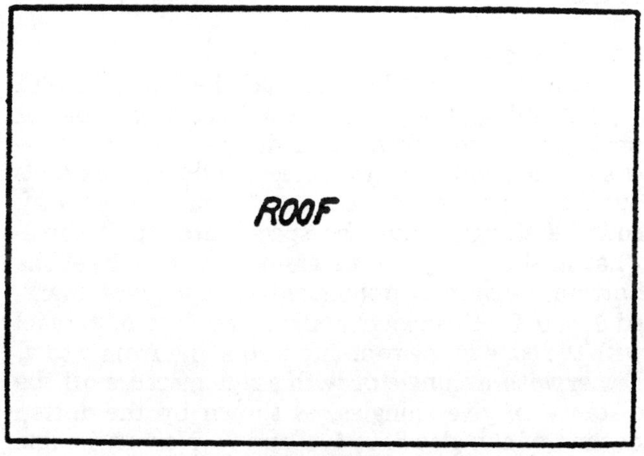

Fig. 505

The shingling is then started, not at the bottom as the custom is, but at the left of the top division, as shown by Fig. 507, section *1*, and working toward the right as indicated. When section *1* is shingled, remove the toe-holt in section *2* and begin shingling, starting at the point indicated by the arrow and working toward the right. The last course of each section following must be nailed under the first course of each preceding section.

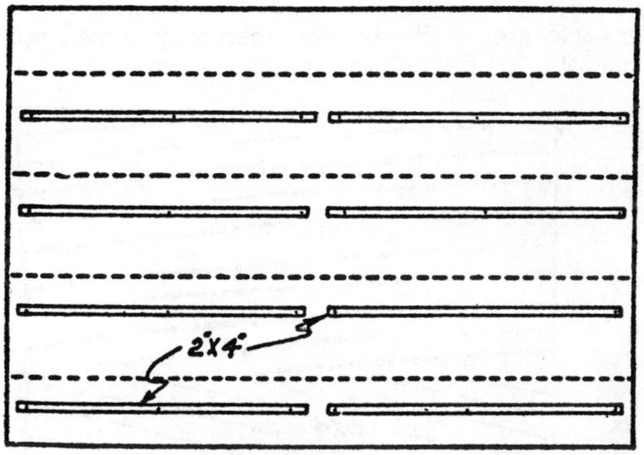

Fig. 506

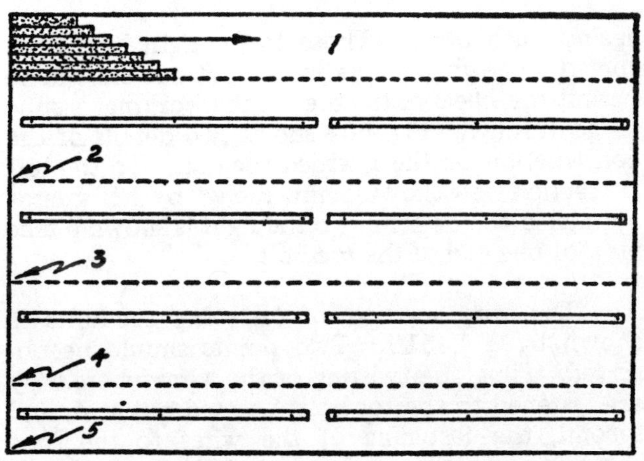

Fig. 507

This makes it necessary to lift the shingles of the first course of one section while the shingles of the last course of the next section are nailed. Section

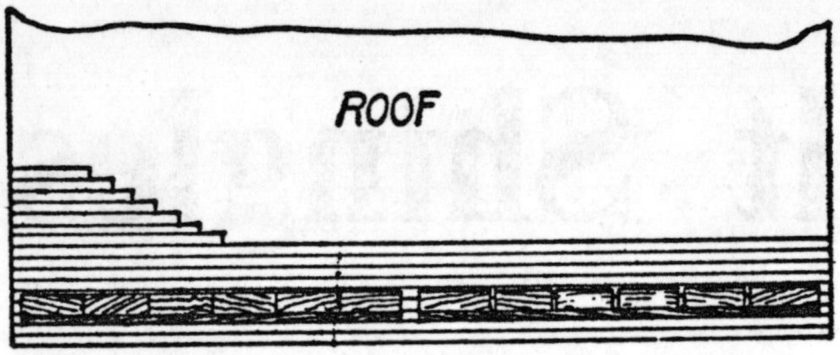

Fig. 508

2 shingled, shingle section *3* and section *4*. Section *5* will be the last to be shingled, which must be done from the scaffold. This method, as can be seen, prevents the workman from walking on the hot asphalt shingles and thus prevents shoe damage to them.

Another method of protecting asphalt shingles from shoe damage is illustrated by Figs. 508 and 509. The first of these figures shows the bottom part of a roof partly shingled. The first toe-holt is in place. It provided with guards made of the covers that are on the shingles when they come onto the job. To the left the guards are shown butting

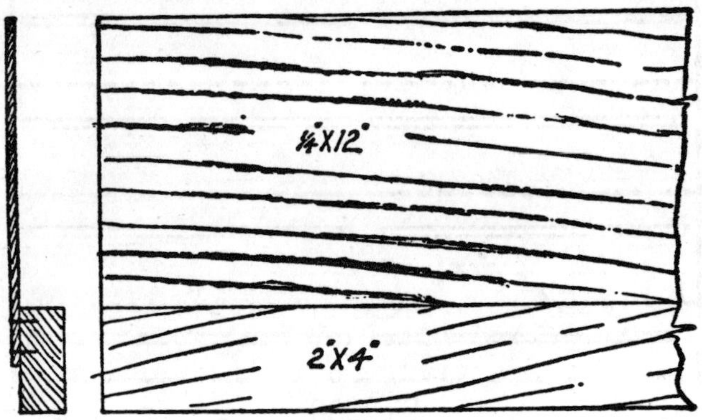

Fig. 509

against each other. Those to the right have been spread somewhat at the joints, but not enough to permit the heels or the toes of the workman's shoe to go through. Fig. 509 shows two details of the construction of the guarded toe-holt. To the left is a section view showing how the ¼" by 12" guards are nailed to the 2x4. To the right is shown a face view of one end of the toe-holt.

The wire method of fastening toe-holts is shown by Fig. 510. Two points should be observed. First, the twisting of the wires where they are fastened to the toe-holt, pointed out at *A*, and second, the fastening of the wires to the roof sheathing. This is pointed out at *B*. To the left is

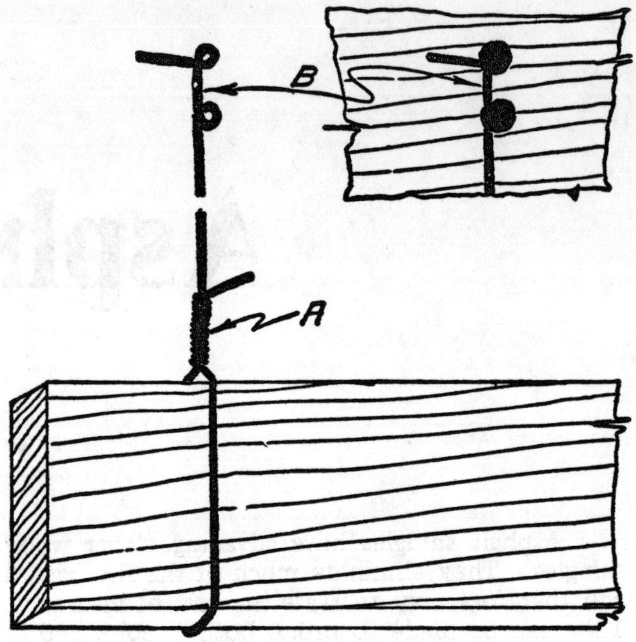

Fig. 510

shown how the wires should be looped, and to the right is shown the manner in which the wires are nailed to the roof sheathing.

Permanent Shingles

Neither wood nor asphalt shingles make a permanent roof. Both are perishable and therefore the life of such roofs come to an end through deterioration. In a general way we refer to such roofs as permanent, but from a standpoint of fact, both wood shingles and asphalt shingles must be classed as temporary roofing materials. There are three roofing materials that last for more than a lifetime, excluding metals. They are tile, slate and asbestos. These materials will not deteriorate by reason of decay and will, if they are without flaw in the beginning, last as long as the structure that holds them. While these materials are more expensive than wood or asphalt, in the long run they are a great deal cheaper. Only the best quality and most permanent buildings are covered with slate, asbestos or tile.

Ashpalt Problems

Fig. 511 shows how to start the first course of asphalt shingles when the roof has some sort of obstruction (for instance, a dormer that must be worked around). In this case the dormer extends out from the main building at the eave of the roof, and the shingles must be spaced around it somewhat as shown by the diagram. To the left, at the bottom, we show two shingles tacked in place, marked *1* and *2*. Strike a chalkline from *a* to *b*, parallel with the edge of the roof and also a line from *c* to *d*. Either with a shingle or with a rule measure off the distance of five shingles, as shown by the dotted outline of shingles *3*, *4*, *5*, *6* and *7*. Now strike line *e-f* parallel with the edge of the roof and crossing

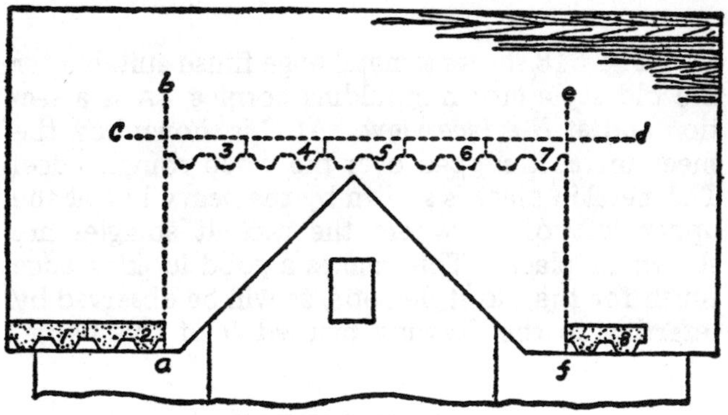

Fig. 511

the right edge of outline number 7. Then tack in place shingle number 8, and go on with the shingling. Shingles of the alternate courses must be kept relatively the same where they contact the lines *a-b* and *e-f*.

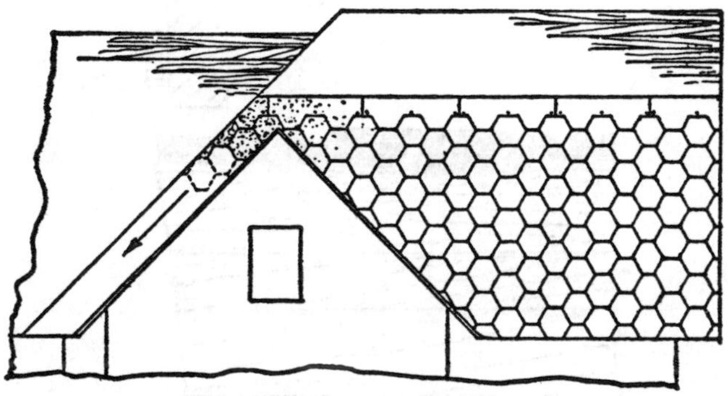

Fig. 512

A similar dormer located close to a hip is shown by Fig. 512. The dotted outlines of two crippled shingles and the arrow show a good way to shingle crippled spaces of roofs so that the shingles will meet properly at the comb of the dormer. While it is possible to shingle such spaces from the eave up, it is much safer and may save labor to shingle them from the top down. Of course, in doing this, the shingles in the last course must be lifted so the shingles in the next course can be nailed.

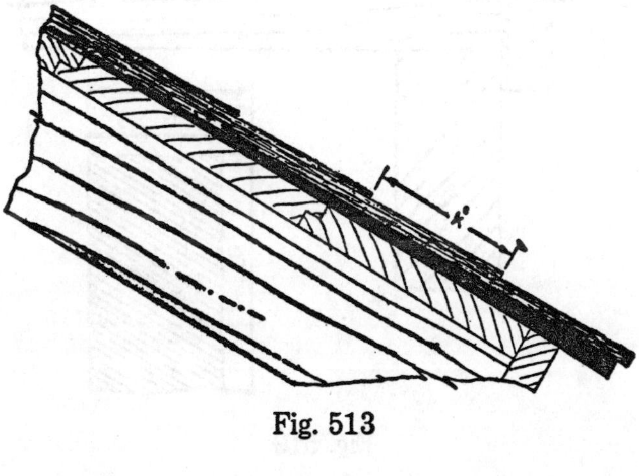

Fig. 513

Fig. 513 shows one of the best methods of supporting the drip of asphalt shingles at the eave when an inch or more drip is desired. The support is represented by the heavy shading, showing an edge view of a wood shingle which projects about 1½ inches beyond the fascia. A course of wood shingles put on in this way will not only support the drip but will block up the first courses of the asphalt shingles so that they will not have a dubbed-off effect. The asphalt shingle should project not less than ½ of an inch.

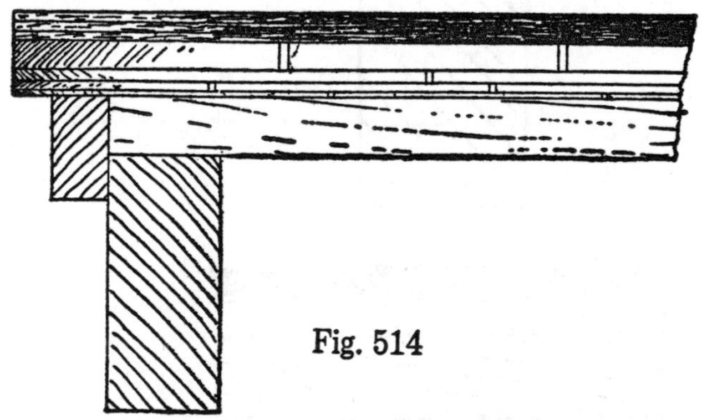

Fig. 514

A section of an edge of a roof having a bargerafter finish shingled with asphalt shingles over wood shingles is shown by Fig. 514. A face view of this edge is shown by Fig. 515. These illustrations show a common method of finishing the edges of roofs when asphalt shingles are put on over wood shingles. (Notice the open spaces marked

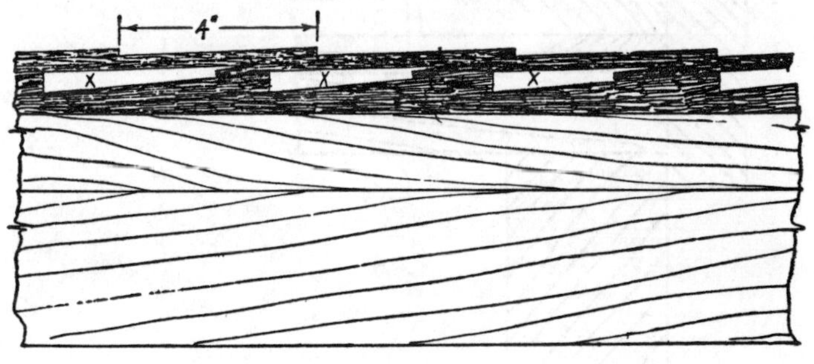

Fig. 515

with X's.) This is probably the most unsightly edge finish that can be used with asphalt shingles. A much better edge finish is shown by Fig. 516, where *A* shows a section and *B* a face view of the edge. Here enough of the wood shingles are removed at the edge to permit a 1x4 edge-board to be put on as shown by the two drawings.

A good edge finish is shown by Fig. 517, where *A* is a section and *B* a face view. In this finish the wood shingles have been trimmed off in line with the bargerafter and a sort of fascia has been nailed over the edges of the shingles and ends of sheathing with a little lap over onto the bargerafter.

Fig. 518 shows a metal edge finish suitable for the old style crown-moulding cornice. *A* is a section and at *B* a face view. At *C* is shown how the sheet metal is slipped over the wood shingle edge. The metal in place is shown by the heavy line at the upper left of *A*, where the asphalt shingles are shown in place. This makes a good looking edge finish for the old-style roof, as will be observed by referring to the drawing marked *B* of this figure.

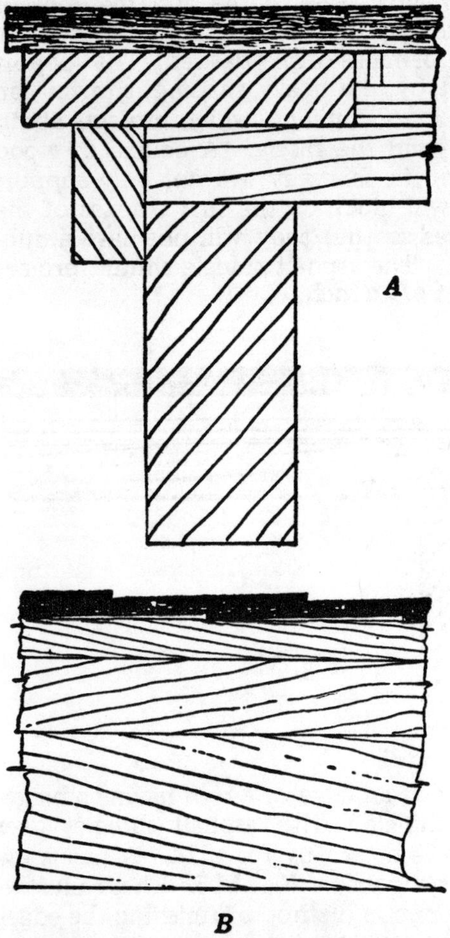

Fig. 516

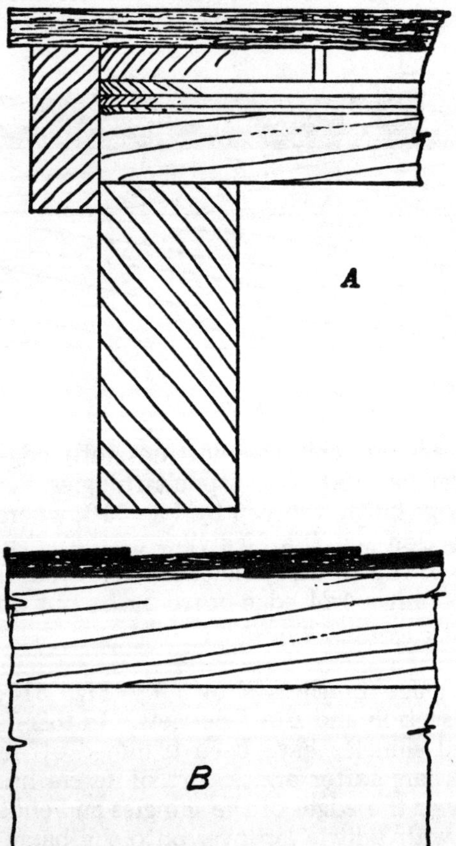

Fig. 517

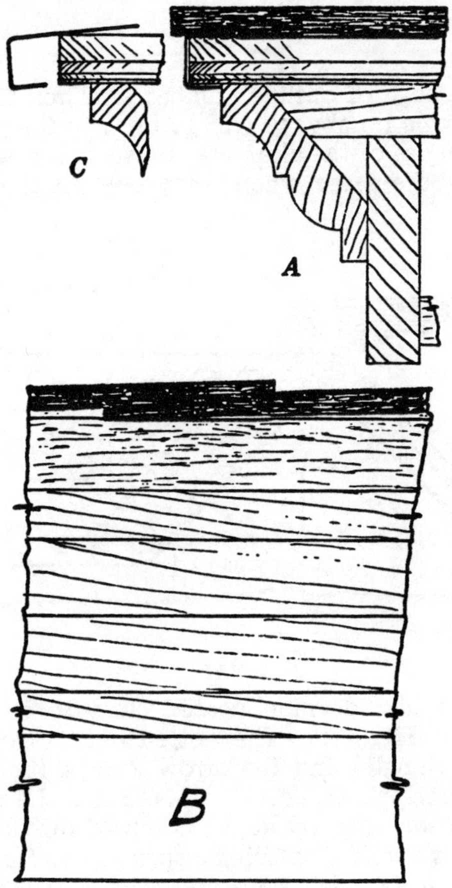

Fig. 518

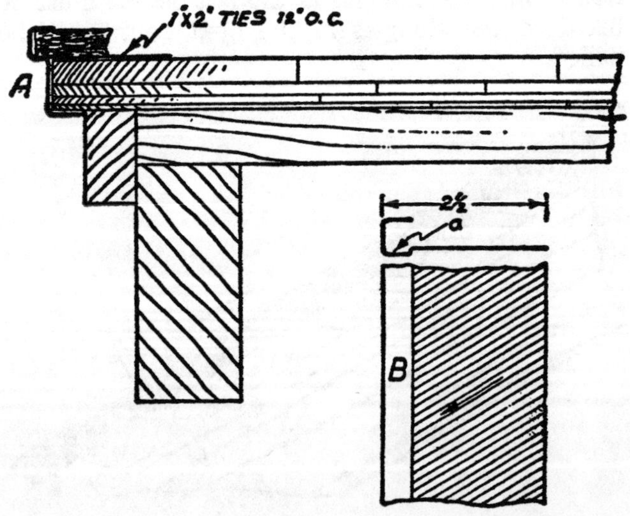

Fig. 519

Two metal edge finishes are shown by Fig. 519. At *A* the heavy lines give an end view of the metal. Metal ties 1x2, 12 inches on center are pointed out. These ties are soldered to the main strip as shown and fasten the edge finish to the roof. At *B* we show a top view of a metal edge finish suitable for use on a new roof. The shaded area represents corrugations and the arrow pointing downward to the left indicates how these corrugations carry the water to the edge where a trough carries it off, as pointed out by the section at *a*. The shoulder shown, caused by the trough at *a*, is held against the edge of the crown moulding or crown moulding substitute when the metal finish is in place.

35 Roll Roofing

Roll roofing is made in 1, 2 and 3-ply thicknesses. Some of these roofings have a slate covered top. That makes them a little more expensive, but gives the roof a better appearance and, as a rule, makes it last a little longer.

The 1-ply roofing is acceptable for roofs that are to serve for only a season, or perhaps two. The 2-ply will give a little longer service, and the 3 ply, if it is kept well painted, will last for many years unless damaged by hail or some foreign object.

Roofing Problems

Fig. 520 shows a side of a pitch roof covered with roll roofing. The arrow at No. *1* points out how the comb covering should be cut into not-too-long pieces. At *2* the piece of comb covering is too long. The pieces of roofing numbered *3* are too long. The end joints shown numbered *4* are bad

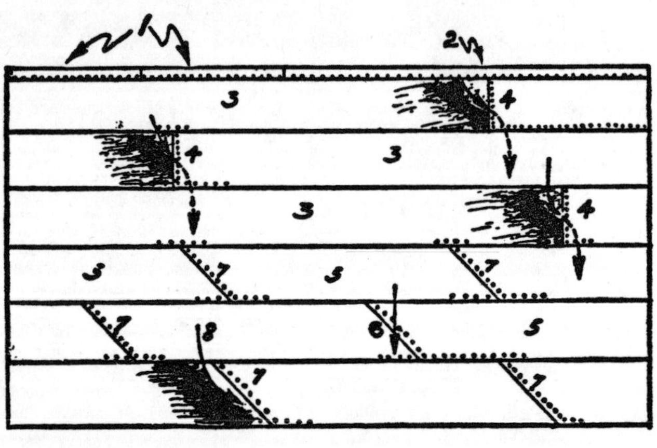

Fig. 520

for, as the arrows indicate, if the roofing buckles, as it often does, the water will be carried through any open joint resulting in a leak. The pieces numbered *5* are cut as they should be. The end joints numbered *7* are cut on a 45-degree angle, making a water-tight joint. At *6* the arrow shows how the water will run over the joint and *8* shows that even if the roofing buckles, the water will not be carried through the joint.

Four edge finishes for roll roofing are shown by Fig. 521. At *a* is a good finish. This fits snuggly to the sheathing and projects a little below the edge, which provides a drip. At *b* is a similar edge finish reinforced by a strip of wood. The edge finishes shown at *c* and *d* are not good. Even if they should give the required service they do not provide a drip edge to keep runoff away from the side of the structure.

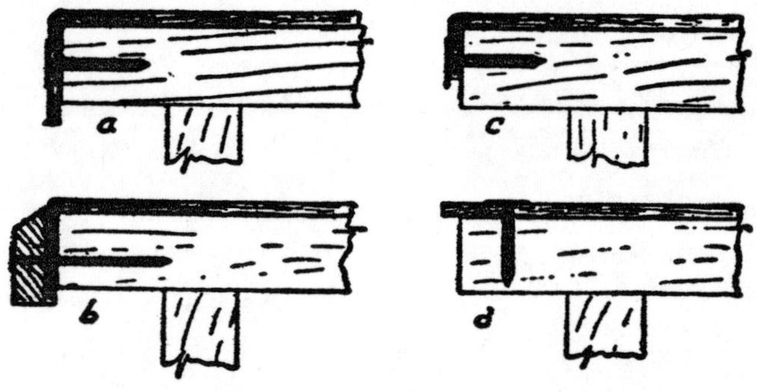

Fig. 521

Three laps are shown by Fig. 522. *1* is a good lap. Number *2* and *3* are both bad. Fig. 523 shows other laps and nailings. Number *1* is a good lap onto a valley. Note how it was nailed. Number *2* and number *3* indicate gross carelessness, and numbers *4* and *5* are good examples of what happens when the workman tries to do more nailing than he has time for.

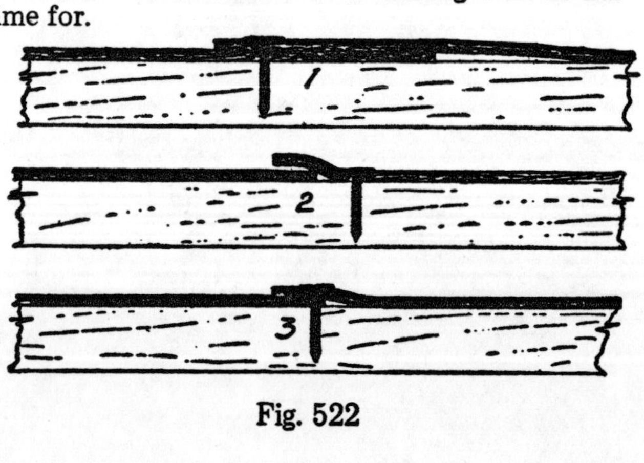

Fig. 522

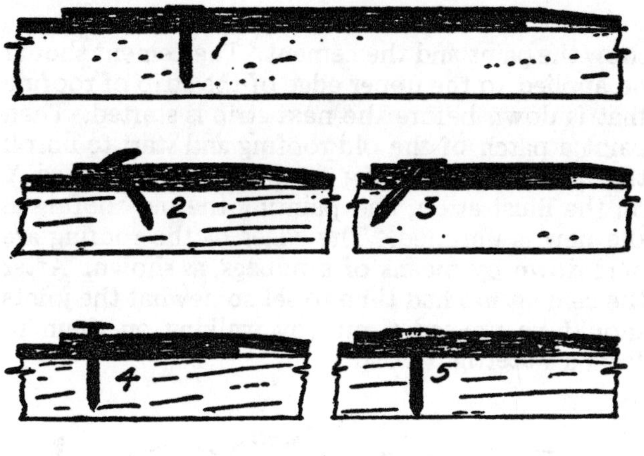

Fig. 523

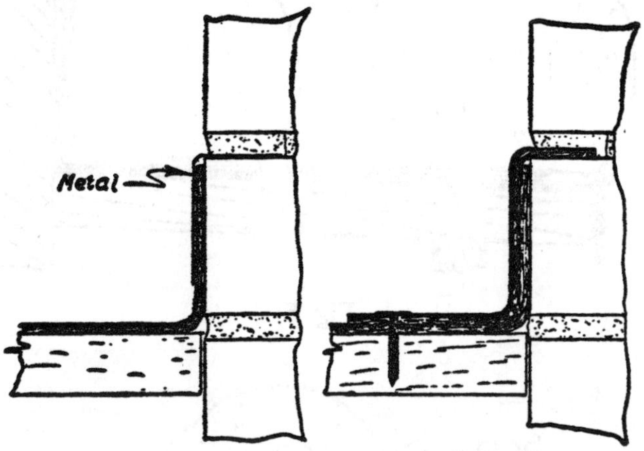

Fig. 524

Two methods of flashing where the roofing joins brick or stone work are shown by Fig. 524. The illustration to the left represents a metal flashing and the drawing to the right shows a flashing made of roofing. This should have a coating of plastic or fibered cement on the under side before

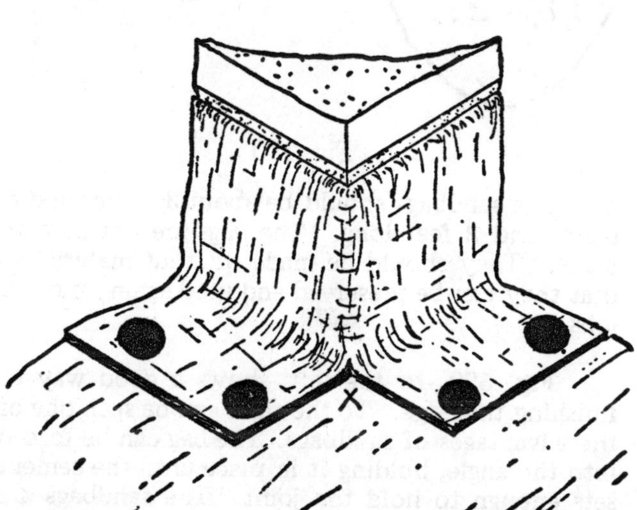

Fig. 525

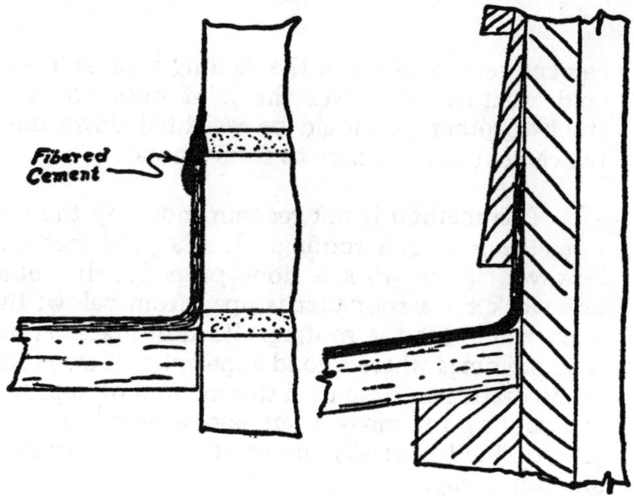

Fig. 526

it is put into place. How the corner is made is shown by the perspective view given by Fig. 525. The area marked *X* should be covered with plastic cement, somewhat on the order shown by Fig. 527. The flashing should be well nailed as shown.

Two ways of joining roofing to a wall without counterflashing are shown by Fig. 526. To the left we show how the end of the roofing is lapped onto a brick wall and sealed with plastic cement. To the right we have the roofing slipped behind the first piece of siding, this makes a water-tight job. A good corner finish for any of these flashing designs is shown by Fig. 527.

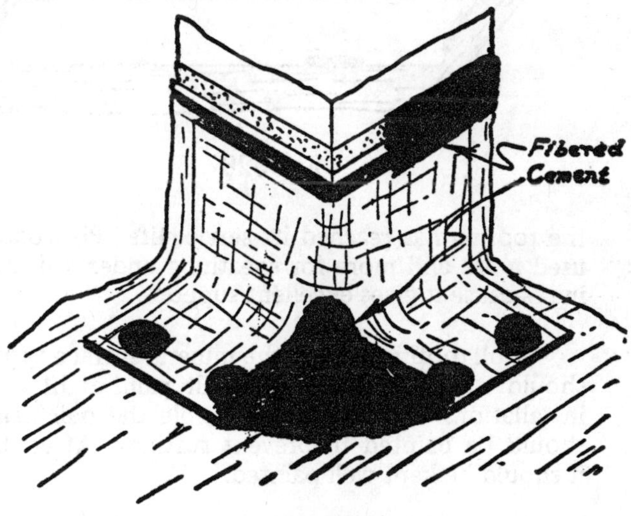

Fig. 527

Limited Nailing

There is a way of putting on roll roofing and keeping the nailing concealed. The strips of roofing are put down in the customary way but the upper edges only are nailed to the sheathing just enough to keep the roofing from slipping while it is put on. The bottom edge of the roofing is sealed to the strip below with plastic cement. This method works especially well in hot weather for then the cement sets readily and the roofing is plyable. In cold weather, whenever the joint does not stay stuck together, it should be weighted down until the cement has had time to form a bond.

This method is not recommended by the manufacturers of roll roofing. It is a good method, however, if the work is done properly. It would not work on a roof that is open from below; the wind would lift the roofing. Its most practical use is on buildings where a good appearance is essential. Some craftsmen have used this method of applying roll roofing for many years and wouldn't put on roll roofing (especially the graveled-top roofing) in any other way.

The sheathing for roll roofing should receive special care for the quality of the sheathing will determine how long the roof will last. Wide cracks should be avoided and holes should be covered with stiff sheet metal. The sheathing should be well seasoned and dry when the roofing is put down. Warping, shrinking and swelling of the sheathing are often the cause of leaks in roll roofing long before

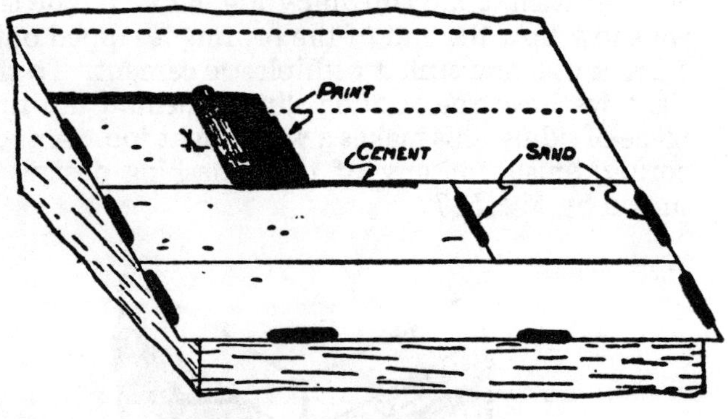

Fig. 528

the roofing has reached its service life. Plywood is used more and more for sheathing under roll roofing because it gives excellent support.

Roll roofing (except the mineral surface types) should receive a coat of asphalt paint right after installation. If that is impossible the nail heads should be painted to prevent rusting. After that it should be kept well painted.

Nailless Methods

Roll roofing can be put on over old roofing without the use of nails. This is the best way to put roll roofing over old roofing. How this is done is illustrated by Fig. 528. This figure shows a shed roof where two strips of roofing (shown somewhat shaded) are in place and a third is partly on. The dotted lines show joints of the old roofing. Begin by painting a section of the old roof where you want to start and then unroll the roofing, making the joints where necessary. This is shown by the partly-applied strip of roofing in Fig. 528. Arrows show the paint and the cement. The cement should be applied to the upper edge of the strip of roofing that is down before the next strip is started. Then paint a patch of the old roofing and start to unroll the next strip, working from the point marked *X* on the illustration, and painting the old roofing as the new is unrolled. The edges of the roofing are held down by means of sandbags, as shown. After the cement has had time to set somewhat the joints should be pressed firmly by walking on them to form a watertight bond.

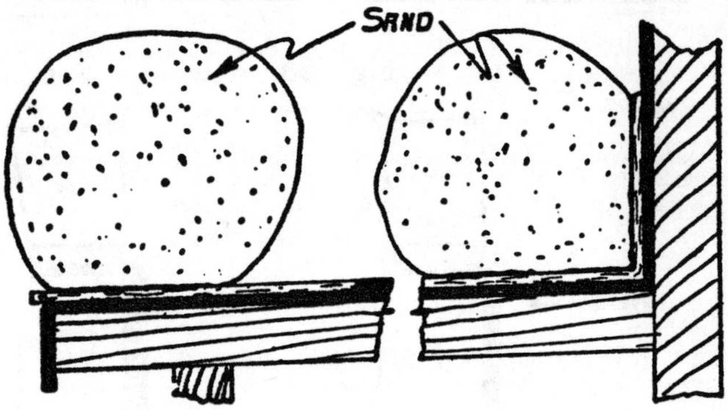

Fig. 529

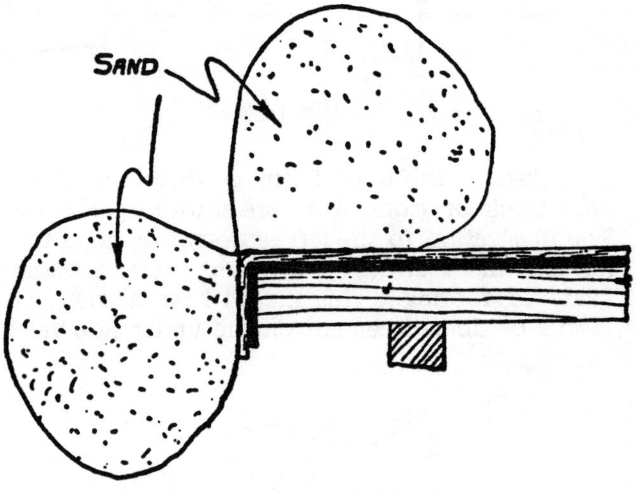

Fig. 530

The sandbags should be about 3 inches in diameter and 2 feet long. The bags are not hard to make. They should be made of stout material so that they can be preserved and used again on other jobs.

Fig. 529, to the left, shows a good way of finishing the edge. To the right can be seen one of the advantages of sandbags. The bag can be forced into the angle, holding it in place until the cement sets enough to hold the joint. The sandbags are shown exaggerated in size.

Another edge finish is shown by Fig. 530, where the new roofing is bent over the edge and a little beyond the old edge finish to give the roof a drip edge. A double sandbag is used to hold the joint until the cement is set. Sometimes the upper part of a double sandbag must be reinforced with extra weight in order to keep it from slipping off the roof.

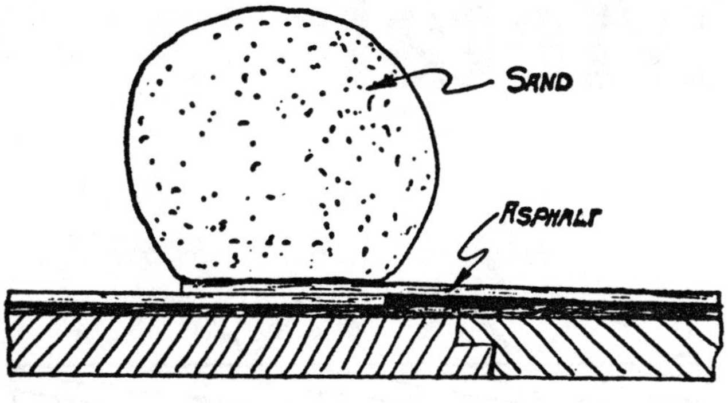

Fig. 531

A lap joint that is nailless and is held temporarily in place by a sandbag is shown by Fig. 531. To the right of the joint there is a pocket under the new roofing which has been filled with plastic cement to keep the surface smooth. It should be noted that in all of these illustrations we indicate the old roofing with heavy shading and the new with light shading.

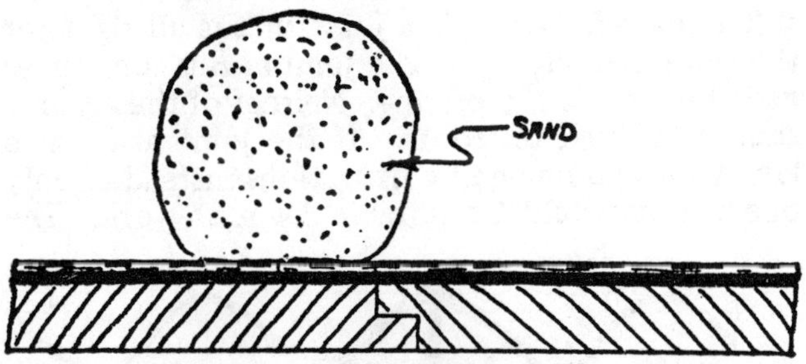

Fig. 532

A butt-joint is shown by Fig. 532. This is perhaps the best joint possible (if carefully made) where the old roofing is still rather good. A coating of fibered cement should be placed directly under the joint, and, when the roof is on, the whole surface should be painted with asphalt paint.

Fig. 533, at the top, shows a common fault of roll roofing – a buckle. Before laying new roofings, such buckles should be cut as indicated. The results are shown at the bottom. Roll roofing ought

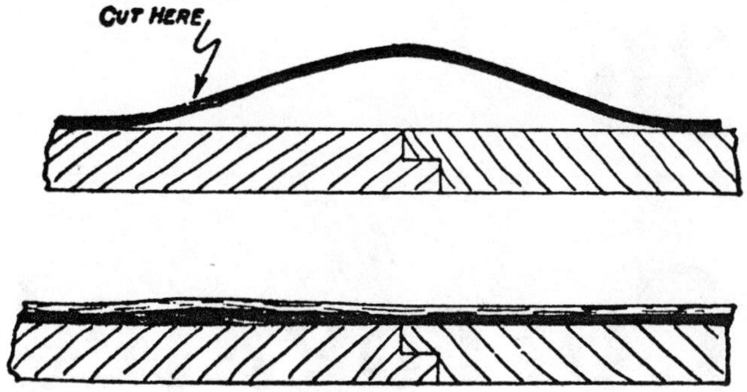

Fig. 533

to be put down when the weather is warm. This is especially true when the method we are dealing with here is used. The roofing should be warm enough so as to render it plyable. When it is put on in cold weather, it does not have enough flexibility to bend without breaking, nor will it fit itself to the roofing support or angles, as it should.

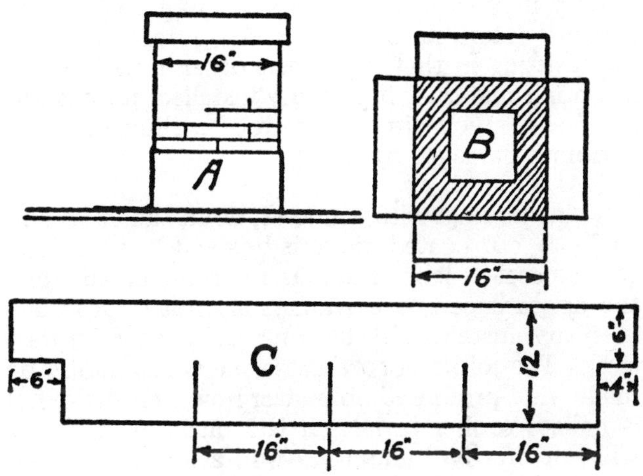

Fig. 534

Fig. 534 shows, *A*, an elevation of a chimney on a flat roof, *B*, a plan of the chimney, and *C* how to cut the flashing for this chimney. The figures given on the drawing give all the information needed. When the flashing is cut to the shape shown, give the chimney and the roof where the flashing is to go a good coat of roofing cement and then stick the flashing in place. The flashing should be weighted down until the cement has had time to set. All the edges should be carefully examined before leaving the job to be sure that the joints are all waterproof.

36

Corrugated Metal

Corrugated metal has a very practical place among the building materials and is used more extensively in rural areas. It is used both as siding and roofing. Galvanized metal is durable and relatively inexpensive, and does not require paint to prevent deterioration. It is fire-proof itself and fire-resisting in that it protects other materials. As a roofing material, if properly installed, it does not require as much pitch in the roof as shingles and it seldom is damaged by hail.

In buildings with windows, doors, dormers and so forth, corrugated metal is less and less suitable, for the cost of labor and material is increased while the appearance and usefulness are not improved -- in many instances it becomes altogether impractical. The joints of corrugated metal can hardly be made wind-proof and no matter how well the metal is nailed, sooner or later it will rattle or grate. It will not brace a building as well as lumber or plywood siding. With corrugated metal siding the frame must be thoroughly braced if it is to hold its shape. The last disadvantage of this material is that when it has deteriorated there is little that can be done to salvage it.

A sheet of corrugated metal 6 feet long and 26 inches wide is shown in Fig. 535. It is sold in sheets 26 inches wide and 6, 8, 10 and 12 feet long. These lengths make possible a great deal of variation in the sizes of the buildings. Little cutting should have to be done since cutting is rather difficult.

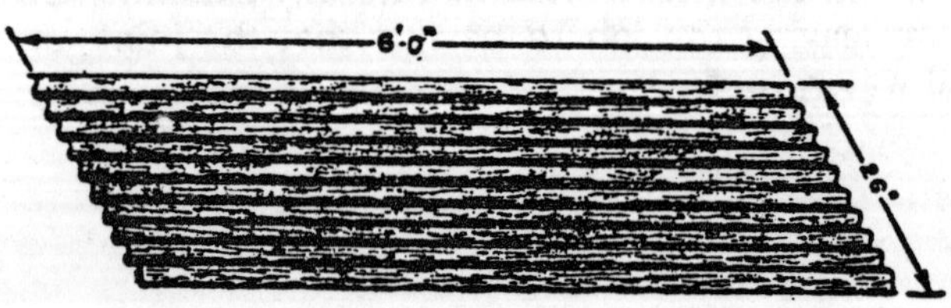

Fig. 535

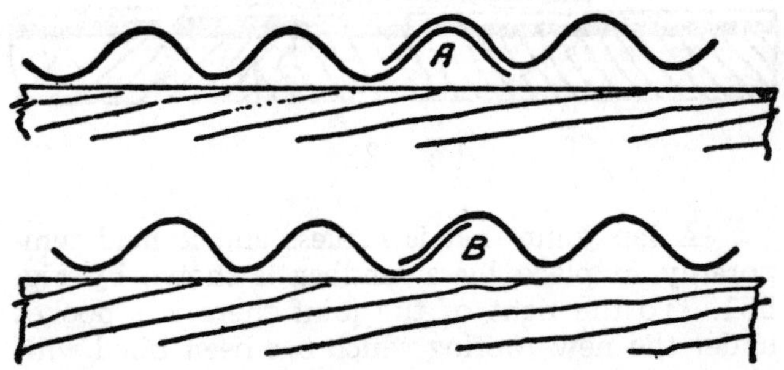

Fig. 536

Use of Corrugated Metal

Fig. 536 shows cross sections of two joints still unnailed. At *A* is a joint that is all right for the sides of buildings. The joint at *B* is sometimes used but is not a good one. Neither of these joints should be used on roofs. If the joint at *B* has a lap of one and one-half corrugations instead of only one-half it would be suitable for roof work. The

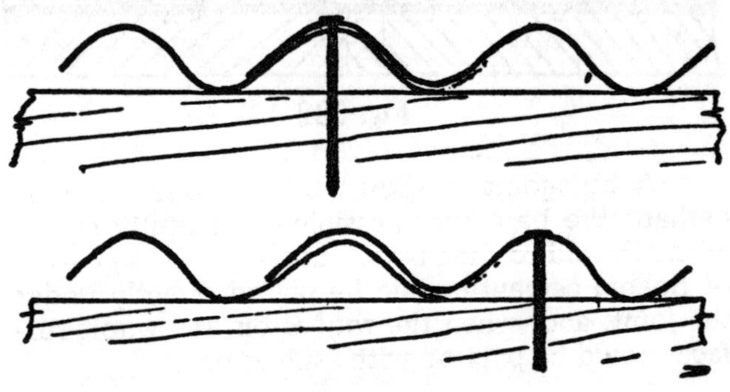

Fig. 537

lap shown in Fig. 537 is right for use on corrugated metal roofs. The right and wrong ways of nailing the joints are also shown in this figure. The upper drawing shows the right way and the bottom one the wrong way. The top of Fig. 538 shows (to the left) the right way to nail between joints and (to the right) the wrong way. The bottom drawing shows what happens when the nailing is carelessly done.

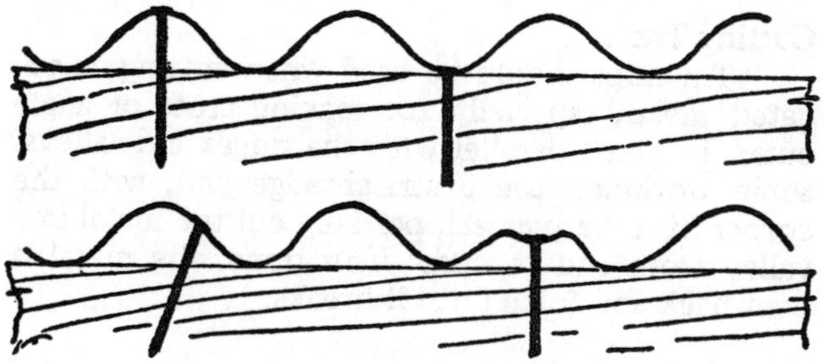

Fig. 538

Fig. 539 shows a good comb construction, both for the metal itself and for the support and nailing provisions. Instead of using two nailing strips, a single nailing strip can be used so that it

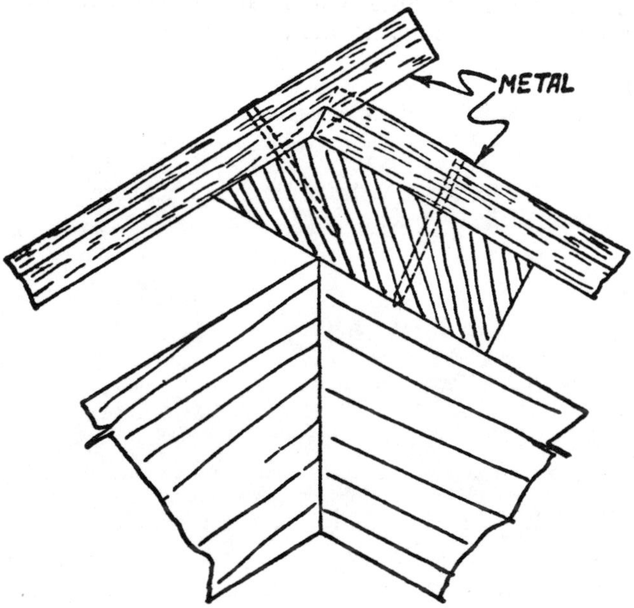

Fig. 539

will give all the service required. Fig. 540 shows a good method of eave construction. Here we show the plate, the foot of the rafter and the bottom

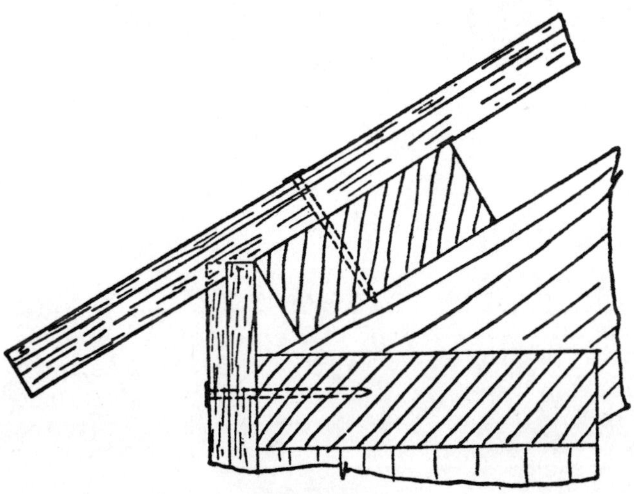

Fig. 540

nailing strip in addition to the metal as it is nailed on. To obtain the best results, nailing strips should not be smaller than 2x4's spaced 3 or 4 feet on center. Use the same dimensions for the rafter spacing.

Fig. 541 shows a rake construction where both the sides of the building and the roof are covered with corrugated metal. This method does not leave

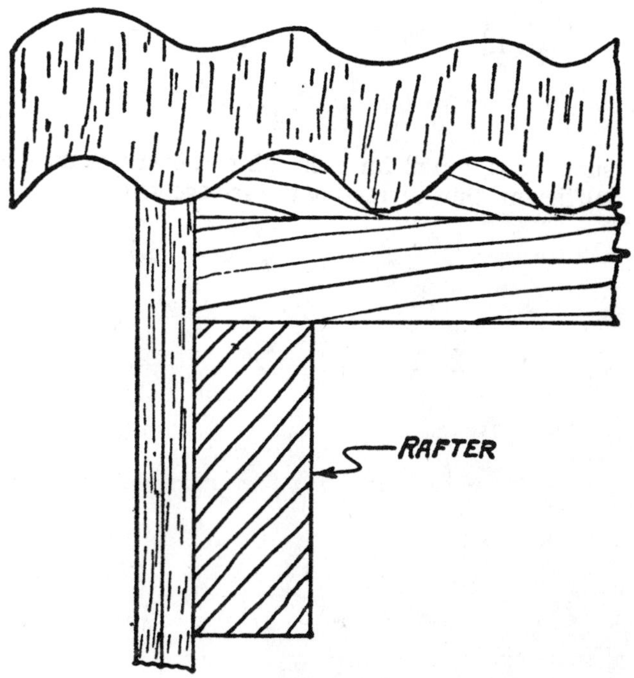

Fig. 541

any wood exposed and for most purposes will give the required service. If, though, a more conventional rake is desired, it can be used with corrugated metal roofing.

Another method of framing the skeleton roof for corrugated metal is shown by Fig. 542. The upper drawing is a section view showing the ends of two rafters with headers for supporting the roofing between the rafters. The rafters are spaced 2 feet on center. The bottom drawing shows, from left to right, the construction of the eaves, the comb, and the relative position of the header to the rafter.

Nailing Problems

A word should be said about nailing. A good rule is to nail every third ridge to every bearing or nailing strip. If, however, the nailing strips are rather closely spaced or the building is to serve for only a short time, then the workman should use his judgment in the matter of nailing. Where it is known that a building is to be dismantled in a short time, the fewer nails used the better, keeping in mind a relative substantial construction. In most

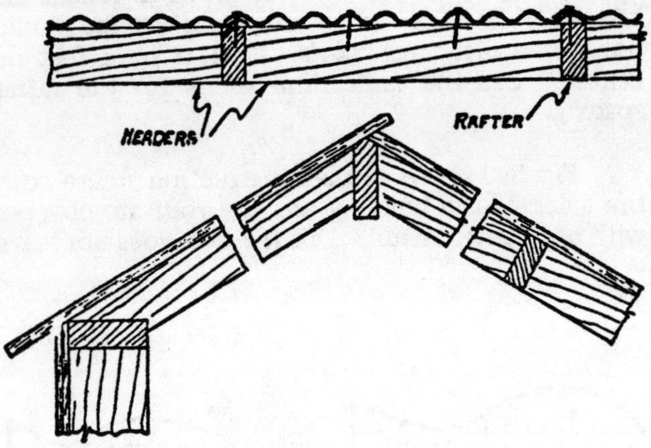

Fig. 542

such cases, a nail for each joint at every bearing is sufficient, thus increasing the salvage value.

Cutting Tools

Tin snips should be used when cutting corrugated metal, especially for making cross or angle cuts. For cuts parallel with the ridges and valleys, some workmen use a straightedge and, with the corner of a sharp chisel, partially cut the metal in a valley from end to end. This done, the metal is bent back and forth until it breaks.

Where a corrugated metal roof joins the side of a building covered with corrugated metal, flashing is difficult. The best method is to use tin flashing and fit it as closely as possible and then seal it with plastic cement to the side of the buildings. The edge of the flashing that comes onto the roof should not be sealed, for expansion and contraction of the metal in time will break the cement from the side of the building and cause leaks.

37

Door Jacks and Butts

Hanging Doors

The carpenter who fits and hangs an ordinary door in one hour can be considered a good average door hanger. Of course, there are many conditions that enter into a door-hanging job. The door itself has much to do with it. If the door is made of hard or tough material and requires much dressing off, it will require more time than if the door is made of soft material that dresses easily. A heavy door is harder to hang than a light one.

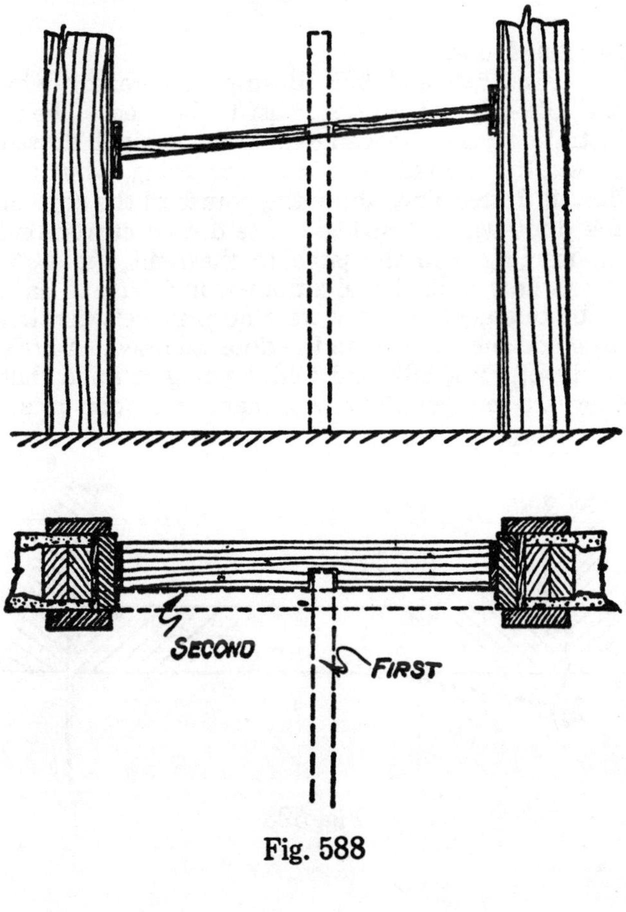

Fig. 588

A pretty good job of door fitting can be done with a jack plane, and many carpenters do not use anything else. Other carpenters use the fore plane, but for the better class of work, the door fitting should be finished with a jointer. Of course the jambs will have to be set perfectly plumb and true in every respect. If the jambs are warped and out of shape, a jack plane will in most cases give the best results. This, perhaps, is the reason why so many door hangers use the jack plane exclusively for fitting doors.

Door Jacks

Fig. 588 shows in elevation and plan views one of the first door jacks this writer used. It consists

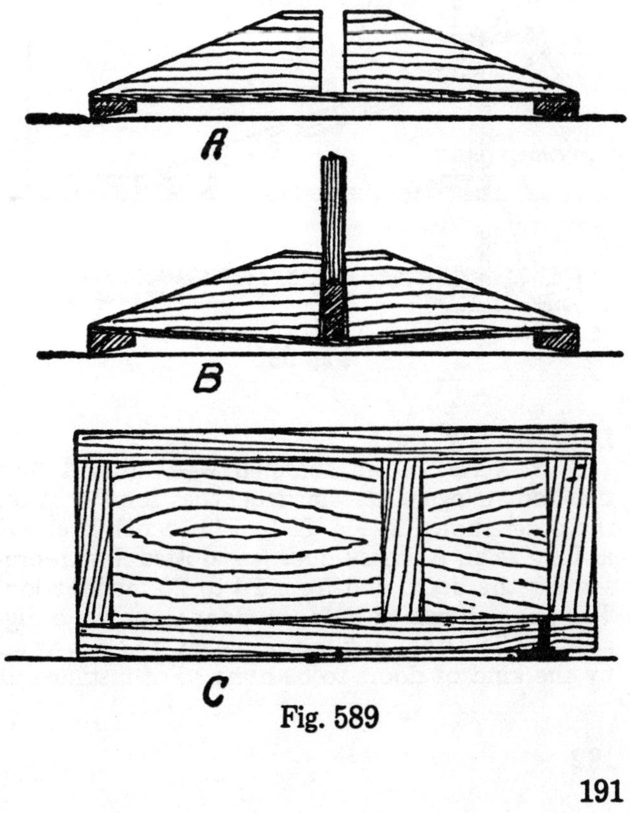

Fig. 589

of a piece of board with a block nailed at either end and a little longer than the distance between the jambs so that it will stick between them. A notch cut at about the center holds the door on edge and in an upright position. This is indicated by the dotted lines on both the elevation and the plan. On the plan it is marked "First," and shows the position of the door when it is being dressed. the dotted lines marked "Second" show how the jack keeps the door from going too far back when it is set up in the jamb for trial, making the jack serve two purposes. The second service, however, is eliminated when the jambs are rabbeted.

Another door jack in three positions, is shown by Fig. 589. At *A* is a side view giving the greater part of the construction. The jaws are made of either 2x4 or 2x6 material. The springboard is made of tough springy wood about 3/8 of an inch thick. At *B* is shown how the weight of the door springs the board so that the jaws pinch the door and hold it in an upright position. At *C* a side of the door is shown and to the right, shaded, an end view of the door jack in position at a much-reduced scale.

Perhaps the most practical door jack that is used by carpenters, is shown by Fig. 590. This jack is also used for fitting window sash. The designs shown are not the only ones in use. In fact almost every carpenter has some special feature that he likes and therefore he adds to, or takes away, until he has a door jack that is strictly of his own design. The material available when the door jack is made also influences the style. It is interesting to observe the door jacks in use on a large job where a number of carpenters are at work hanging doors.

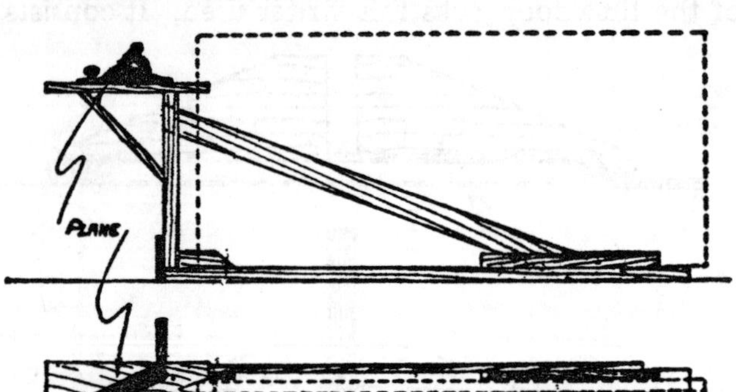

Fig. 590

Door Hanging Bench

We have known foremen who insisted that the door hangers use small benches somewhat on the order shown in Fig. 591. This bench is about as long as an ordinary door (see dotted lines representing the door) and from 18 to 20 inches wide. The hook shown to hold the door is merely a suggestion. The door holding device must be governed by the kind of doors to be hung. For instance, if

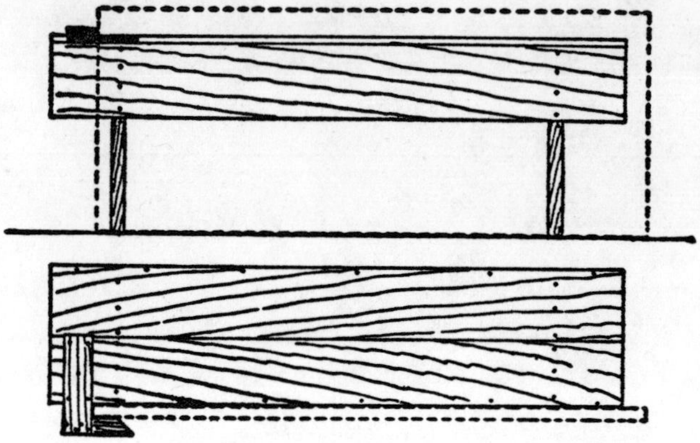

Fig. 591

small cabinet doors are to be fitted and hung, the carpenter must figure out some way of holding the doors so he can dress them. There is no jack equal to the bench arrangement for fitting small doors.

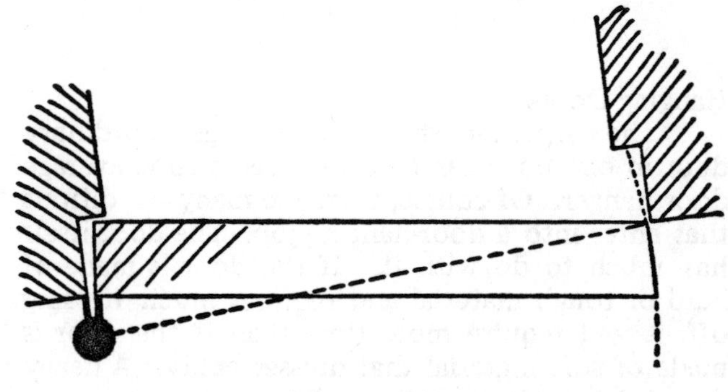

Fig. 592

Beveling Doors

Figs. 592 and 593 illustrate the reason why the lock-edge of a door must be beveled. The illustrations are exaggerated. The straight dotted lines show the radius of the door swing, while the curved dotted lines show the course of the edge of the door when it swings. The dotted curved line intersecting with the jamb to the right, Fig. 593, shows how much bevel a door would have to have to be operated, if there were no play between the door and the jamb when the door is closed. A little study of Figs. 592 and 593, bearing in mind that they are exaggerations, will make this point clear.

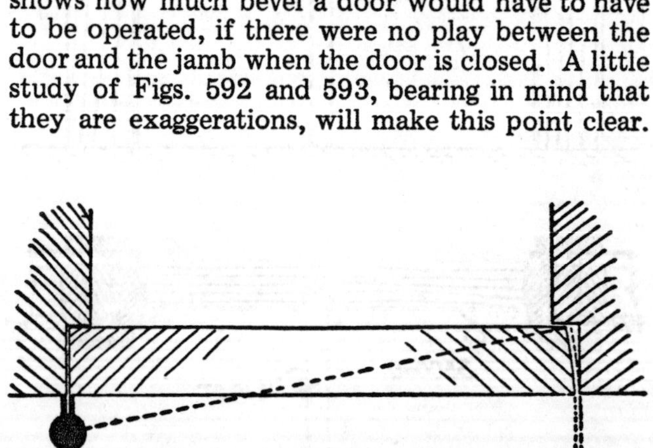

Fig. 593

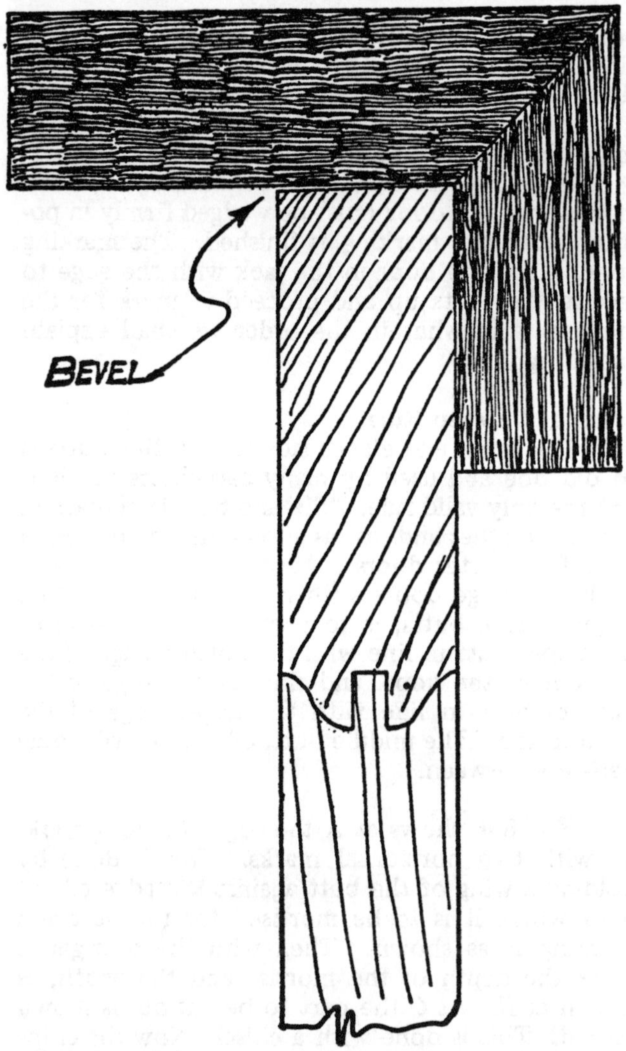

Fig. 594

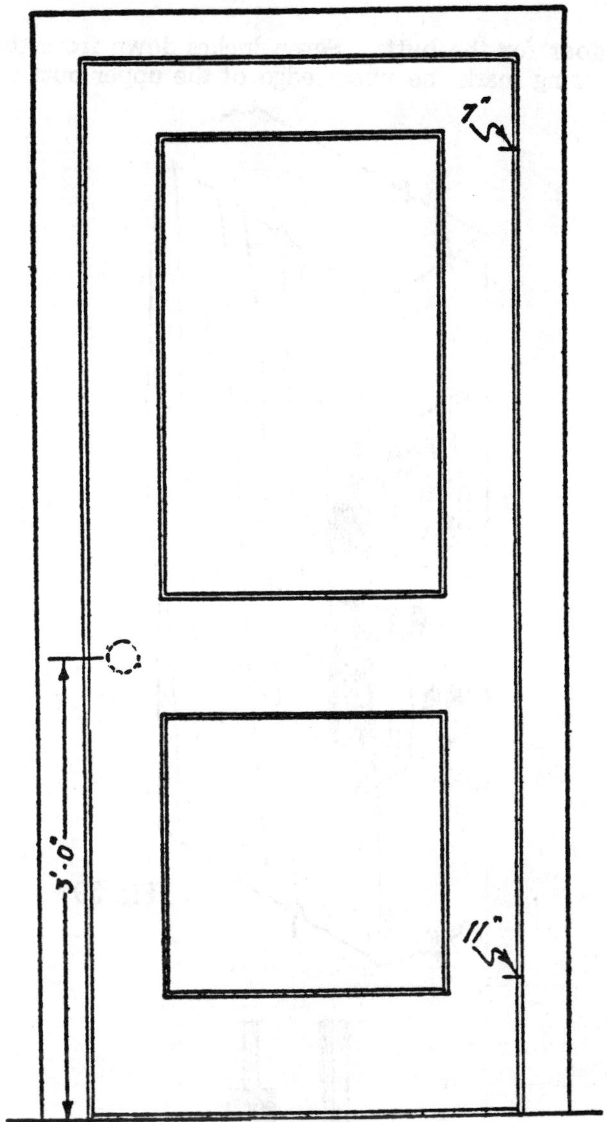

Fig. 595

Fig. 594 shows a cross section of a door with a tri-square between the bevel of the butt-edge of a door and the lock-edge. The butt-edge of a door should never have any more bevel than what is shown and could have less, while the lock-edge should never have any less, but could have more, depending on circumstances. For example, a wide, thin door with the pivot of the butts close to the surface of the door would need little beveling on the lock-edge and less on the butt-edge. But if a door is thick and narrow and the pivot of the butts projects out from the door, then it is obvious that the lock-edge of the door will have to be beveled so that the door can be operated and still not show too much play between the jamb and the door when it is closed.

The following figures show some of the first principles of fitting butts and some of the problems the workman will have when he hangs doors. Fig. 595 shows an elevation of a door that is ready for the butts. It is shown in place. To the left is shown by dotted lines and figures the location of the lock, and to the right with figures, how to mark a

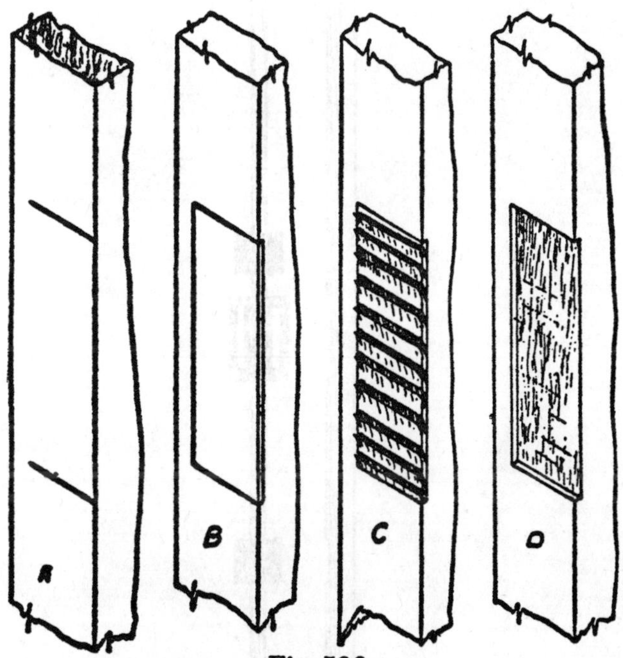

Fig. 596

door for the butts. Seven inches down from the casing mark the upper edge of the upper butt on

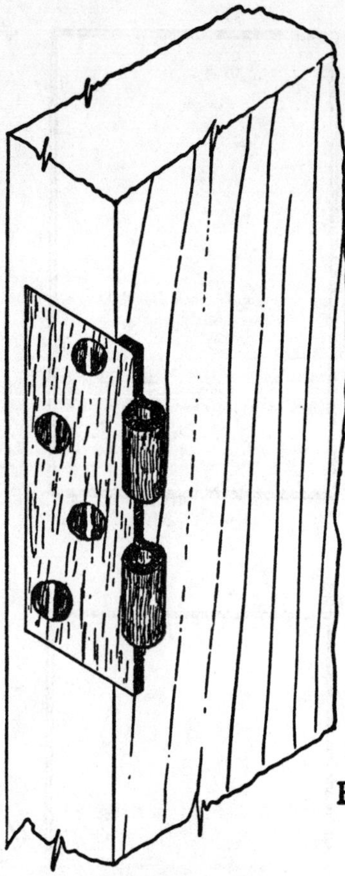

Fig. 597

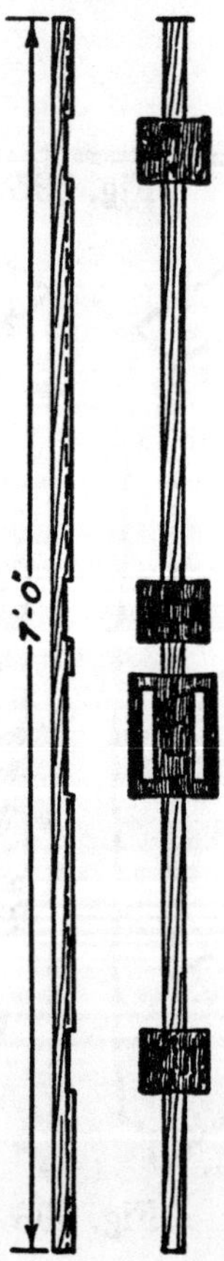

Fig. 598

both the door and the door jamb with a sharp pencil, chisel or knife. Then eleven inches from the finish floor mark the bottom edge of the bottom butt in the same way. If there are three butts, mark the upper edge of the center butt also. It should be kept in mind that in marking for butts in this way the door must be wedged firmly in position until the marking is finished. The marking done, place the door in the jack with the edge to receive the butts up and proceed to mark for the mortises somewhat in the order we shall explain under Fig. 596.

Seven and Eleven Rule

The seven and eleven rule for locating butts is an old one and used by many carpenters but it is not the only valid rule. "Six and ten" is perhaps as old as the other and just as good -- maybe better, at least for shorter doors. Another good rule, especially for large doors with rather wide stiles that require three butts, is to keep the upper edge of the upper butt in line with the bottom edge of the top rail of the door, and the bottom edge of the bottom butt in line with the upper edge of the bottom rail. The middle butt, of course, will come half-way between.

Fig. 596 shows at *A* the edge of a door marked with two horizontal marks. This is done by holding a wing of the butt against the edge of the door where it is to be mortised for the butt and marking it as shown. Then with the butt-gauge mark the depth of the mortise and the width, as shown at *B*. At *C* the part to be cut out is shown scored. This is done with a chisel. Now the chips are cut out and the mortise is trimmed up, giving it the shape shown at *D*. Next the wing of the butt is inserted and fastened to the door with screws, as shown by Fig. 597. The other mortises are done in the same way.

Butt-Marking Templet

Some carpenters use templets for marking the mortises for both the butts and the locks. The templet shown in Fig. 598 is a good one. It provides not only for three butts, but for the lock also. This templet is made for 7-foot doors. To the left is an edge view, where the notches for the sheet metal wings can be seen. To the right is a face view of the templet pole with the wings attached.

Fig. 599 has details of the top construction and also details of the wings for marking the butt mortises of the templet in Fig. 598. The wings are made of sheet metal and cut exactly the size of the part of the butt that is to be mortised into the door and into the jamb. These wings are sunk into the templet pole about ¼-inch and fastened as shown by the details. Two wings are necessary, one on either edge of the pole so that both the door and the jamb can be marked with the same templet.

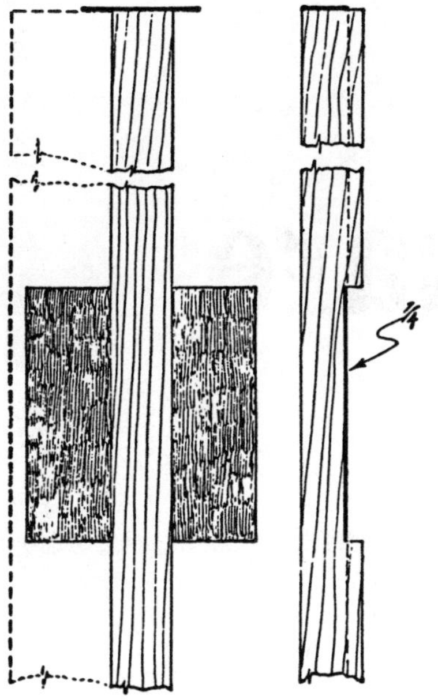

Fig. 599

On the top of the pole, set ¼-inch back, a piece of sheet metal is fastened, as shown, which gauges the marking so that the door will have the proper play at the top. The dotted lines to the left of the details represent the edge of a door and give a good idea of how the templet is applied when the marking is done. When the jamb is marked, the metal at the top of the pole is held against the rabbet of the head jamb.

Faulty Installations of Butts

Fig. 600 shows four faults that always cause binding. At *A*, the hinge is properly mortised to the door but not to the jamb. This prevents the door from closing freely, as indicated by the dotted line. At *B*, the hinge is poorly mortised both to the door and to the jamb, so it binds. At *C*, the binding is caused by the uneven mortise in the door. At *D*, the mortises are too deep and the door binds.

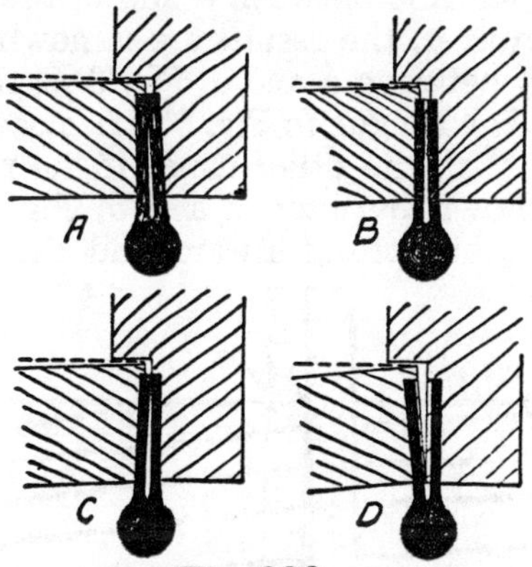

Fig. 600

Fig. 601 shows four more faults. At *A* the mortise of the door is too narrow. At *B* the mortise of the jamb is too wide, and at *C* it is too narrow. At *D* the mortise is too wide. The depths of all these mortises are right.

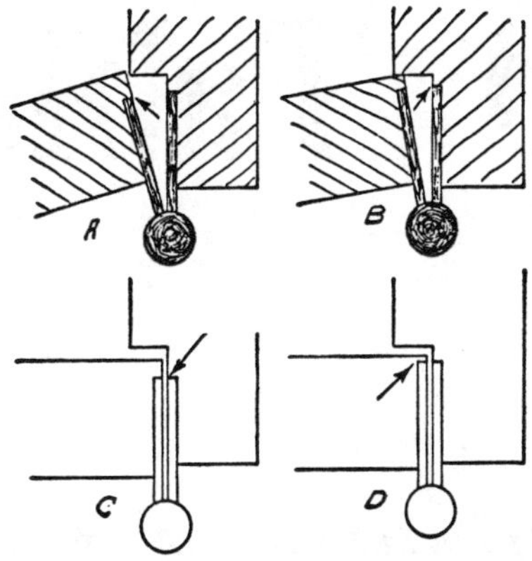

Fig. 601

In Fig. 602, *A* and *B* show the same fault, one on the door and the other on the jamb. The dotted lines and arrows point out the faults. At *C*, we have faulty mortising that has been corrected by padding the mortises with paper. Sandpaper is usually used, and in most cases, if it is carefully done is unnoticeable. At *D* is a hinge that is properly installed.

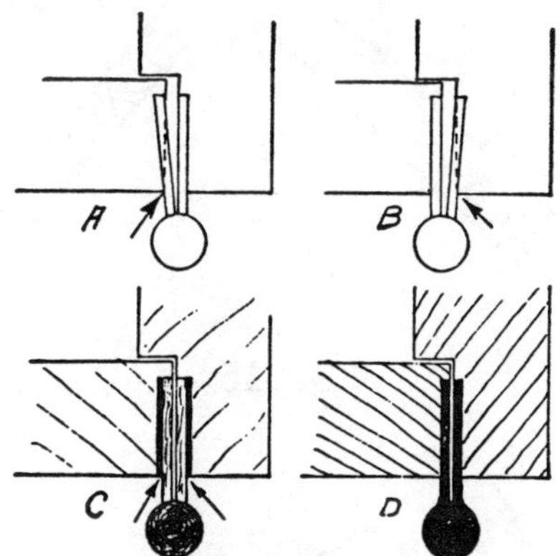

Fig. 602

38 The Steel Square and Tables

The Steel Square

Almost everyone knows what a steel square is: It has a large arm and a smaller one branching off from it at a right angle. The large arm is 2 inches wide and 24 inches long, while the smaller one is only 1½ inches wide and 16 inches long. The large arm is called the body, and the small one is called the tongue, see Fig. 615. The intersection of the outside edges of the body and the tongue is called the heel. The face side of a steel square is the side on which the manufacturer's name is stamped. The square in Fig. 615 has the face side up so that if the body of the square is held in the left hand the tongue will be in the right. Fig. 616 shows the square with the back side up. The tables on a square are usually on the back.

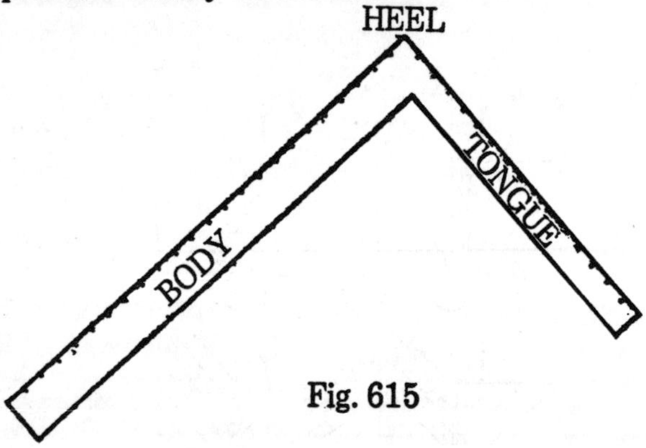

Fig. 615

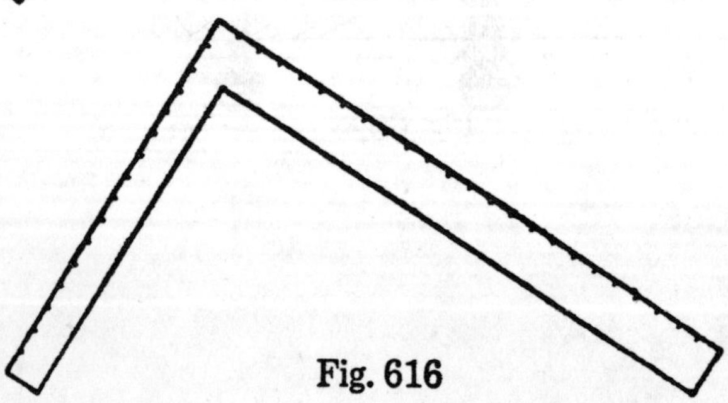

Fig. 616

Testing Squares

Fig. 617 shows how to test a steel square to see if it is true. Take a board about 12 inches wide and joint one edge so that is is perfectly straight. Then apply the square as shown at *A*. With a sharp pencil draw a line from *a* to *b* along the outside edge of the tongue. Reverse the square to the

Fig. 617

position indicated at *B*, shown by dotted lines. If the outside edge of the tongue coincides perfectly with the pencil line between *a* and *b*, the square is true. If, however, the results are somewhat on the order shown between *a* and *b*, Fig. 618, the square is not true. The tongue in Fig. 618 has been sprung outward. Or, on the other hand, if the results are like those shown between *a* and *b*, Fig. 619, the tongue has been sprung inward and the square is not true.

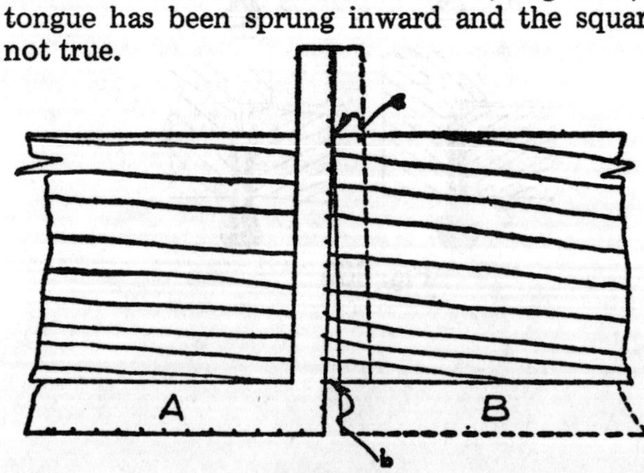

Fig. 618

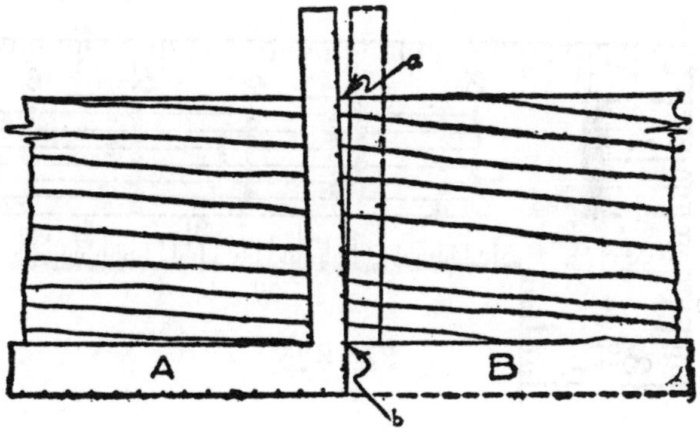

Fig. 619

Testing Half-Rounds

Fig. 620 shows how the square can be used to test a half-round. If the edges are kept in contact with the points indicated at *a* and *b*, the square (shown shaded) will describe a true half-circle with the heel, indicated at *c*. Compare the position of the shaded square with the position of the square marked 1 (shown by the dotted lines). The position of the square marked 2 shows how the other half of the circle can be described by keeping the edges in contact with *a* and *b*, and moving the heel from one point to the other.

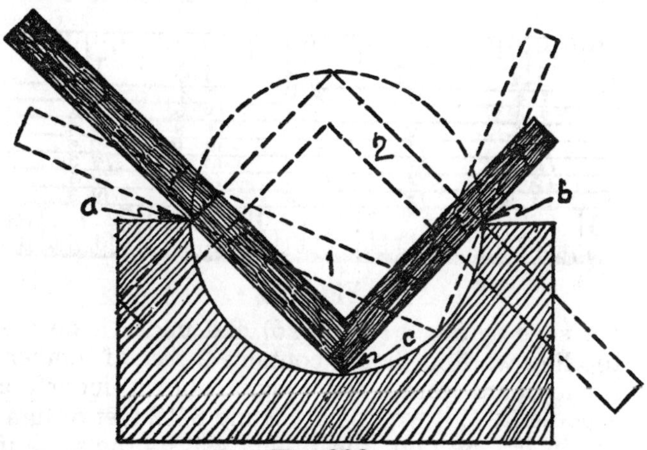

Fig. 620

Finding Center of a Circle

Fig. 621 shows how the center of a circle can be found by means of the steel square. Draw two lines *A-B* and *a-b* at any convenient position. Establish the center of the two lines, *(c, c,)* and place the steel square in the positions shown. The point at *C* is the center of the circle.

Another method of finding the center of a circle with the steel square is illustrated by Fig. 622. Place the square in two positions, making the heel come in contact with the circle, and draw lines along the outside edge of the square. Then join *a* to *b* and *c* to *d*, as shown with dotted lines; the point where the two lines cross is the center of the circle. One of the two lines just mentioned can be omitted and a line drawn between *e* and *f*. This will also establish the center at the crossing.

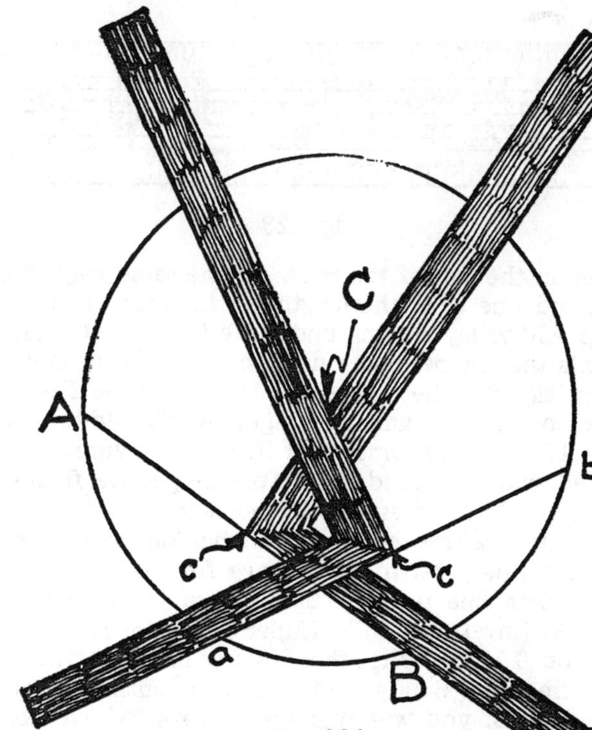

Fig. 621

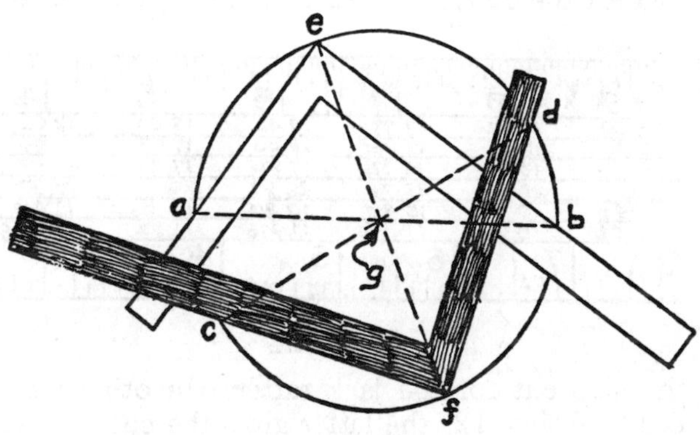

Fig. 622

Different Kinds of Square

The square illustrated above is the standard steel square with a 24-inch body and a 16-inch tongue. There are different kinds of squares on the market. Some carpenters like to carry, besides a full-sized square, a small square with a 12-inch body and an 8-inch tongue. Such a square has several advantages over the full-sized square, especially where space is limited.

Roof Framing Tables

Few roofs are framed without the use of the steel square. The square that will be described here is the R100C. If the student will lay this square before him with the face up, he will find to the left of the body, top line, these words: "Length of main rafters per foot run," just as shown in Fig. 623. Then, if he will go on to the figure 18, he will find under it 21.63, which means that the length of the main or common rafter per foot run for an 18-inch rise is 21 and 63/100th inches. To find the length of the rafter, multiply 21 63/100 by the number of

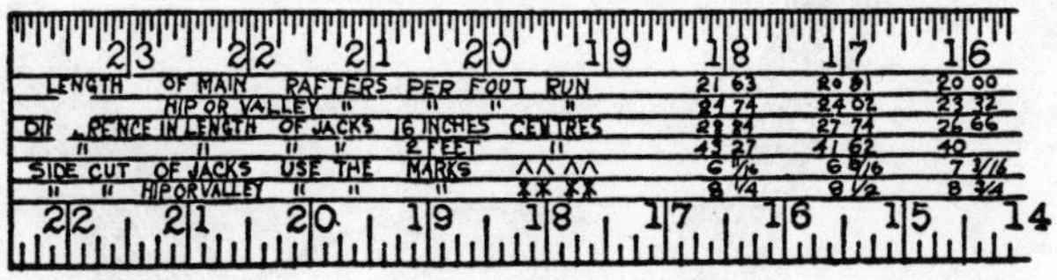

Fig. 623

feet in the run of the roof. In the same way, the second line gives the length per foot run for both hip and valley rafters, under the figure that represents the rise per foot run of the roof. If the rise is less, say 8 inches per foot run, pass up all figur excepting those under the figure 8. Should the rise be 10 inches or some other figure, the figures to be used will be found under the respective figures. The third line gives the difference in the lengths of the jack rafters, spaced 16 inches on center; the fourth line gives the difference for 2-foot spaces. The fifth line reads: "Side cut of jacks, use the marks (inverted V's)". Under the figure 18 will be found 6 11/16. With these figures in mind, pass on to the figure 6 (Fig. 624) and on counting eleven-sixteenths, you will find the V-mark, which indicates the graduating mark to be used on the outside edge of the body, with 12 on the tongue, to obtain

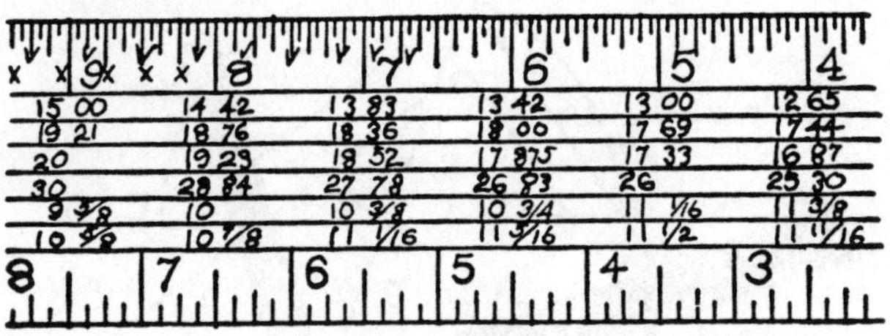

Fig. 624

the side cut for the jack rafter. In other words, 6 11/16 and 12, the latter gives the cut, will give the bevel for the side cut of the jack rafters for a roof having an 18-inch rise per foot run.

The sixth line gives the figures to be used with the figure 12 to obtain the side cut for hips and valleys. The graduating marks are indicated with X's. Under the figure 18, on this line, we find 8¼.

Now refer to Fig. 624 and you will find the X mark, at the graduating mark, representing 8¼ inches . . . Application: 8¼ and 12 (the latter gives the cut) will give the bevel for the side cut for hips and valleys on roofs having an 18-inch rise per foot run. We are using the 18-inch rise for convenience. In practice, the rise is usually below 12. Study Figs. 623 and 624.

Lumber Table

Now lay the square before you with the back to the top and you will find on the back of the body what is known as the lumber table; Figs. 625 and 626 give two sections of the back of the square showing enough of the lumber table to serve our purpose. The base figure for this table is 12. Now,

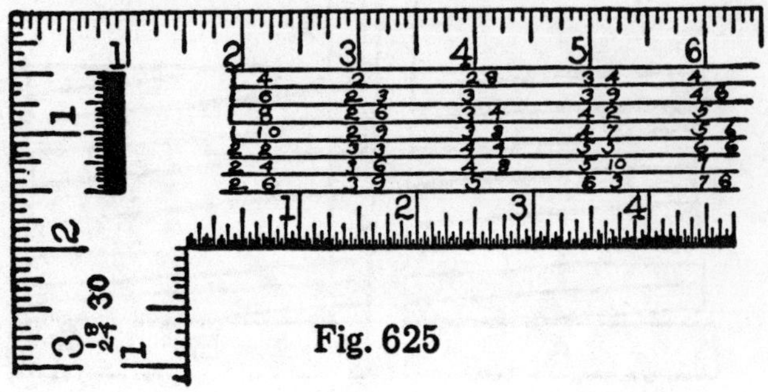

Fig. 625

if you will look to Fig. 626, you will find under the base figure these figures: On the first line 8; on the second line, 9; on the third line, 10, and so on down to 15, omitting 12. The table can be used in two ways. First, let the figures on the edge of the square represent the length of the board, and the figures under the base figure, the width. Then a board 12 feet long, 8 inches wide has 8 feet of lumber in it; if the board is 13 feet long, it has 8 and 8/12 feet of lumber in it; if 14 feet long, it has 9 and 4/12 feet of lumber in it. These figures are found in Fig. 626, on the part of the square shown, and on the upper line. For instance, we have a board 8 inches wide and 3 feet long. Turn back to

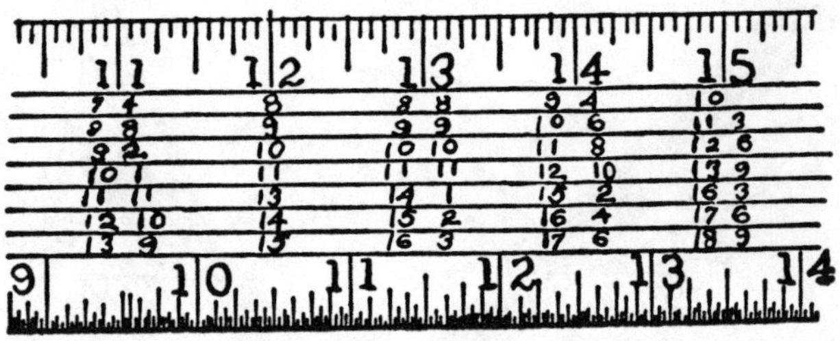

Fig. 626

the edge-figure, 3 (Fig. 625) and under it on the top line we find that it contains 2 feet of lumber. If the board were 4 feet long, as shown under the edge-figure 4, it would have 2 and 8/12 feet of lumber in it. Second, the results will be the same if you let the edge-figure represent the width and the figures under the base figure the length. For example, a board 8 feet long, 13 inches wide, has, as we find under the figure 13, 8 and 8/12 feet of lumber in it. If it were 8 feet long and only 3 inches wide, it would have, as we find under the figure 3, 2 feet of lumber in it.

The lumber table is very simple, although it is necessary to understand it thoroughly before it can be used with certainty. To be thorough, study is necessary. Study the table in the light of the foregoing explanations.

Diagonal Scale

Some squares have what is called a "diagonal scale," like the one shown in Fig. 627. The measuring here is also done with a compass, as in the case just explained, but the count is somewhat different. The diagonal lines, as well as the horizontal lines, are exactly 10/100ths of an inch apart. From

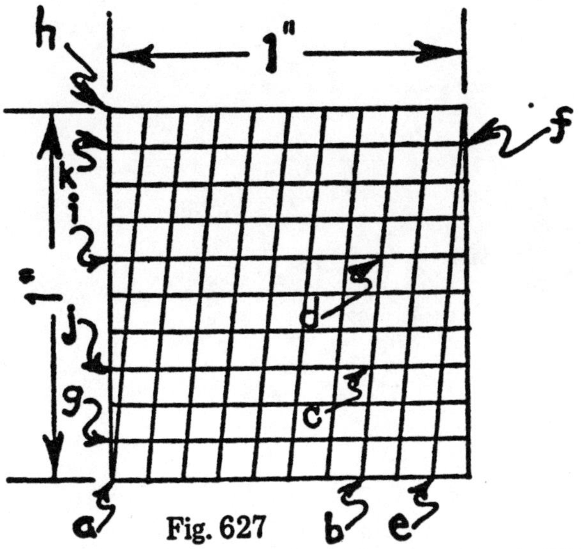

Fig. 627

left to right, as from *a* to *b*, each space counts 10/100ths of an inch, or in all, 70/100ths of an inch. But from *j* to *c*, 3 spaces up, the distance is 73/100ths of an inch, and from *i* to *d*, 6 spaces up, the distance is 76/100ths of an inch. Now from *a* to *e* the distance is 90/100ths of an inch, but from *k* to *f*, 9 spaces up, the distance is 99/100ths of an inch, or as shown, 1 inch. The space between the two lines at *g*, 1 space up, is 1/100th of an inch. A little study will make clear the application of the diagonal scale.

Brace Tables

Fig. 628, at *A*, shows one inch divided into one-hundred parts, which is necessary in getting the different lengths of braces. Directly under the edge figure 3 will be found these figures: 18 over 24, and to the right 30, which means that a brace joining the frame 18 inches one way from the angle, and 24 inches the other way, will have to be 30 inches long. This is illustrated by Fig. 629. Under

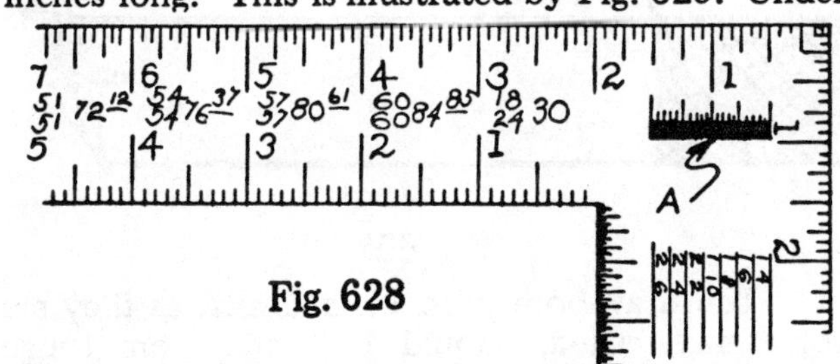

Fig. 628

the edge figure 4 (Fig. 628) you will find 60 over 60, and to the right 84.85, which means that a brace joining the frame 60 inches each way from the angle will have to be 84.85 inches long.

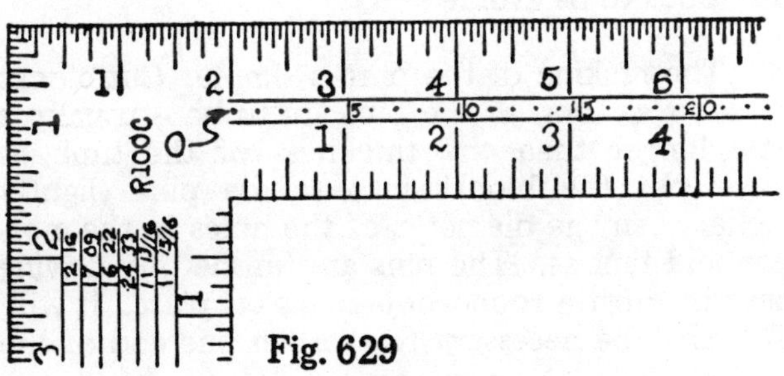

Fig. 629

Octagon Table

Fig. 630 shows a part of the octagon table on the face side of the tongue. To describe an octagon, take as many spaces shown by dots in the table, as there are inches in one side of the square to be changed to an octagon. For example, if you want to make an octagon out of an 8x8, start at the dot pointed out at O, count eight spaces to the right, and set the compass as shown by the drawing. Then set one leg of the compass at the center of one side, as at *C*, Fig. 631, and mark points *1* and 2, as indicated by the dotted lines from *1* to *3* and from *2* to *4*, and from these four points mark off the four corners on a 45-degree angle, as shown by Fig. 631. That completes the laying out of the octagon.

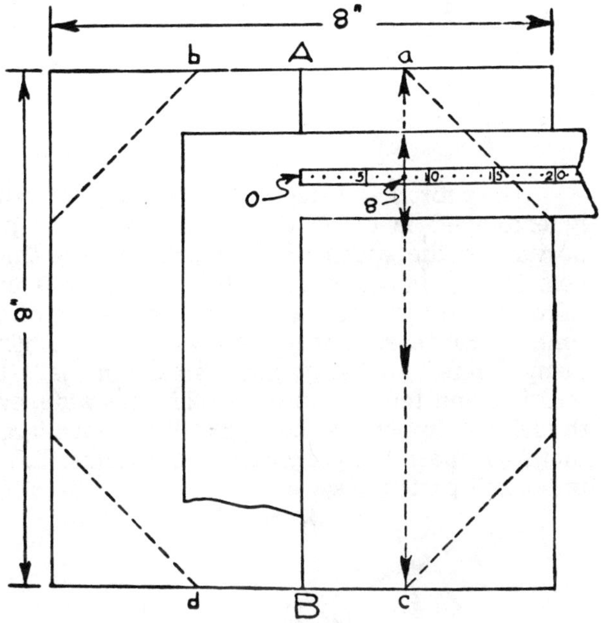

Fig. 630

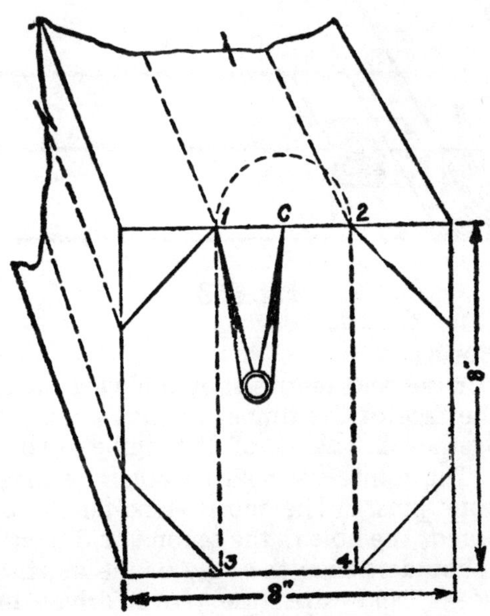

Fig. 631

39 Mortise-And-Tenon Joints

The mortise-and-tenon joint is the joint usually employed in heavy timber framing. The proportion of the width of the mortise to the thickness of the timber into which it is worked is one-third, and the mortise is cut into the center of the timber. However, this is not always true. A great many framers of heavy timbers seldom make the mortises and tenons more than 2 inches wide, even though the timbers are 8x8's, 8x10's, or even larger. In this respect the judgment of the framer has an important part to play.

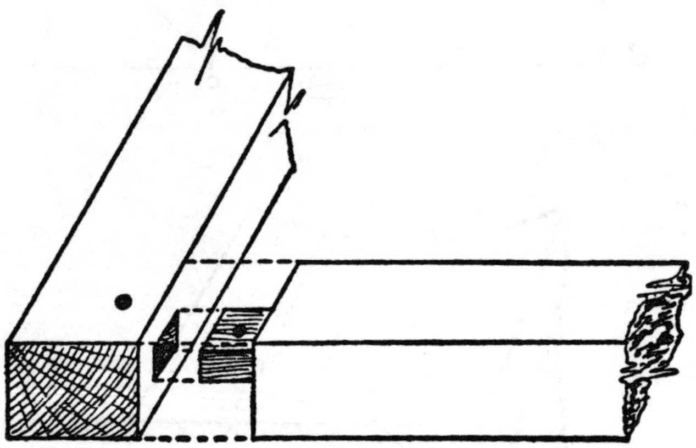

Fig. 656

Draw Boring

Mortises and tenons should always be laid off from the face of the timber so that when they are put together the faces of the timbers will come flush. The joints are held together by means of draw-bore pins. The mortise-and-tenon joint is draw-bored; the hole in the tenon is 1/8-inch closer to the shoulders than the hole of the mortise is to the cheeks, and when the pin is driven in, the shoulders of the tenon will be tight against the cheeks of the mortise. A mortise-and-tenon joint ready to go together is shown by Fig. 656. The holes for the draw-bore pin are heavily shaded. The dotted lines show how the two timbers are to come together. Fig. 657 shows the joint completed.

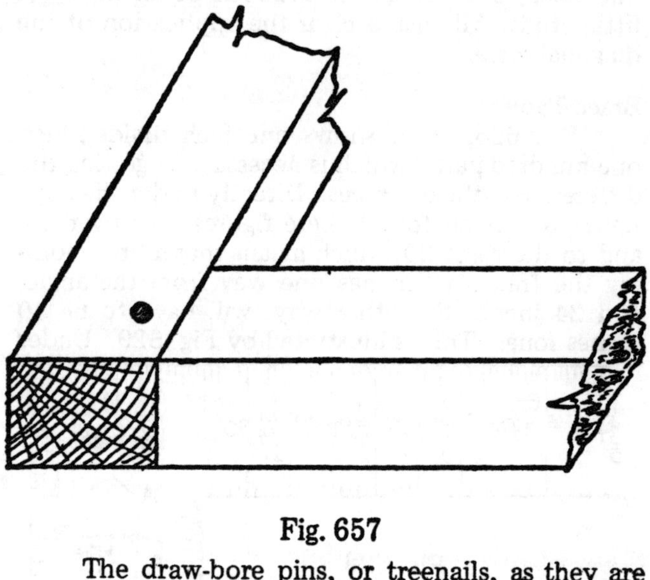

Fig. 657

The draw-bore pins, or treenails, as they are sometimes called, should be made from tough straight-grained wood, such as oak or locust. Wood that is liable to be worm eaten, even though it may be tough and straight-grained, should be avoided for draw-bore pins. Hickory is the best example of the woods to be avoided.

The making of the pins is simple. Cut blocks of wood as long as the pins are to be – usually a little longer than the thickness of the timbers. Then split the blocks into square pins slightly thicker than the diameter of the holes in the mortises and tenons. The pins are finished by driving them through a round hole in a steel plate. It will, of course, be necessary to sharpen one end of the

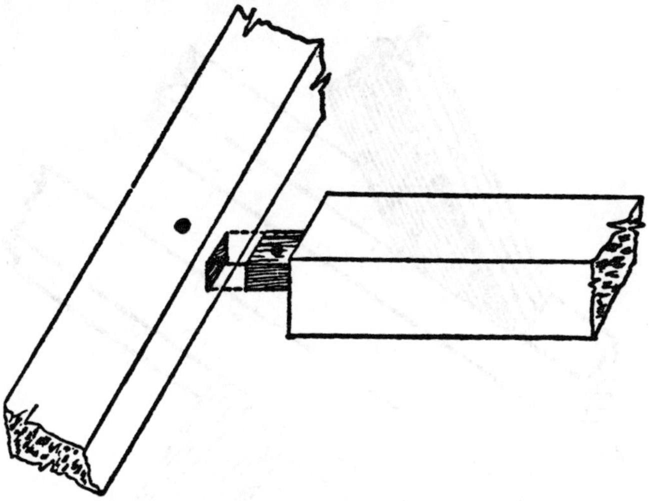

Fig. 658

square pins in order to start them. The hole in the steel plate must be slightly larger than the holes in the mortises and tenons because the pins, in order to fit tightly, must be a little larger than the drawbore holes. Pins from ¾ to 1 inch in diameter will answer for most purposes.

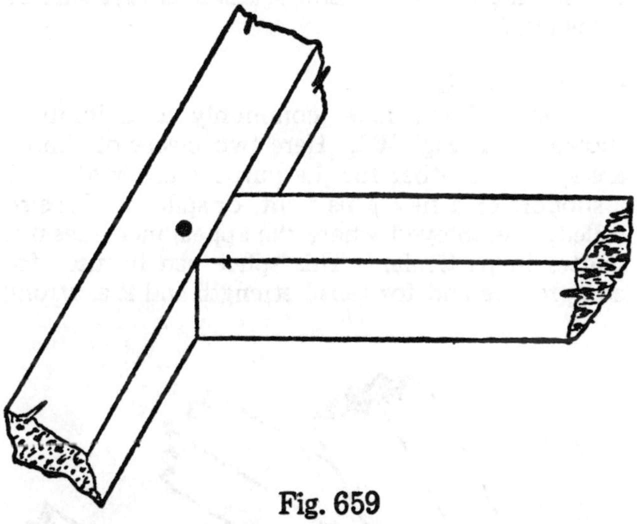

Fig. 659

A T-mortise-and-tenon joint about to go together is shown by Fig. 658. The T-joint completed, is shown by Fig. 659. A mortise-and-tenon joint for a corner post joining a sill is shown by Fig. 660. The joint for the plate where the corner post joins it is exactly the same as the one shown here except that the mortise comes above the tenon.

A mortise and a tenon for a girt joining a corner post is shown by Fig. 661. Fig. 662 shows a mortise-and-tenon joint for a brace. Notice that the tenon of the brace is cut square at the bottom. This not only gives the brace a bearing, but it lets the joint come together easily.

Mortises and tenons must be well executed. Otherwise trouble will be the result when the frame-

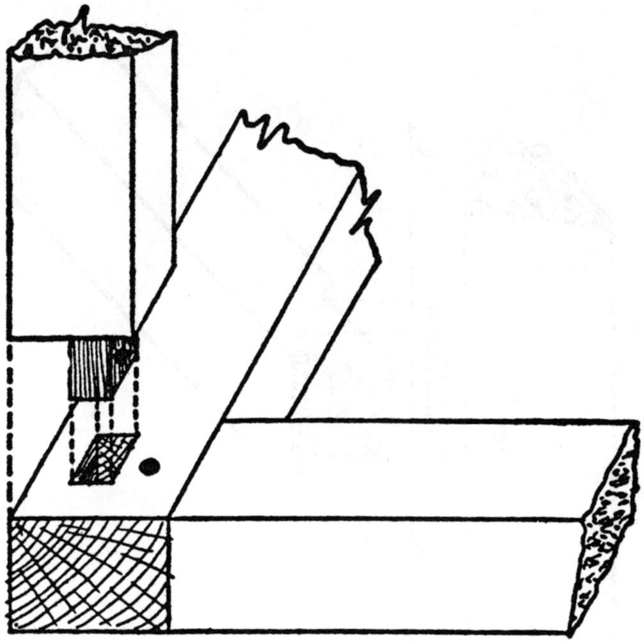

Fig.660

work is put together. The tenons should be made to fit snugly into the mortises – not too large and not too small. The sharp corners of the end of the tenons should be dressed off with a plane so that they will slip into the mortises with ease. The shoulders of the tenons must be perfectly square and the cheeks of the mortises must be even in order to insure a uniformly tight fit in every joint. The depths of the mortises should be slightly deeper than the

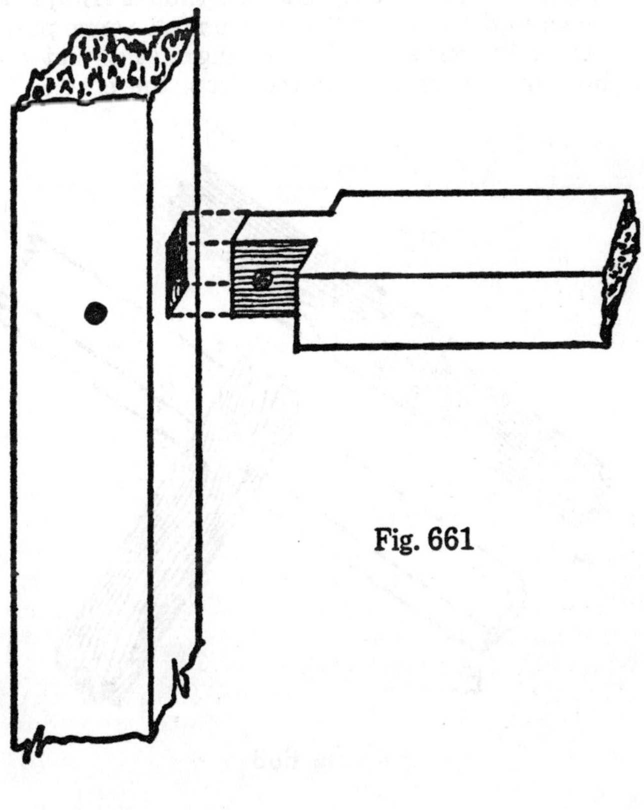

Fig. 661

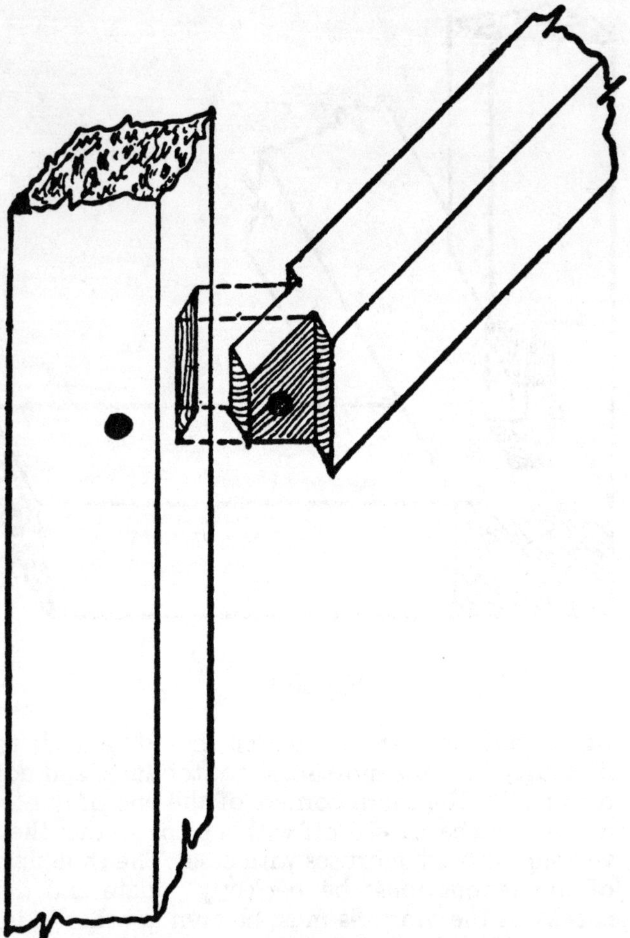

Fig. 662

length of the tenons, and care should always be taken in draw-boring that it will not become push-boring by mistake. Push-boring means that the hole in the tenon is farther from the cheeks; in

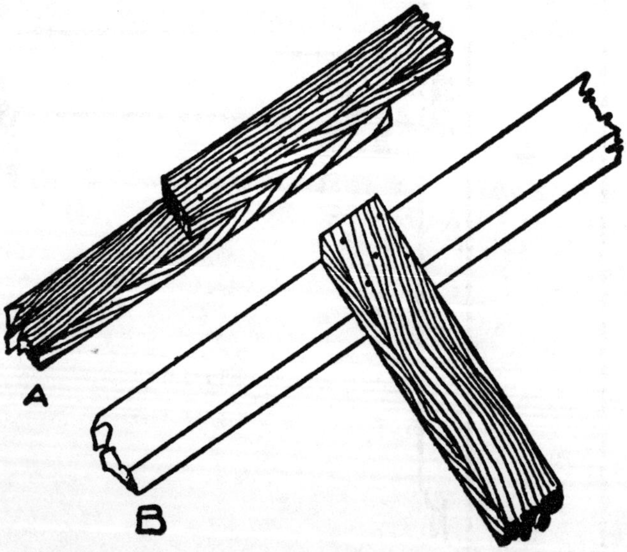

Fig. 663

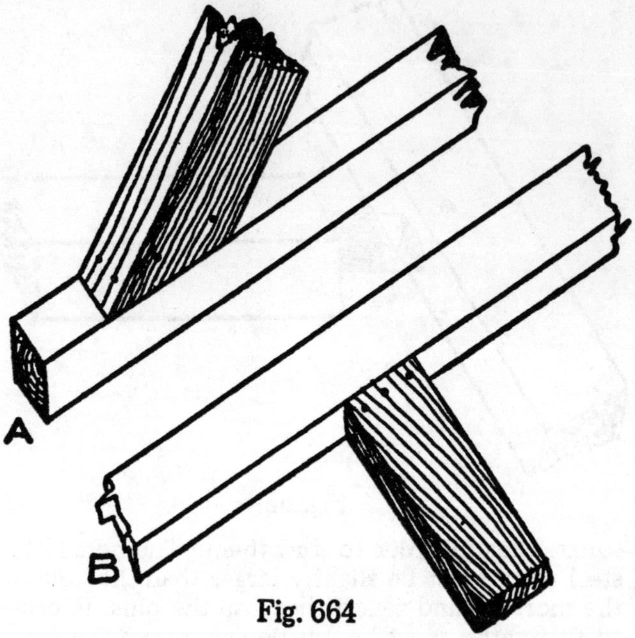

Fig. 664

which case the pin would open the joint on driving it, instead of drawing it together. Measurements for locating mortises and tenons should be made painstakingly. A mistake in this is always sure to bring grief.

Common Joints

One of the most commonly used joints is shown at *A*, Fig. 663. Here two pieces of timber are spiked together for the purpose of lengthening a support or a tie. This joint, or splice as it is also called, is employed where the appearance does not matter in particular. This splice can be used for compressive and for tensil strength and is as strong

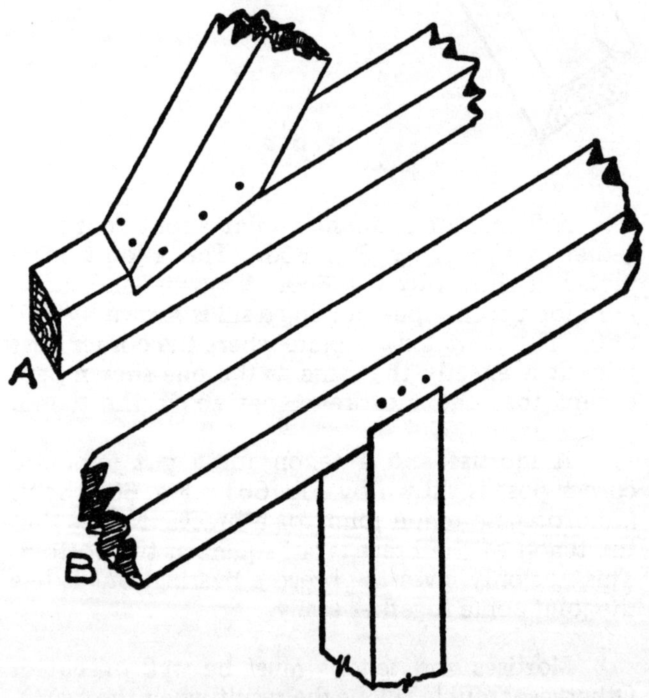

Fig. 665

one way as it is the other. It will also resist a great deal of shearing stress. At *B*, Fig. 663, is another very common joint, used mostly in rough and temporary work. Its strength, as in the joint shown at *A*, depends entirely on the nails used in fastening the two timbers together.

A beveled butt joint is shown at *A*, Fig. 664. This joint is held together by toe-nailing. This joint should always be used cautiously and should never be used where a great deal of strength is required. There are, though, many legitimate uses for this joint, both in temporary and in permanent constructions. The builder's judgment must determine when to, and when not to use it, unless the architect specifies the kind of joint to use. A plain T-butt joint is shown at *B*, Fig. 664. The joint is fastened with toe nails.

A butt joint that is suggestive of the mortise-and-tenon joint is shown at *A*, Fig. 665. The notch cut into the horizontal piece prevents the slanting timber from slipping. Another mortise-and-tenon suggesting joint is shown at *B*, Fig. 665. The notch here, as in the first case, prevents slipping of the joint. The joints shown in *A* and *B* of this figure are fastened together with toe nails, but the strength is not altogether dependent on nails. These joints can be called the dividing line between butt joints and mortise-and-tenon joints.

40 Miscellaneous Joints

Making Joints

It is one thing to make a joint and make it right, but it is quite a different thing to be able to select the most suitable joint for each connection that must be made in constructing a building. There are many factors that must be considered when selecting the right joint and no hard-and-fast rule will apply. The strength necessary to carry the load is the principal consideration. The carrying capacity of the wood used is another factor. The toughness, hardness and durability of the wood all enter into the matter of selecting and making a joint. The appearance of a joint and its harmony with other joints used on a job are matters that should not be overlooked in selecting a joint. Even if a cheaper joint would answer all other purposes, it should not be used if it will mar the effectiveness of the appearance of the job as a whole. On the other hand, when a simple inexpensive joint will answer the purpose, especially in concealed work, there is no reason why it should not be used. Again, the reader should remember that joints can easily be changed, so far as their forms are concerned, to meet the various conditions that may exist.

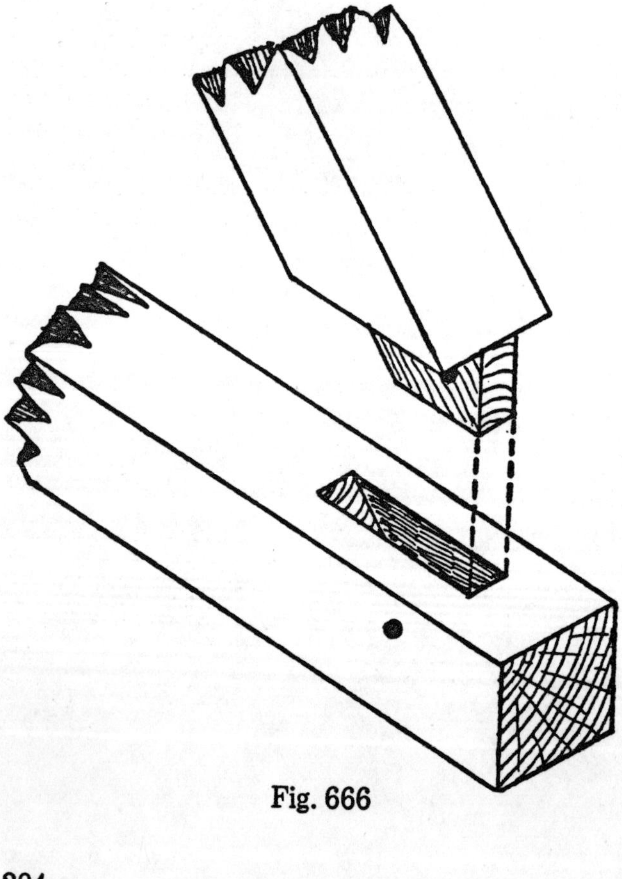

Fig. 666

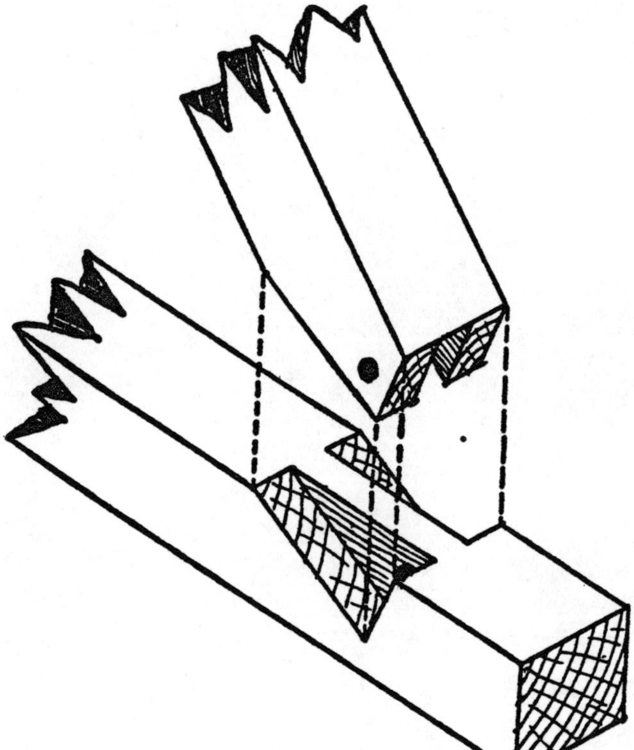

Fig. 667

Bridge Joint

Fig. 666 is what is called a "bridge" joint. This joint is a modification of the mortise-and-tenon joint and is frequently so called. It is used where one of the timbers holds an inclined position, as braces or rafters of trusses. Another modification of a mortise-and-tenon joint is shown by Fig. 667. This joint is called a "straddle" joint. It is also used where one of the timbers hold an inclined position, such as braces and truss-rafters. In cases where such joints must resist a great deal of pressure they should be reinforced with strap-iron clamps.

Gained Joint

What is known as a "gained joint" is shown by Fig. 668. This joint is designed for such timbers as floor joists where they join onto the sills or onto beams. The purpose of the gain is to provide additional bearing for the joist, which is often necessary where heavy timber sills are used. But in modern construction the heavv timber sill is losing favor and the box-sill is taking its place. The box sill provides ample bearing for the joists. This is especially true in residential work. The part marked

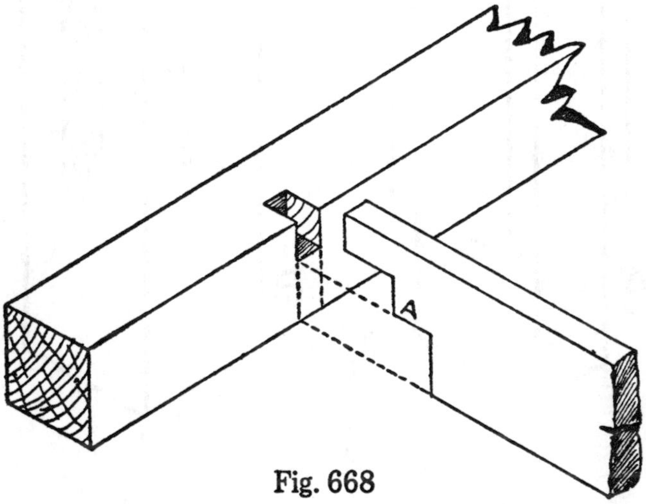

Fig. 668

A is sometimes called the tenon, and the gain is called a mortise.

Tusk-and-Tenon-Joint

Two forms of tenon-and-tusk joints are shown by Fig. 669. The one shown at *A* is held together

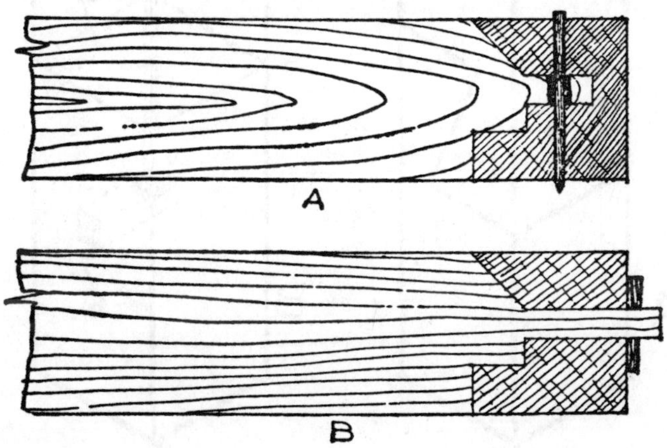

Fig. 669

with a pin, while the one shown at *B* is held in place with a peg. A bolted tenon-and-tusk joint is shown at *C*, Fig. 670, and a double tenon joint at *D*. These joints are usually employed in heavy timber framing, and, because of that, are not as extensively used as they were when heavy timbers were popular as building materials.

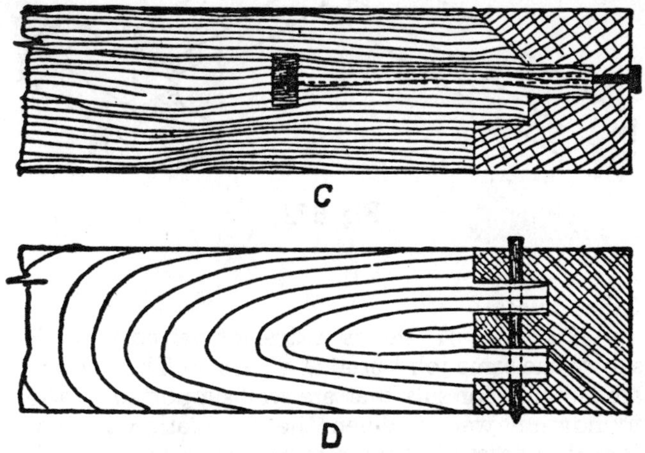

Fig. 670

Notched Joints

Two footing joints for truss rafters are shown in Fig. 671. The one shown at *A* shows the single notch joint, while the one shown at *B* shows the double notch. These joints are subject to modification, as are all of the joints shown in this work. In cases where the depth of the notches has a tendency to weaken the tie-beam to the extent that its strength will be below the safety point, the number of notches should be increased, and the depth of them reduced. In other words, where the number of notches is increased, the shoulder-surface of each notch can be decreased proportionately. This, of course, will reduce the depth of the notches. Moreover, where the joint must resist a great deal of pressure, it should be reinforced with a strap-iron

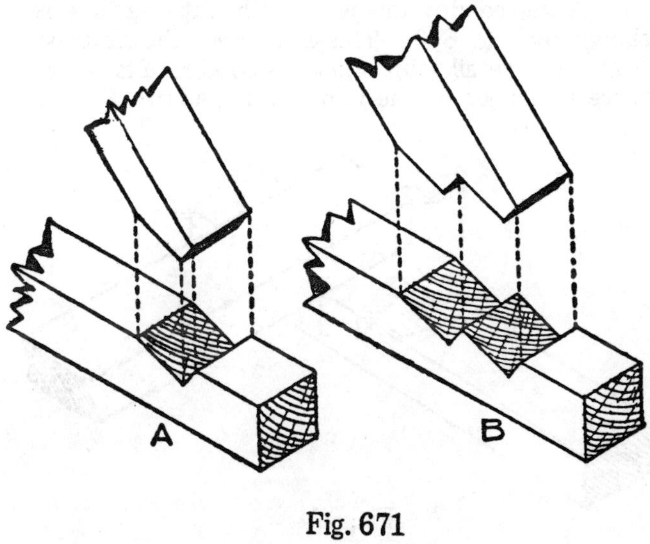

Fig. 671

clamps or bolts. Where the pressure is light, the fastening may be done with nails. It is always good policy in framing joints of this type to follow the architect's specifications.

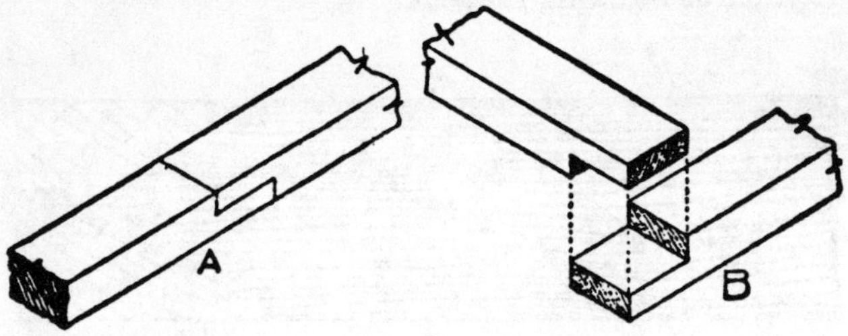

Fig. 672

Other Joints

Fig. 672, *A*, shows a straight halved joint; *B* shows the same joint used at a corner. These joints are suitable for sills that are well supported by the foundation walls. When the foundation is reinforced concrete, a simple joint such as we are dealing with here will be adequate in most cases. This joint is also suitable for plates but should come directly over a bearing.

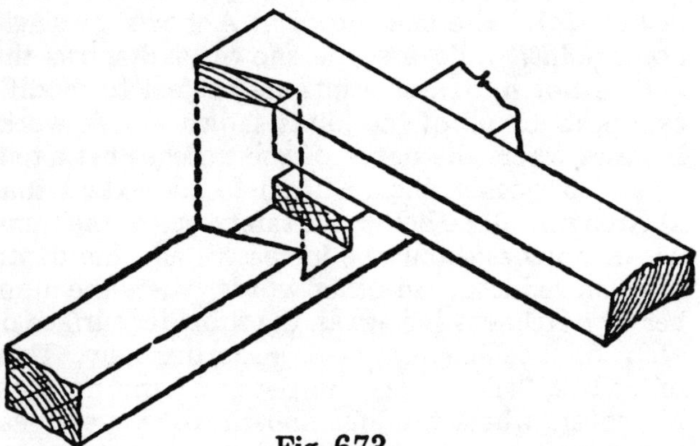

Fig. 673

A halved dovetail joint with but one flare is shown by Fig. 673. This is perhaps the simplest form of dovetail joint and gives good results where there is danger of the joint pulling apart. Such a

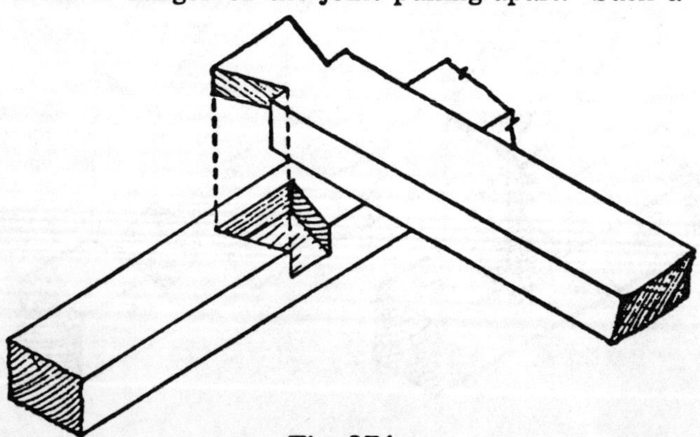

Fig. 674

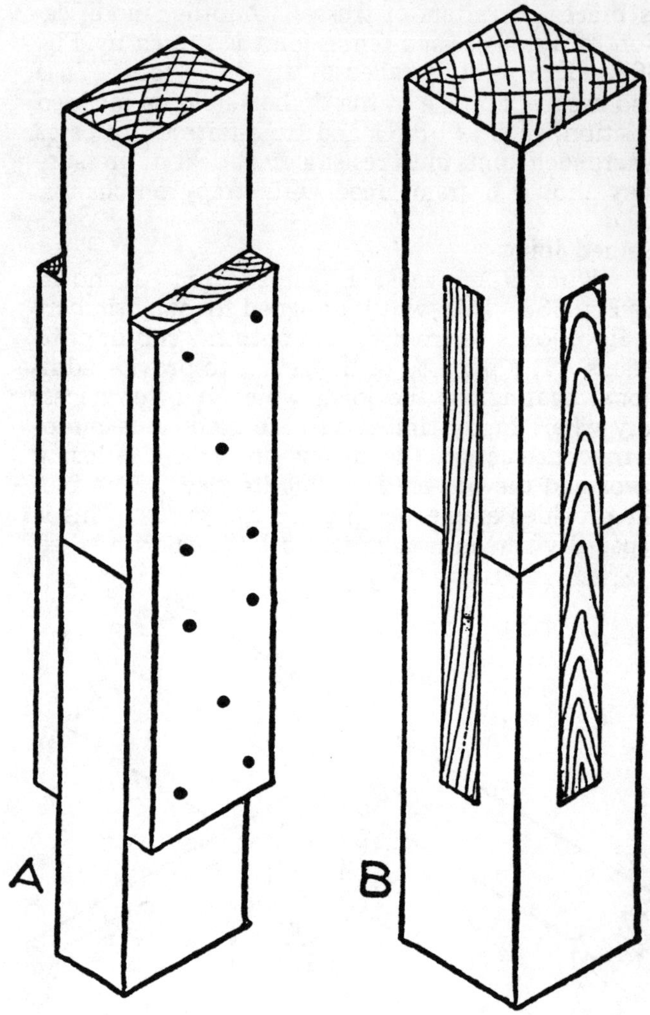

Fig. 675

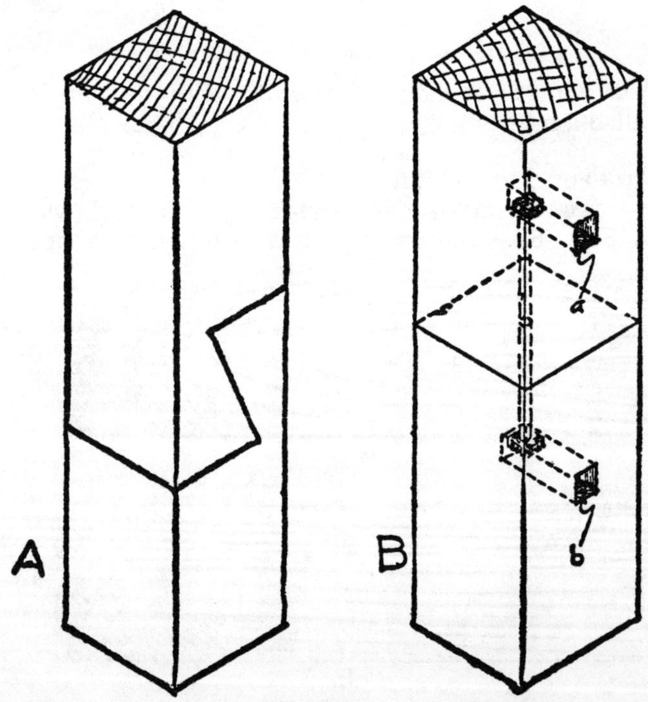

Fig. 676

joint should fit closely and may be pinned or nailed into place. Another form of dovetail joint is shown at Fig. 674. This joint has both flares but does not extend all the way through the timber. It answers the same purposes as the joint shown under Fig. 673 but leaves the main timber with more strength. It can be fastened with pins or nails, whichever the framer may think most suitable.

Two joints suitable for resisting compression are shown by Fig. 675 at *A* and *B*. The first one is a plain butt joint except that it has two short pieces of wood nailed over the joint on opposite sides, as shown. These pieces of wood are called "fish plates" and consequently the joint is called a fished joint. The second joint is also a butt joint with dowels set into four sides, as shown. This is known as a "doweled joint." The dowels may be either metal or wood and the size can be governed by the circumstances.

Two joints for resisting compression only (suitable for short posts, etc.) are shown at *A* and *B*, Fig. 676. The joint at *A* is called a "beveled joint" and may be toe-nailed or bolted, whichever the case may require, while the joint shown at *B* is called a bolted butt joint. The two ends are sawed perfectly square and a hole bored at the center of each, as shown by dotted lines. Through this hole is slipped a bolt threaded at both ends. The nuts are put onto the bolt through the hole shown at *a* and *b* and tightened the joint.

41

Shoring, Needling and Bracing

The problems covered in this chapter are emergency problems, and for that reason, temporary. They all belong to carpentry, and carpenters should be prepared to solve any or all of them when the emergency arises.

Fig. 724 shows how to shore up floors when the outside wall is to be taken down and rebuilt. The indicators *a, a, a, a* point to the side of the shores that has been boarded up. This is usually done where the building is occupied during the process of reconstructing the wall. In these instances, the shoring is much on the order of a partition wall, having a plate at the top and one at the bottom, with studding between, spaced so as to support the weight above. When boarding up is not necessary the shoring can be done with heavier timbers. However, the shores should then be braced in order to prevent buckling. The sizes of the timbers, of course, must depend on the amount of weight to be carried.

Fig. 725 shows a method of shoring a brick, stone or cement wall either for repairs, or for underpinning. A T-bearing of the upper ends of the shores is shown by Fig. 726. This is not necessary except where the wall is rather soft or otherwise unreliable.

Another method of fastening the shores to wall is shown by Fig. 727. Details of one of the bearings are shown by Fig. 728, *A* and *B*. The indicators at *A* show: *a*, the reinforcing block; *b*, the needle; *c*, the end of the shore and *d*, the needled plank. At *B* is a face view of the bearing.

Still another method of shoring a wall is shown by Fig. 729. Here the shores are parallel to each other. Notice the bracing and the built-up base.

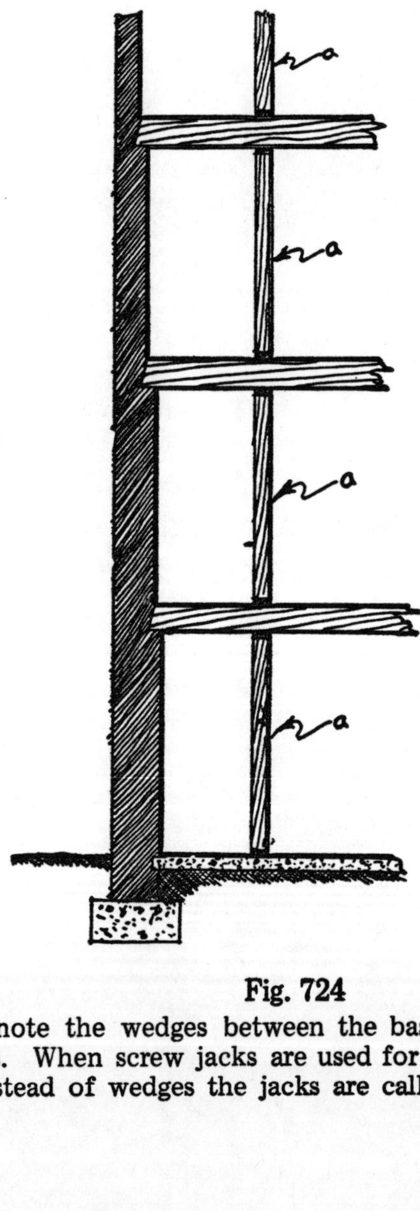

Fig. 724

Also note the wedges between the base and the shores. When screw jacks are used for tightening up instead of wedges the jacks are called pumps.

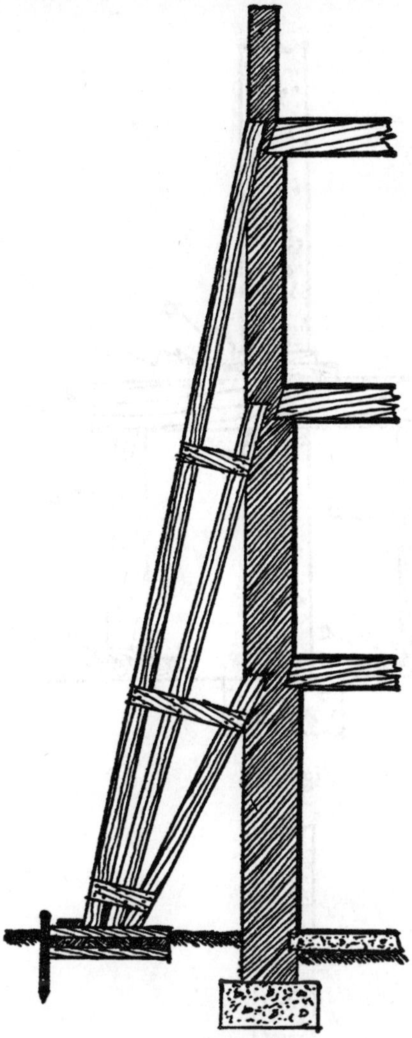

Fig. 725

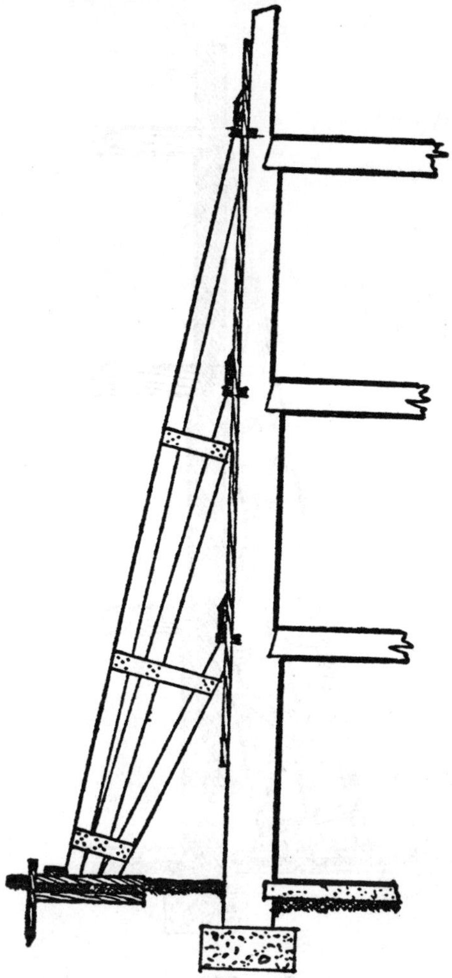

Fig. 727

Fig. 726

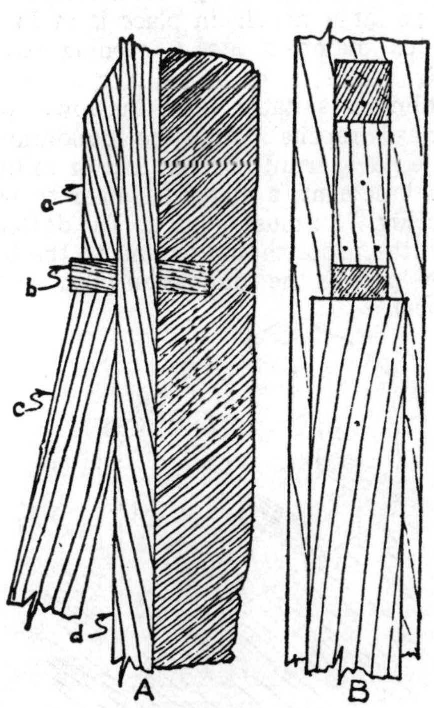

Fig. 728

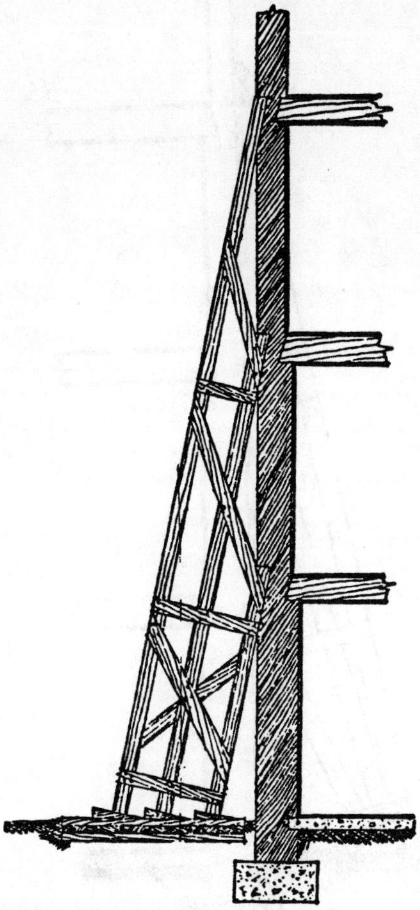

Fig. 729

Needling

A wall with needles *a*, *b* and *c*, in place ready for the props is shown in Fig. 730. The positions of the props are shown by the dotted lines. A clearer view of a needle in place is in Fig. 731. The various members are *a*, evening block; *b*, wedges; *c*, needle; *d*, *d*, props; *c*, *c*, base blocks. This needling is suitable when the lower part of the wall needs repairs. When the foundation is to be extended deeper into the ground or to the bottom of a basement, a method similar to what is shown by Fig. 732 must be used. The dotted lines show how the props should extend to the bottom of the new part of the foundation (also shown by dotted lines).

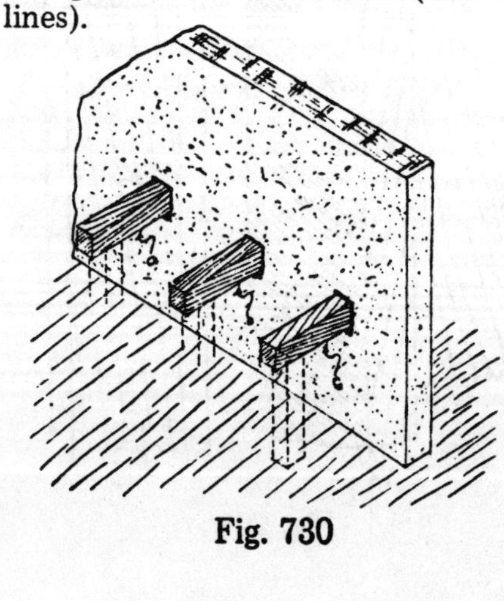

Fig. 730

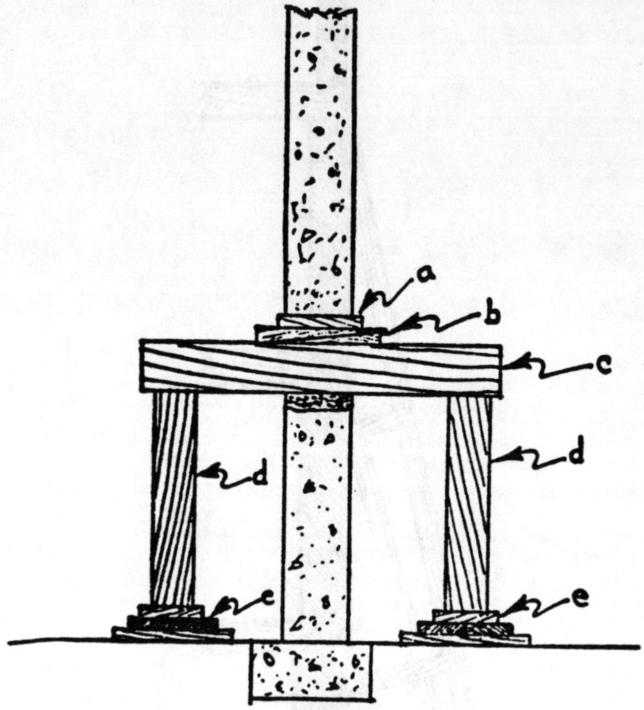

Fig. 731

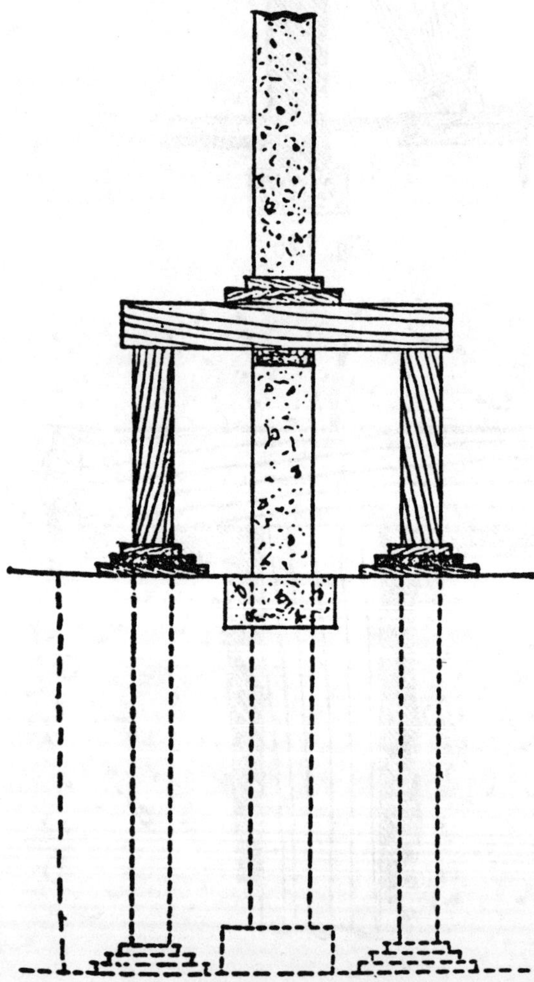

Fig. 732

Fig. 733 is a detail of the base block. In this case the needle is tightened by means of the wedges shown directly under the bottom end of the prop. When the tightening of the needle is done by means of a jack it will be somewhat as shown by Fig. 734. Fig. 735 shows a screw jack used for the same purposes. Notice how the threaded part of the jack runs up into the bottom end of the prop. This is indicated by the dotted lines.

All of the methods of tightening the needles shown are good, but extremely heavy loads should not be placed onto light screw jacks for it is possible to break the threads of the screw. Remember that the ground onto which the base blocks are placed should be examined to make sure that it will carry the weight of the wall.

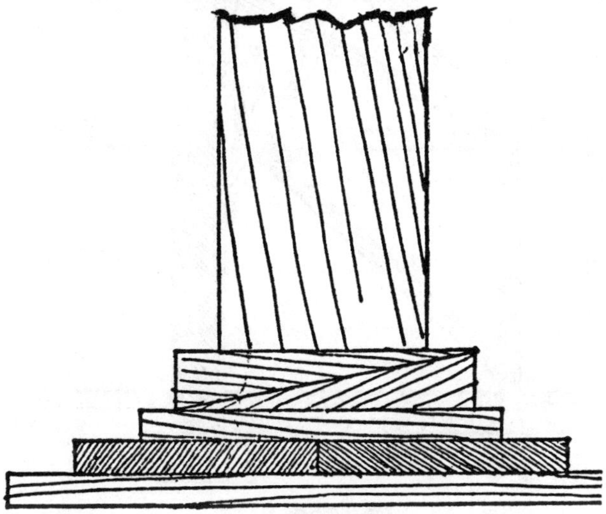

Fig. 733

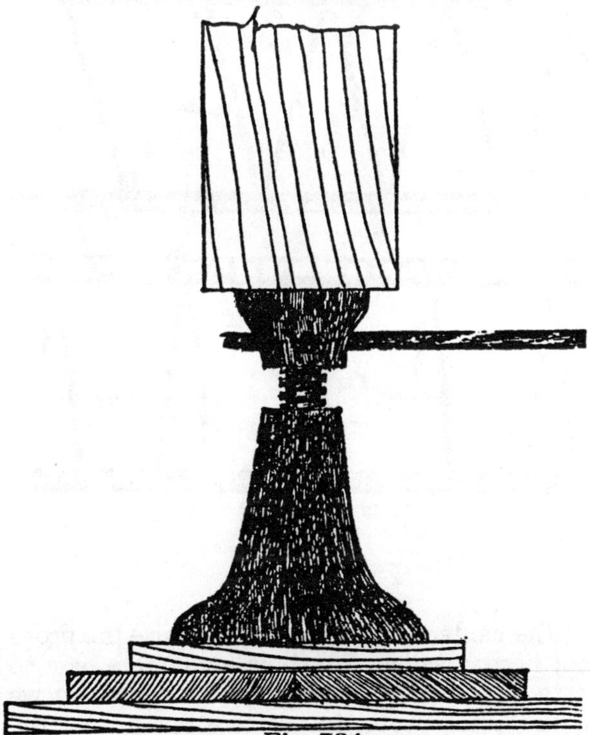

Fig. 734

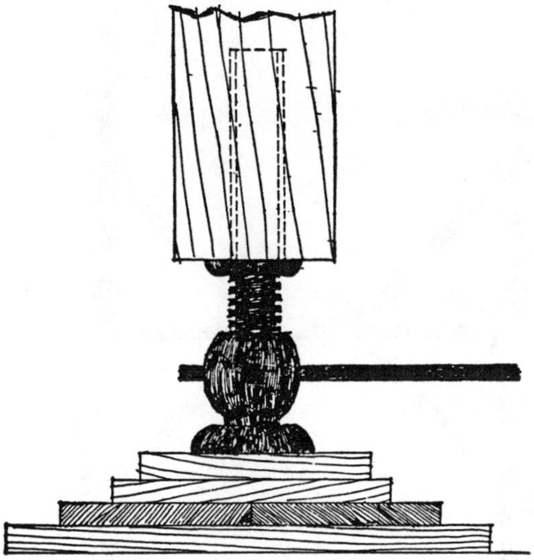

Fig. 735

Heavy Timber Bracing

Fig. 736 shows a method of bracing walls, either for repairs or for preventing further damage. This method is not the only one, nor the best one, but it is, perhaps, the quickest. We will assume that the two walls are twenty-five feet apart. The bracing is done with timbers twenty-six feet long beginning at the bottom and working up. The upright braces are used for preventing buckling. The dotted lines show how the braces of the next section should be placed.

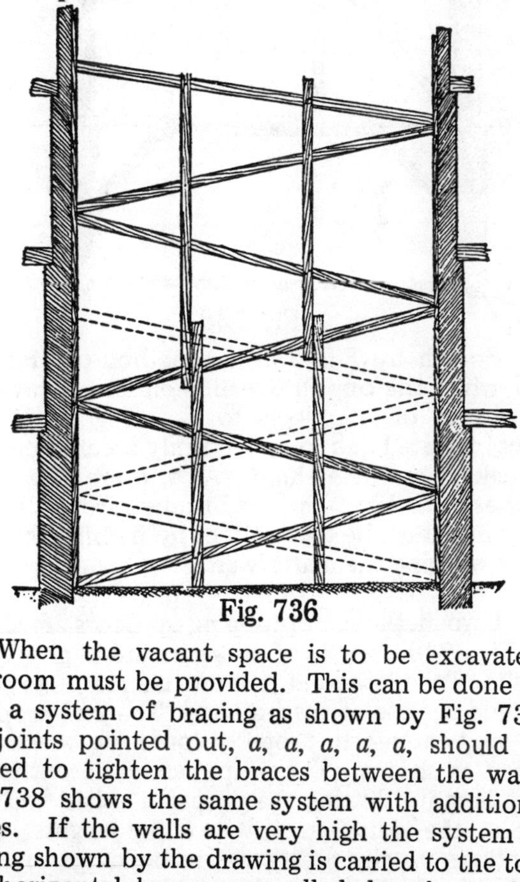

Fig. 736

When the vacant space is to be excavated, headroom must be provided. This can be done by using a system of bracing as shown by Fig. 737. The joints pointed out, *a*, *a*, *a*, *a*, *a*, should be wedged to tighten the braces between the walls. Fig. 738 shows the same system with additional braces. If the walls are very high the system of bracing shown by the drawing is carried to the top. The horizontal braces are called dog shores, but taken as a whole, this bracing is called a system of flying shores.

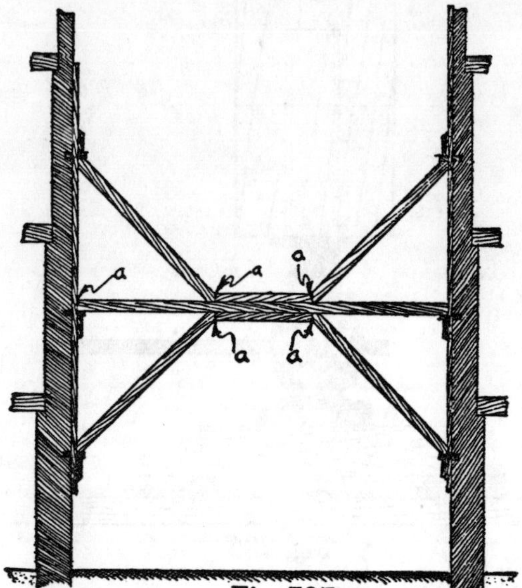

Fig. 737

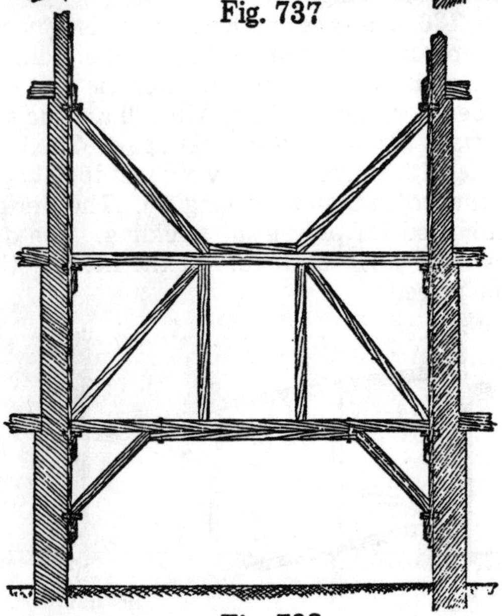

Fig. 738

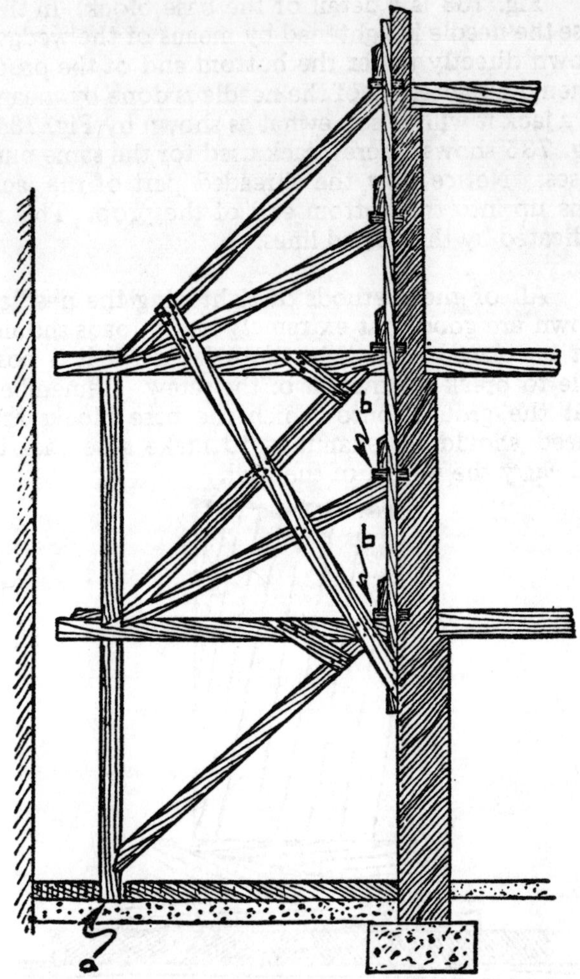

Fig. 739

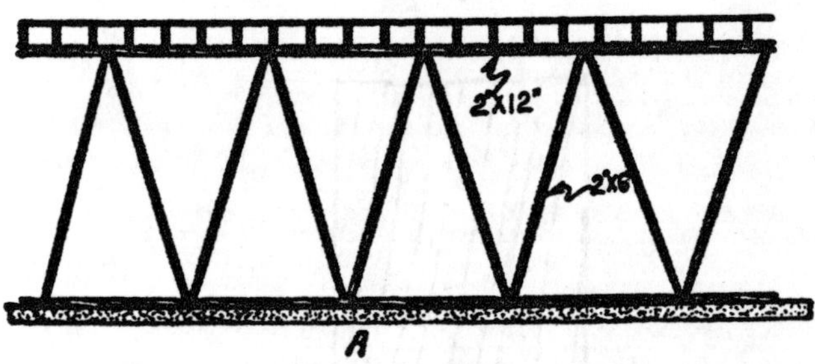

Fig. 740

Shown by Fig. 739 is a method of bracing a wall when the opposite wall is either too weak or the owner of it objects to its being used. The point indicated at *a* must be firmly secured in order to make the bracing hold. At *b*, *b* are two of the needles that keep the wall-timber from slipping. All joints must be tight, either by framing them so, or by wedging them afterward.

Two methods of propping up floors are shown by Fig. 740. Both of these methods give quick results. The method at *A* shows a 2x12 plate at the top and a plate at the bottom. The 2x6 props are cut long enough to properly space the bearings, if placed as shown. The props can all be cut to the same length. Setting the props on a slant will take care of the variations in the ceiling heights. Such variations usually are found in buildings to be remodeled. At *B* is a similar system except that the props are set more nearly upright. If the weight above is very great, stay braces should be applied about the center of the props, otherwise the props do not need bracing. We show a concrete base to indicate that the base must be firm. While we show 2x6's for props, in practice the size of the props is governed by circumstances.

Index

A

B

C

D

E

F

S

T

V

W

NUMBER OF WOOD JOISTS REQUIRED FOR ANY FLOOR AND SPACING

Length of Span	Spacing of Joists 12"	16"	20"	24"	30"	36"	42"	48"	54"	60"
6	7	6	5	4	3	3	3	3	2	2
7	8	6	5	5	4	4	3	3	3	2
8	9	7	6	5	4	4	3	3	3	3
9	10	8	6	6	5	4	4	3	3	3
10	11	9	7	6	5	4	4	4	3	3
11	12	9	8	7	5	5	4	4	3	3
12	13	10	8	7	6	5	4	4	4	3
13	14	11	9	8	6	5	5	4	4	4
14	15	12	9	8	7	6	5	5	4	4
15	16	12	10	9	7	6	5	5	4	4
16	17	13	11	9	7	6	6	5	5	4
17	18	14	11	10	8	7	6	5	5	4
18	19	15	12	10	8	7	6	6	5	4
19	20	15	12	11	9	7	6	6	5	5
20	21	16	13	11	9	8	7	6	5	5
21	22	17	14	12	9	8	7	6	6	5
22	23	18	14	12	10	8	7	7	6	5
23	24	18	15	13	10	9	8	7	6	6
24	25	19	15	13	11	9	8	7	6	6
25	26	20	16	14	11	9	8	7	7	6
26	27	21	17	14	11	10	8	8	7	6
27	28	21	17	15	12	10	9	8	7	6
28	29	22	18	15	12	10	9	8	7	7
29	30	23	18	16	13	11	9	8	7	7
30	31	24	19	16	13	11	10	9	8	7
31	32	24	20	17	13	11	10	9	8	7
32	33	25	20	17	14	12	10	9	8	7
33	34	26	21	18	14	12	10	9	8	8
34	35	27	21	18	15	12	11	10	9	8
35	36	27	22	19	15	13	11	10	9	8
36	37	28	23	19	15	13	11	10	9	8
37	38	29	23	20	16	13	12	10	9	8
38	39	30	24	20	16	14	12	11	9	9
39	40	30	24	21	17	14	12	11	10	9
40	41	31	25	21	17	14	12	11	10	9

One joist has been added to each of the above quantities to take care of extra joist required at end of span.

Add for doubling joists under all partitions.

NUMBER OF FEET OF LUMBER B.M. REQUIRED PER 100 SQ. FT. OF SURFACE WHEN USED FOR STUDS, JOISTS, RAFTERS, WALL AND FLOOR FURRING STRIPS, ETC.

The following table does not include any allowance for waste in cutting, doubling joists under partitions or around stair wells, extra joists at end of each span, top or bottom plates, etc. These items vary with each job. Add as required.

Size of Lumber	12-Inch Centers	16-Inch Centers	20-Inch Centers	24-Inch Centers
1" x 2"	16-2/3	12-1/2	10	8-1/3
2" x 2"	33-1/3	25	20	16-2/3
2" x 4"	66-2/3	50	40	33-1/3
2" x 5"	83-1/3	62-1/2	50	41-2/3
2" x 6"	100	75	60	50
2" x 8"	133-1/3	100	80	66-2/3
2" x 10"	166-2/3	125	100	83-1/3
2" x 12"	200	150	120	100
2" x 14"	233-1/3	175	140	116-2/3
3" x 6"	150	112-1/2	90	75
3" x 8"	200	133-1/3	120	100
3" x 10"	250	187-1/2	150	125
3" x 12"	300	225	180	150
3" x 14"	350	262-1/2	210	175

FURRING

Size of Strips	Spacing on Centers	Board Feet per Square Feet of Area	Lbs. Nails per 1000 Board Feet
1 x 2	12" 16" 20" 24"	.18 .14 .11 .10	55
1 x 3	12" 16" 20" 24"	.28 .21 .17 .14	37
1 x 4	12" 16" 20" 24"	.36 .28 .22 .20	30

PLYWOOD THICKNESSES, SPANS AND NAILING RECOMMENDATIONS

(Plywood Continuous Over 2 or More Spans; Grain of Face Plys Across Supports)

FIR PLYWOOD FLOOR CONSTRUCTION

Application	Recommended Thickness	Maximum Spac. of Supports (C. to C.)	Nail Size and Type	Nail Spacing: Panel Edges	Intermediate
Subflooring	1/2" (a)	16" (b)	6d Common (c)	6"	10"
	5/8" (a)	20"	8d Common (c)	6"	10"
	3/4" (a)	24"	8d Common (c)	6"	10"
	2.4.1	48"	8d Ring Shank (c)	6"	6"
Underlayment	3/8" (d) 5/8"		6d Ring Shank or Cement Coated 8d Flathead	6"	8" Ea. Way

(a) Provide blocking at panel edges for carpet, tile, linoleum or other non-structural flooring. No blocking required for 25/32" strip flooring.
(b) If strip flooring is perpendicular to supports 1/2" can be used on 24" span.
(c) If resilient flooring is to be applied without underlayment, set nails 1/16".
(d) FHA accepts 1/4" plywood.

If supports are not well seasoned, use ring-shank nails.

FIR PLYWOOD ROOF SHEATHING

Recommended Thickness	Max. Spacing of Supports, (C. to C.) 20 PSF	30 PSF	40 PSF	Nail Size and Type	Nail Spacing: Panel Edge	Intermediate
5/16"	20" (b)	20"	20"	6d Common	6"	12"
3/8"	24" (b)	24"	24"	6d Common	6"	12"
1/2" (a)	32" (b)	32"	30"	6d Common	6"	12"
5/8" (a)	42" (b)	42"	39"	8d Common	6"	12"
3/4" (a)	48" (b)	47"	42"	8d Common	6"	12"

(a) Provide blocking or other means of suitable edge support when span exceeds 28" for 1/2"; 32" for 5/8" and 36" for 3/4".
(b) For special case of two span continuous beams, plywood spans can be increased 6-1/2% except as roof indicated by (b) in chart.

BUILT-UP GIRDERS

Size of Girder	Bd. Ft. per Lin. Ft.	Nails per 1000 Bd. Ft.
4 x 6	2.15	53
4 x 8	2.85	40
4 x 10	3.58	32
4 x 12	4.28	26
6 x 6	3.21	43
6 x 8	4.28	32
6 x 10	5.35	26
6 x 12	6.42	22
8 x 8	5.71	30
8 x 10	7.13	24
8 x 12	8.56	20

PARTITION STUDS

(Studs including top and bottom plates)

Size of Studs	Spacing on Centers	Bd. Ft. per Sq. Ft. of Area	Lbs. Nails per 1000 Bd. Ft.
2" x 3"	12" 16" 24"	.91 .83 .76	25
2" x 4"	12" 16" 24"	1.22 1.12 1.02	19
2" x 6"	16" 24"	1.48 1.22	16

PARTITION STUDS

Number of Feet of Lumber Required Per Sq. Ft. of Wood Stud Partition Using 2" x 4" Studs.

Studs spaced 16" on centers, with single top and bottom plates.

Length Partition in Feet	No. Studs Required	Ceiling Heights in Feet			
		8'-0"	9'-0"	10'-0"	12'-0"
2	3	1.25	1.167	1.13	1.13
3	3	0.833	.812	.80	.80
4	4	0.833	.812	.80	.80
5	5	0.833	.812	.80	.80
6	6	0.833	.812	.80	.80
7	6	0.833	.75	.75	.80
8	7	0.75	.75	.75	.70
9	8	0.75	.75	.75	.70
10	9	0.75	.75	.75	.70
11	9	0.75	.70	.70	.67
12	10	0.75	.70	.70	.67
13	11	0.75	.70	.70	.67
14	12	0.75	.70	.70	.67
15	12	0.70	.70	.70	.67
16	13	0.70	.70	.70	.67
17	14	0.70	.70	.70	.67
18	15	0.70	.70	.67	.67
19	15	0.70	.70	.67	.67
20	16	0.70	.70	.67	.67
For dbl. plate, add per sq. ft.		0.13	.11	.10	.083

For 2"x8" studs, double above quantities.
For 2"x6" studs, increase above quantities 50%.

Example: Find the number of feet of lumber, b.m. required for a stud partition 18'-0" long and 9'-0" high. This partition would contain 18x9 = 162 sq. ft. The table gives 0.70 ft. of lumber, b.m. per sq. ft. of partition. Multiply 162 by 0.70 equals 113.4 ft. b.m.

RAFTERS

	BOARD FEET REQUIRED 100 Square Feet Surface Area			NAILS
	12" O.C.	16" O.C.	24" O.C.	Per 1000 Board Feet
2 x 4	89	71	53	17
2 x 6	129	102	75	12
2 x 8	171	134	112	9
2 x 10	212	197	121	7
2 x 12	252	197	143	6

NOTE: Includes common, hip and valley rafters, ridge boards and collar beams.

EXTERIOR WALL STUDS

(Studs Including Corner Bracing.)

Size of Studs	Spacing on Centers	Bd. Ft. per Sq. Ft. of Area	Lbs. Nails per 1000 Bd. Ft.
2" x 3"	12" 16" 20" 24"	.83 .78 .74 .71	30
2" x 4"	12" 16" 20" 24"	1.09 1.05 .98 .94	22
2" x 6"	12" 16" 20" 24"	1.66 1.51 1.44 1.38	15

SHEATHING AND SUBFLOORING

(HORIZONTAL APPLICATION)

Type	Size	Bd. Ft. per Sq. Ft. of Area	LBS. NAILS PER 1000 BD. FT. — Spacing of Framing Members			
			12"	16"	20"	24"
T & G	1 x 4	1.32	66	52	44	36
	1 x 6	1.23	43	33	28	23
	1 x 8	1.19	32	24	21	17
	1 x 10	1.17	37	29	24	20
Shiplap	1 x 4	1.38	69	55	46	38
	1 x 6	1.26	44	34	29	24
	1 x 8	1.21	32	25	21	17
	1 x 10	1.18	37	29	25	20
S4S	1 x 4	1.19	60	47	40	33
	1 x 6	1.15	40	31	26	22
	1 x 8	1.15	30	23	20	17
	1 x 10	1.14	36	28	24	19

ESTIMATING DATA ON ALUMINUM NAILS

No. Nails per Lb.	Size and Type of Nail	No. Nails per Box	Coverage
604	6d Wood Siding — Sinker Hd.	575	500 bd. ft. 1/2"x8" Bevel Sdg
468	7d Wood Siding — Sinker Hd.	575	500 bd. ft. 3/4"x8" Bevel Sdg
319	8d Wood Siding — Sinker Hd.	575	500 bd. ft. 3/4"x8" Bevel Sdg.
185	10d Wood Siding — Sinker Hd.	290	250 bd. ft. 3/4"x8" Bevel Sdg.
604	6d Wood Siding — Casing Hd.	575	500 bd. ft. 1/2"x8" Bevel Sdg.
468	7d Wood Siding — Casing Hd.	575	500 bd. ft. 3/4"x8" Bevel Sdg.
319	8d Wood Siding — Casing Hd.	575	500 bd. ft. 3/4"x8" Bevel Sdg.
185	10d Wood Siding — Casing Hd.		
1230	1-1/4" Asbestos Siding	885	5 sqs. Asb. Sdg. Face Nailing
720	1-3/4" Asbestos Siding	885	5 sqs. Asb. Sdg. Face Nailing
785	1-1/4" Asbestos Shingle	885	5 sqs. Asb. Sdg. Conc. Nailing
659	1-1/2" Asbestos Shingle	885	5 sqs. Asb. Sdg. Conc. Nailing
544	1-3/4" Asbestos Shingle	885	5 sqs. Asb. Sdg. Conc. Nailing
1300	1-1/4" Cedar Shake	1680	3 sqs. Single Course
724	1-3/4" Cedar Shake	1680	3 sqs. Double Course
1480	3d Cedar Shingle	3150	3 sqs. with 5" Exposure
1313	7/8" Standard Shingle	2600	General Purpose
1009	3d Standard Shingle	2000	Barn Battens, Joist Lining, etc.
1048	1-3/8" Dri-Wall	1530	1000 sq. ft. 3/8" Sheet Rock
900	1-5/8" Dri-Wall	1530	1000 sq. ft. 1/2" Sheet Rock
988	1-1/8" Rock Lath	2666	35 square yards
939	1-1/4" Rock Lath	2666	35 square yards
725	1-1/2" Rock Lath	1900	25 square yards
495	2" Insulated Siding	1680	5 squares
295	2-1/2" Insulated Siding	600	60 Buttress Corners
663	7/8" Roofing	860	500 sq. ft. Roll Roofing
605	1" Roofing	980	3 squares Sq. Tab. Shingles
491	1-1/4" Roofing	980	3 squares Sq. Tab. Shingles
017	1-1/2" Roofing	980	3 squares Sq. Tab. Shingles
368	1-3/4" Roofing	980	3 squares Sq. Tab. Shingles
336	2" Roofing	980	3 squares Sq. Tab. Shingles
274	2-1/2" Roofing	650	2 squares Sq. Tab. Shingles
318	1-3/4" Roofing w/w/attached	1050	10 sqs. Aluminum Roofing
285	2" Roofing w/w/attached	1050	10 sqs. Aluminum Roofing
242	2-1/2" Roofing w/w/attached	1050	10 sqs. Aluminum Roofing

WIRE NAILS

Size of Nail	Length of Nails Inches	Gauge Number	Approximate Number to Pound
2d	1	15	876
3d	1-1/4	14	568
4d	1-1/2	12-1/2	316
5d	1-3/4	12-1/2	271
6d	2	11-1/2	181
7d	2-1/4	11-1/2	161
8d	2-1/2	10-1/4	106
9d	2-3/4	10-1/4	96
10d	3	9	69
12d	3-1/4	9	63
16d	3-1/2	8	49
20d	4	6	31
30d	4-1/2	5	24
40d	5	4	18
50d	5-1/2	3	14
60d	6	2	11

FLOORING NAILS

Width Flooring	Joist Spacing 12" on Centers	Joist Spacing 16" on Centers
2"	40 lbs. 8d. flg.	30 lbs. 8d. flg.
3"	30 lbs. 8d. flg.	23 lbs. 8d. flg.
4"	22 lbs. 8d. flg.	17 lbs. 8d. flg.
6"	24 lbs. 10d. com.	18 lbs. 10d. com.
8"	17 lbs. 10d. com.	13 lbs. 10d. com.

NOTE: Quantities determined for 1000 board feet of material.

STANDARD NAIL REQUIREMENTS

Description of Material	Unit of Measure	Size and Kind of Nail	Number of Nails Required	Pounds of Nails Required
Wood Shingles	1,000	3d Common	2,560	4 lbs.
Individual Asphalt Shingles ...	100 sq. ft.	7/8" Roofing	848	4 lbs.
Three in One Asphalt Shingles .	100 sq. ft.	7/8" Roofing	320	1 lb.
Wood Lath	1,000	3d Fine	4,000	6 lbs.
Wood Lath	1,000	2d Fine	4,000	4 lbs.
Bevel or Lap Siding, 1/2" x 4".	1,000'	6d Coated	2,250	*15 lbs.
Bevel or Lap Siding, 1/2" x 6".	1,000'	6d Coated	1,500	*10 lbs.
Byrkit Lath, 1" x 6"	1,000'	6d Common	2,400	15 lbs.
Drop Siding, 1" x 6"	1,000'	8d Common	3,000	25 lbs.
3/8" Hardwood Flooring	1,000'	4d Finish	9,300	16 lbs.
13/16" Hardwood Flooring ...	1,000'	8d Casing	9,300	64 lbs.
Softwood Flooring, 1" x 3" ...	1,000'	8d Casing	3,350	23 lbs.
Softwood Flooring, 1" x 4" ...	1,000'	8d Casing	2,500	17 lbs.
Softwood Flooring, 1" x 6" ...	1,000'	8d Casing	2,600	18 lbs.
Ceiling, 5/8" x 4"	1,000'	6d Casing	2,250	10 lbs.
Sheathing Boards, 1" x 4"	1,000'	8d Common	4,500	40 lbs.
Sheathing Boards, 1" x 6"	1,000'	8d Common	3,000	25 lbs.
Sheathing Boards, 1" x 8"	1,000'	8d Common	2,250	20 lbs.
Sheathing Boards, 1" x 10" ...	1,000'	8d Common	1,800	15 lbs.
Sheathing Boards, 1" x 12" ...	1,000'	8d Common	1,500	12-1/2 lbs.
Studding, 2" x 4"	1,000'	16d Common	500	10 lbs.
Joist, 2" x 6"	1,000'	16d Common	332	7 lbs.
Joist, 2" x 8"	1,000'	16d Common	252	5 lbs.
Joist, 2" x 10"	1,000'	16d Common	200	4 lbs.
Joist, 2" x 12"	1,000'	16d Common	168	3-1/2 lbs.
Interior Trim, 5/8" thick	1,000'	6d Finish	2,250	7 lbs.
Interior Trim, 3/4" thick	1,000'	8d Finish	3,000	14 lbs.
5/8" Trim where nailed to jamb	1,000'	4d Finish	2,250	3 lbs.
1" x 2" Furring or Bridging ...	1,000'	6d Common	2,400	15 lbs.
1" x 1" Grounds	1,000'	6d Common	4,800	30 lbs.

* NOTE: Cement Coated Nails sold as two-thirds of Pound equals 1 Pound of Common Nails.

NOTE: Quantities determined for 1000 board feet of material or 100 sq. ft. of Surface.

FLOOR JOIST

SIZE OF JOIST	MATERIAL — Board Feet Required for 100 Sq. Ft. of Surface Area				NAILS — Per 1000 Bd. Ft.
	12" O.C.	16" O.C.	20" O.C.	24" O.C.	Pounds
2 x 6	128	102	88	78	10
2 x 8	171	136	117	103	8
2 x 10	214	171	148	130	6
2 x 12	256	205	177	156	5

CEILING JOIST

SIZE OF JOIST	MATERIAL — Board Feet Required for 100 Sq. Ft. of Surface Area				NAILS — Per 1000 Bd. Ft.
	12" O.C.	16" O.C.	20" O.C.	24" O.C.	Pounds
2 x 4	78	59	48	42	19
2 x 6	115	88	72	63	13
2 x 8	153	117	96	84	9
2 x 10	194	147	121	104	7
2 x 12	230	176	144	126	6

BEVEL SIDING

SIZE	MATERIAL — Siding for 100 Square Foot Wall			NAILS — Per 100 Square Feet
	Exposed to Weather	Add for Lap	BM per 100 Sq. Ft.	
1/2 x 4	2-3/4	46%	151	1-1/2 Pounds
1/2 x 5	3-3/4	33%	138	1-1/2 Pounds
1/2 x 6	4-3/4	26%	131	1 Pound
1/2 x 8	6-3/4	18%	123	3/4 Pound
5/8 x 8	6-3/4	18%	123	3/4 Pound
3/4 x 8	6-3/4	18%	123	3/4 Pound
5/8 x 10	8-3/4	14%	119	1/2 Pound
3/4 x 10	8-3/4	14%	119	1/2 Pound
3/4 x 12	10-3/4	12%	117	1/2 Pound

NOTE: Quantities include 5% for endcutting and waste. Deduct for all openings over ten square feet.

DROP SIDING

SIZE	MATERIAL — Siding for 100 Square Foot Wall			NAILS — Per 100 Square Feet
	Exposed to Weather	Add for Lap	BM per 100 Sq. Ft.	
1 x 6	5-1/4	14%	119	2-1/2 Pounds
1 x 8	7-1/4	10%	115	2 Pounds

NOTE: Quantities include 5% for encutting and waste. Deduct for all openings over ten square feet.

SQUARE-EDGED BOARDS

Following data is based on lumber surfaced 1 or 2 sides and 1 edge. The waste allowance shown includes width lost in dressing plus 5 percent waste in end-cutting. If laid diagonally, add 5 percent additional waste.

Measured Size, Inches	Finished Width, Inches	Add for Waste %	Quantity Lbr. Required, Multiply Area by	Feet of Lumber Required, 100 Sq. Ft. Surface
1 x 3....	2-1/2	25	1.25	125
1 x 4...	3-1/2	20	1.20	120
1 x 6...	5-1/2	14	1.14	114
1 x 8...	7-1/2	12	1.12	112
1 x 10...	9-1/2	10	1.10	110
1 x 12...	11-1/2	9-1/2	1.095	109-1/2
2 x 4...	3-1/2	20	2.40	240
2 x 6...	5-1/2	14	2.28	228
2 x 8...	7-1/2	12	2.25	225
2 x 10...	9-1/2	10	2.20	220
2 x 12...	11-1/2	9-1/2	2.19	219
3 x 6...	5-1/2	14	3.43	343
3 x 8...	7-1/2	12	3.375	337-1/2
3 x 10...	9-1/2	10	3.30	330
3 x 12...	11-1/2	9-1/2	3.29	329

T&G, SHIPLAP BOARDS

Following data applicable on most D&M lumber. Waste allowance shown includes width lost in dressing and lapping plus 5 percent waste in end-cutting and matching.

Measured Size, Inches	Finished Width, Inches	Add for Waste %	Quantity Lbr. Required, Multiply Area by	Feet of Lumber Required, 100 Sq. Ft. Surface
1 x 2	1-3/8	50	1.50	150
1 x 2-3/4	2	42-1/2	1.425	142-1/2
1 x 3	2-1/4	38-1/3	1.383	138
1 x 4	3-1/4	28	1.28	128
1 x 6	5-1/4	20	1.20	120
1 x 8	7-1/4	16	1.15	115
1-1/4 x 3	2-1/4	38-1/3	1.73	173
1-1/4 x 4	3-1/4	28	1.60	160
1-1/4 x 6	5-1/4	20	1.50	150
1-1/2 x 3	2-1/4	38-1/3	2.08	208
1-1/2 x 4	3-1/4	28	1.92	192
1-1/2 x 6	5-1/4	20	1.80	180
2 x 4	3-1/4	28	2.60	260
2 x 6	5-1/4	20	2.40	240
2 x 8	7-1/4	16	2.32	232
2 x 10	9-1/4	13	2.25	225
2 x 12	11-1/4	12	2.24	224
3 x 6	5-1/4	20	3.60	360
3 x 8	7-1/4	16	3.48	348
3 x 10	9-1/4	13	3.39	339
3 x 12	11-1/4	12	3.36	336

TABLE OF SQUARE FEET PER SHEET OF CORRUGATED STRUCTURAL PLASTIC PANELS

Length of Sheet	26" Width	27-1/2" Width	33" Width	33-3/4" Width	35" Width	36" Width	40" Width	42" Width
* 6"	1.08	1.15	1.37	1.41	1.46	1.5	1.67	1.75
3'	6.50	6.87	8.25	8.44	8.75	9.0	10.00	10.50
4'	8.67	9.17	11.00	11.25	11.67	12.0	13.33	14.00
5'	10.83	11.46	13.75	14.06	14.58	15.0	16.67	17.50
6'	13.00	13.75	16.50	16.87	17.50	18.0	20.00	21.00
7'	15.17	16.04	19.25	19.69	20.42	21.0	23.33	24.50
8'	17.33	18.33	22.00	22.50	23.33	24.0	26.67	28.00
* 9'	19.50	20.62	24.75	25.31	26.25	27.0	30.00	31.50
10'	21.67	22.91	27.50	28.12	29.17	30.0	33.33	35.00
*11'	23.83	25.21	30.25	30.93	32.08	33.0	36.67	38.50
12'	26.00	27.50	33.00	33.75	35.00	36.0	40.00	42.00
*13'	28.17	29.79	35.75	36.56	37.92	39.0	43.33	45.50
14'	30.34	32.08	38.50	39.37	40.84	42.0	46.66	49.00

* Non-stock lengths.

TABLES OF SQUARE FEET PER SHEET OF FLAT PLASTIC PANELS

Length of Sheet	Width 16"	24"	32"	48"
* 6"	.66	1.00	1.33	2.00
* 1'	1.33	2.00	2.66	4.00
* 2'	2.66	4.00	5.33	8.00
* 3'	4.00	6.00	8.00	12.00
4'	5.33	8.00	10.66	16.00
5'	6.66	10.00	13.33	20.00
6'	8.00	12.00	16.00	24.00
7'	9.33	14.00	18.66	28.00

Length of Sheet	Width 16"	24"	32"	48"
* 8'	10.66	16.00	21.33	32.00
* 9'	12.00	18.00	24.00	36.00
*10'	13.33	20.00	26.66	40.00
*11'	14.66	22.00	29.33	44.00
12'	16.00	24.00	32.00	48.00
13'	17.33	26.00	34.66	52.00
14'	18.66	28.00	37.33	56.00

* Non-stock lengths.

RECTANGLE

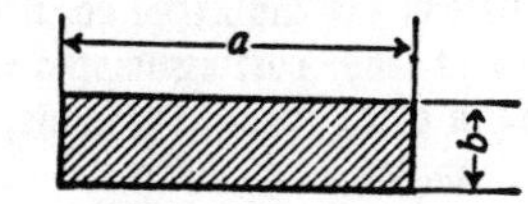

Area = ab

CYLINDER

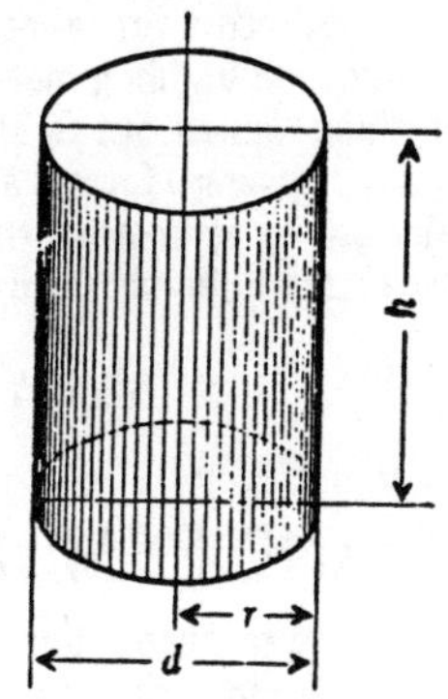

Cylindrical Surface = πdh

Volume = $\pi r^2 h$

TRIANGLE

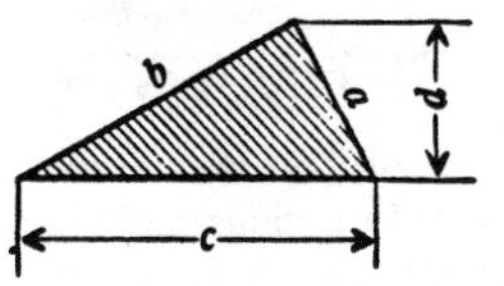

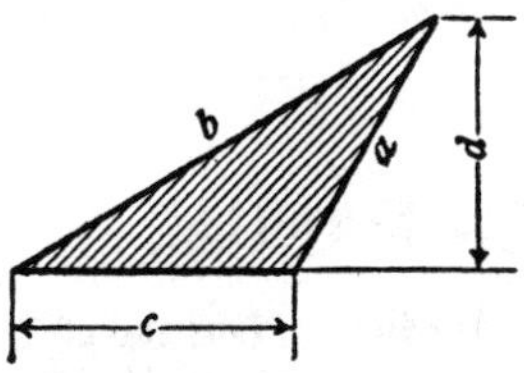

Area = ½ cd

PARALLELOGRAM

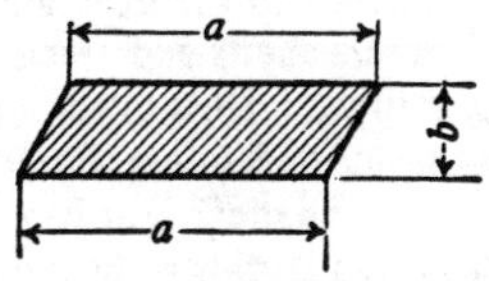

Area = ab

PYRAMID

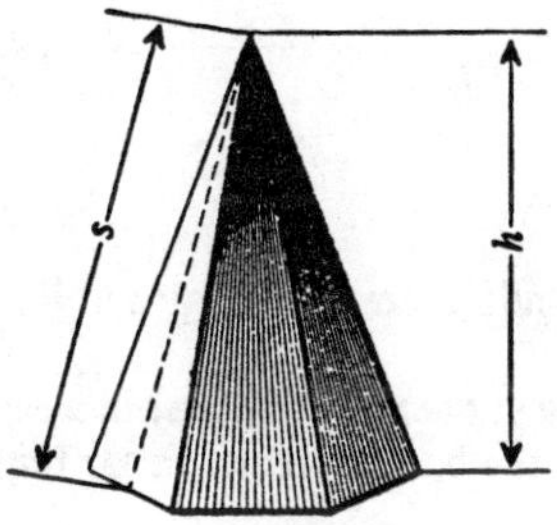

Volume = ⅓ Ah

TRAPEZIUM

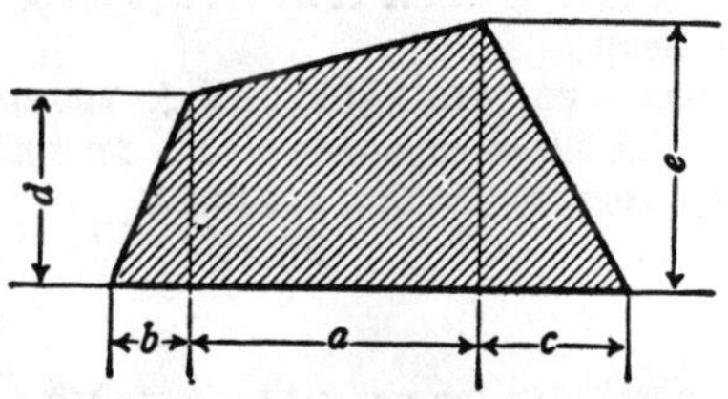

Area = ½ [a (e + d) + bd + ce]

CONE

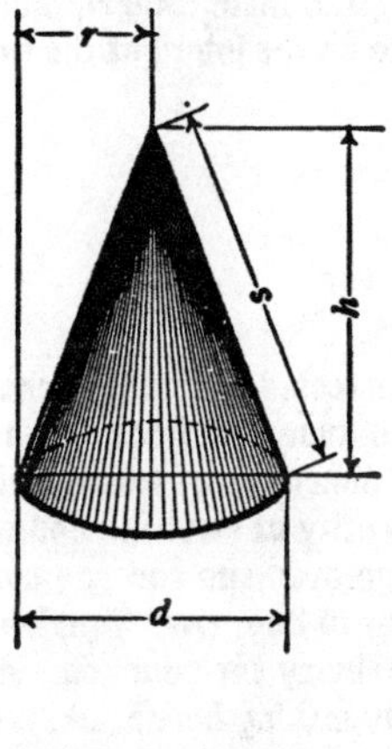

Conical Area = πrs

Volume = ⅓ $\pi r^2 h$ =

REGULAR POLYGON

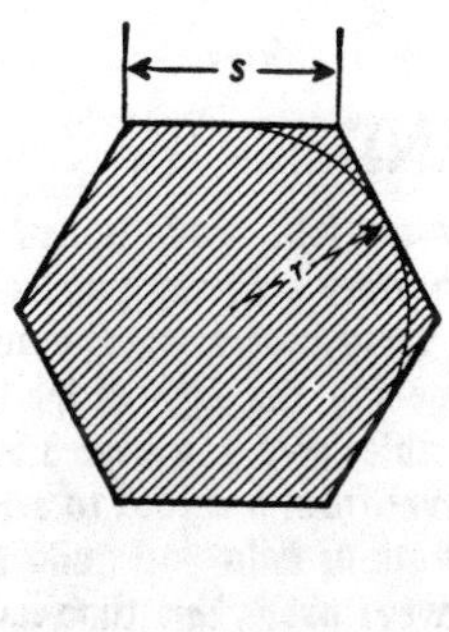

Area = ½ nsr
when n = Number of sides.

Area of a square = length x breadth or height.

Area of a rectangle = length x breadth or height.

Area of a triangle = base x 1/2 altitude.

Area of parallelogram = base x altitude.

Area of trapezoid = altitude x 1/2 the sum of parallel sides.

Area of trapezium = divide into two triangles, total their areas.

Circumference of circle = diameter x 3.1416.

Circumference of circle = radius x 6.283185.

Diameter of circle = circumference x .3183.

Diameter of circle = square root of area x 1.12838.

Radius of a circle = circumference x .0159155.

Area of a circle = half diameter x half circumference.

Area of a circle = square of diameter x .7854.

Area of a circle = square of circumference x .07958.

Area of a sector of circle = length of arc x 1/2 radius.

Area of a segment of circle = area of sector of equal radius — area of a triangle, when the segment is less, and plus area of triangle, when segment is greater than the semi-circle.

Area of circular ring = sum of the diameter of the two circles x difference of the diameter of the two circles and that product x .7854.

Side of square that shall equal area of circle = diameter x .8862.

Side of square that shall equal area of circle = circumference x .2821.

Diameter of circle that shall contain area of a given square = side of square x 1.1284.

Side of inscribed equilateral triangle = diameter x .86

Side of inscribed square = diameter x .7071.

Side of inscribed square = circumference x .225.

Area of ellipse = product of the two diameters x .7854.

Area of a parabola = base x 2/3 of altitude.

Area of a regular polygon = sum of its sides x perpendicular from its center to one of its sides divided by 2.

Surface of cylinder or prism = area of both ends plus length and x circumference.

Surface of sphere = diameter x circumference.

Solidity of sphere = surface x 1/6 diameter.

Solidity of sphere = cube of diameter x .5236.

Solidity of sphere = cube of radius x 4.1888.

Solidity of sphere = cube of circumference x .016887.

Diameter of sphere = cube root of solidity x 1.2407.

Diameter of sphere = square root of surface x .56419.

Circumference of sphere = square root of surface x 1.772454.

Circumference of sphere = cube root of solidity x 3.8978.

Contents of segment of sphere = (height squared plus three times the square of radius of base) x (height x .5236).

Contents of a sphere = diameter x .5236.

Side of inscribed cube of sphere = radius x 1.1547.

Side of inscribed cube of sphere = square root of diameter.

Surface of pyramid or cone = circumference of base x 1/2 of the slant height plus area of base.

Contents of pyramid or cone = area of base x 1/3 altitude.

Contents of frustum of pyramid or cone = sum of circumference at both ends x 1/2 slant height plus area of both ends.

Contents of frustum of pyramid or cone = multiply areas of two ends together and extract square root. Add to this root the two areas and x 1/3 altitude.

Contents of a wedge = area of base x 1/2 altitude.

WOOD FRAME HOUSE CONSTRUCTION

The basic source book and guide to sound wood-frame house construction. Builders will appreciate the modern, practical information on materials and building practice. A useful training aid for apprentices and a standard by which the quality of all housing can be judged.
Everything you need to know about framing, exterior trim, roofing, windows and doors, siding, plumbing and heating, insulation and vapor barrier, finishes, floor covering, millwork, cabinets, chimneys and fireplace, driveways and walks . . . complete step by step "how to" information. Carefully written, well illustrated, full of practical information and worth several times the price.

232 Pages 8 x 10 $2.75

CONSTRUCTION MANUAL: CONCRETE & FORMWORK

The new handbook for the concrete professional . . . The man on the job who needs sure answers to practical problems: What type of mix is best for the job? What admixtures are needed? How deep should the footing be? What is the best way to lay out and design the forms? How much concrete and form material are needed? How can reinforcing and curing be handled most profitably? Nearly 100 pages of step-by-step instructions cover the actual construction and erecting of all forms in common use. Over 200 man hour tables, charts, and clear illustrations make this the most useful single volume for anyone working with concrete formwork.

176 Pages 8½ x 11 $3.75

CONSTRUCTION MANUAL: FINISH CARPENTRY

Money-making know-how for the carpentry "pro". Sure answers to practical carpentry problems. This new handbook has the time-saving methods, inside trade information and proven shortcuts you need to do first class finish carpentry work on any job. It gives you information you can use on every finish carpentry subject: cornices and rakes, gutters and downspouts, wood shingle roofing, asphalt, asbestos and built-up roofing, window frames, sash and trim, prefabricated windows, door bucks and frames, door trim, siding, wallboard, lath and plaster, stairs and railings, cabinets, joinery, and wood flooring. One chapter covers estimating finish carpentry so that you have all the information you need to figure the materials and labor required, lay out the work and cut, fit and install the items required. Best of all, you know the work is done right. Over 350 man-hour tables, charts and clear illustrations make this the practical, step-by step handbook any carpentry "pro" needs and can use. Have the information you need at your fingertips when you need it.

192 Pages 8½ x 11 $4.75

BUILDERS CALCULATOR Not Illustrated

With this handy dual purpose slide chart you make volume calculations for concrete, asphalt, or excavation work in seconds -- with error free accuracy. The reverse side of the chart gives you the board footage per piece for various lengths of lumber for all the common dimensions. A real time saver for take-off work. Made of rugged six-ply paperboard and designed to serve you for years.

8 x 3 $1.50

PLEASE RUSH THESE TO ME:

____	NATIONAL CONSTRUCTION ESTIMATOR	$7.50	____
____	NATIONAL REPAIR AND REMODELING ESTIMATOR	$6.50	____
____	CONSTRUCTION INDUSTRY PRODUCTION MANUAL	$6.00	____
____	PRACTICAL LUMBER COMPUTER	$2.00	____
____	PRACTICAL RAFTER CALCULATOR	$3.00	____
____	STAIR BUILDERS HANDBOOK	$5.95	____
____	WOOD FRAME HOUSE CONSTRUCTION	$2.75	____
____	CONSTRUCTION MANUAL: CONCRETE & FORMWORK	$3.75	____
____	CONSTRUCTION MANUAL: FINISH CARPENTRY	$4.75	____
____	BUILDERS CALCULATOR	$1.50	____

Use your Master Charge or BankAmericard

Charge my BANKAMERICARD ☐ MASTER CHARGE ☐

NO. ____________

EXPIRATION DATE OF CARD ____________

TOTAL ENCLOSED ____________

Subtotal ____________
In Calif. add 6% tax ____________

CRAFTSMAN Book Company of America
SERVING THE CONSTRUCTION INDUSTRY'S NEED FOR INFORMATION SINCE 1950 . . .

542 Stevens Avenue • Solana Beach, California 92075

SATISFACTION GUARANTEED OR YOUR MONEY BACK

Name ____________

Company Name ____________

Address ____________

City ____________ State ____________ Zip ____________